Mastering Skills in Complex Cardiovascular Interventions

Ata Firouzi • Zahra Hosseini • Azin Alizadehasl
Editors

Mastering Skills in Complex Cardiovascular Interventions

A Case-Based Approach

Springer

Editors
Ata Firouzi
Interventional cardiologist
Rajaie Cardiovascular Institute
Tehran, Iran

Zahra Hosseini
Interventional cardiologist
Rajaie Cardiovascular Institute
Tehran, Iran

Azin Alizadehasl
Echo-Cardiologist, Cardio-Oncologist
Rajaie Cardiovascular Center
Tehran, Iran

ISBN 978-3-031-97320-8 ISBN 978-3-031-97321-5 (eBook)
https://doi.org/10.1007/978-3-031-97321-5

This Springer imprint is published by the registered company Springer Nature Switzerland AG
The registered company address is: Gewerbestrasse 11, 6330 Cham, Switzerland

Introduction

In 2020, thanks to the better appreciate the pathophysiology of cardiovascular diseases, continued advances in engineering in interventional equipment, miniaturization of materials, and improvements in knowledge of interventional techniques, percutaneous interventions in complex cardiovascular diseases are growing up. A combination of new innovations in image-guided procedures and high-precision imaging technology has provided clinicians with more confidence in performance of more complex interventions, in such a way that in timeline the guidelines in management of even these complex procedures are going to change.

In the field of coronary heart disease, Percutaneous Coronary Intervention (PCI) was performed until two decades ago just for simple de novo lesions, but within 20 years with improvements in equipment such as stent designs and profile, distinct balloon types, breakthrough in coronary guide-wire designs, lesion modification tools like Rotational Rotablator, shockwave balloons, technical approaches, imaging modalities such as Intra-Vascular Ultrasound (IVUS) and Optical Coherence Topography (OCT), and with concomitant enhancing operators experience, unexpectedly, nowadays, PCI is going to be prefer even for very complex lesions such as Left Main or Chronic Total Occlusion (CTO) ones even in high risk patients to more aggressive alternative approaches, means Coronary Artery Bypass Graft (CABG).

In adult congenital heart disease, in the past, surgical repair was the first option, but recently with new innovations in device types and approaches, trans-catheter procedures had their role first in simple ones but recently even in extremely complex lesions by expert interventionist, advanced imaging modalities and appropriate case selections, these are repaired percutaneously worldwide.

Valvular heart diseases are one of the most reasons that patients are referred to surgeons. Trans-catheter valve technology began with Percutaneous Pulmonary valve Commissurotomy (PTPC) in 1982, and then in 1984, Inoue revolutionized the treatment of rheumatic Mitral Stenosis (MS), who designed the specific dumbbell-shaped balloon which has led to change the management of these patients in such a way that in the suitable patients the first option for treatment of MS is Percutaneous Mitral valve Commissurotomy (PTMC) and finally with percutaneous Balloon Aortic Valvotomy (BAV) in 1986. In long-term follow-ups, although the results of

PTMC and PTPC were comparable to surgical repair, the results of BAV were disappointing. So, for the first time in 1990 to 2000, pulmonary valve implantation in pulmonary conduit by Bonhoeffer and Trans-catheter Aortic Valve implantation (TAVI) by Cribier were performed. In time, with more advancements in prosthetic valves, procedural tactics, and growth in experience in this field, dramatic changes have occurred in valvular treatment, as in the last guidelines, TAVI was classified as class I for prohibited and high-risk patients and Class IIa in moderate risk patients with aortic stenosis. Valve in Valve (ViV) is also another boost in degenerative failed biologic valves in high-risk patients. Another example is Mitral Clips which recently has been approved in primary and secondary MR with prohibited surgical risk and refractory symptoms.

With performance of more aggressive procedures, the complication rates are also increased, especially in the early era of the procedures. But with more intervention experience, performing in high volume centers and may be done in hybrid catheterization room, the complication rates have been decreased, as these days we are bystander the degree of reduction in the rates of major complications.

Cardiovascular Intervention Research Center
Rajaie Cardiovascular Institute
Iran University of Medical Sciences
Tehran, Iran

Ata Firouzi
Zahra Hosseini
Ehsan Khalilipur

Contents

Part I
Percutaneous Coronary Arteries Intervention

Chapter 1
Primary Percutaneous Coronary Intervention

Ata Firouzi, Zahra Hosseini, Ehsan Khalilipur, Armin Elahifar, and Mohammadreza Baay

Abbreviations

ACS	Acute coronary syndrome
CABG	Coronary artery bypass grafting
ECG	Electrocardiography
IABP	Intra-aortic balloon pump
LAD	Left anterior descending artery
LCX	Left circumflex coronary artery
LM	Left main coronary artery
LV	Left ventricle
MCS	Mechanical circulatory support
MI	Myocardial infarction
PCI	Percutaneous coronary intervention
RCA	Right coronary artery
RV	Right ventricle
SCAD	Spontaneous coronary artery dissection
STEMI	ST-segment elevation myocardial infarction
SVG	Saphenous vein graft
TIMI	Thrombolysis in myocardial infarction

A. Firouzi · Z. Hosseini · E. Khalilipur (✉) · A. Elahifar · M. Baay
Cardiovascular Intervention Research Center,
Rajaie Cardiovascular medical and research institute, Tehran, Iran

A. Firouzi et al. (eds.), *Mastering Skills in Complex Cardiovascular Interventions*, https://doi.org/10.1007/978-3-031-97321-5_1

Introduction

The proportion of patients with ACS events presenting with STEMI varies across observational studies but has declined over the past decade. This decline is partly attributable to more sensitive assays for myocardial injury, which have increased the detection of non-STEMI cases relative to STEMI. Between 1999 and 2008, the proportion of patients with STEMI decreased by nearly 50% (Table 1.1) [1].

Primary PCI, defined as percutaneous catheter intervention in the setting of STEMI without previous fibrinolysis, is the preferred reperfusion strategy. It has replaced fibrinolysis in patients with STEMI, provided it can be performed promptly at high-volume PCI centers with experienced operators and 24/7 catheterization laboratory availability.

According to the latest European Society of Cardiology (ESC) guidelines on myocardial revascularization, the optimal time for reperfusion in these high-risk patients is within 120 min of first medical contact for primary percutaneous coronary intervention (PCI) or within 10 min of arrival at the hospital for fibrinolysis, depending on the availability and feasibility of the treatment modality [3].

Table 1.1 The fourth universal definition of myocardial injury and MI [2]

Universal definitions of myocardial injury and MI
Criteria for myocardial injury
The term "myocardial injury" should be used when there is evidence of elevated cardiac troponin (cTn) values, with at least one value above the 99th percentile upper reference limit (URL). Myocardial injury is considered acute if there is a rise and/or fall in cTn values.
Criteria for acute MI (types 1, 2, and 3)
The term "acute MI" should be used when there is an acute myocardial injury with clinical evidence of acute myocardial ischemia, along with the detection of a rise and/or fall in cTn values (with at least one value above the 99th percentile URL) and at least one of the following: • Symptoms of myocardial ischemia; • New ischemic ECG changes; • Development of pathological Q waves; • Imaging evidence of new loss of viable myocardium or new regional wall motion abnormality in a pattern consistent with an ischemic etiology; and • Identification of a coronary thrombus by angiography or autopsy (not for types 2 or 3). Postmortem demonstration of acute atherothrombosis in the artery supplying the infarcted myocardium meets the criteria for type 1 MI. Evidence of an imbalance between myocardial oxygen supply and demand unrelated to acute atherothrombosis fulfills the criteria for type 2 MI. Cardiac death in patients with symptoms suggestive of myocardial ischemia and presumed new ischemic ECG changes, before cTn values become available or abnormal, meets the criteria for type 3 MI.

(continued)

Table 1.1 (continued)

Universal definitions of myocardial injury and MI
Criteria for coronary procedure-related MI (types 4 and 5)
PCI-related MI is termed "type 4a MI." CABG-related MI is termed "type 5 MI." Coronary procedure-related MI occurring ≥48 h after the index procedure is arbitrarily defined by an elevation of cTn values: • ≥5 times the 99th percentile URL for type 4a MI; and • >10 times the 99th percentile URL for type 5 MI in patients with normal baseline values. For patients with elevated preprocedural cTn values, in whom the preprocedural cTn levels are stable (≤20% variation) or falling, the criteria require: • A >five-fold (for type 4a MI) or >ten-fold (for type 5 MI) increase in cTn values; and • A change from the baseline value of >20%. In addition, at least one of the following must be present: • New ischemic ECG changes (this criterion applies to type 4a MI only); • Development of new pathological Q waves; • Imaging evidence of loss of viable myocardium that is presumed to be new and in a pattern consistent with an ischemic etiology; and • Angiographic findings consistent with a procedural flow-limiting complication, such as coronary dissection, occlusion of a major epicardial artery or graft, side-branch occlusion, thrombus, disruption of collateral flow, or distal embolization. The isolated development of new pathological Q waves meets the criteria for type 4a MI or type 5 MI in the context of either revascularization procedure if cTn values are elevated and rising but do not meet the prespecified thresholds for PCI or CABG. Other types of type 4 MI include: • Type 4b MI: Stent thrombosis; and • Type 4c MI: Restenosis, both of which meet the criteria for type 1 MI. Postmortem demonstration of a procedure-related thrombus meets the criteria for type 4a MI or type 4b MI if associated with a stent.
Criteria for prior or silent/unrecognized MI
Any one of the following criteria meets the diagnosis for prior or silent/unrecognized MI: • Abnormal Q waves, with or without symptoms, in the absence of nonischemic causes; • Imaging evidence of loss of viable myocardium in a pattern consistent with ischemic etiology; and • Pathoanatomical findings of a prior MI.

Primary PCI on the RCA and Advanced Cardiogenic Shock Following RV MI as a Consequence of the No-Reflow Phenomenon

Case 1

The patient was a 58-year-old man with a history of chronic cardiac syndrome who had undergone PCI on the LAD and the posterior descending artery 8 years earlier. He was referred to a PCI-capable hospital due to severe, ongoing retrosternal chest pain accompanied by vomiting.

On physical examination, the patient's hemodynamic status was stable. No murmurs or abnormal lung sounds were detected on auscultation. ECG findings were normal. Transthoracic echocardiography revealed normal LV size, mild LV systolic dysfunction, severe hypokinesia in the inferior and inferoseptal segments, normal RV size and function, and no evidence of valvular heart disease. Due to persistent chest pain, the patient was scheduled for primary PCI.

As demonstrated in the angiography (Fig. 1.1), the culprit lesion was located in the mid-segment of the RCA with a TIMI flow grade II. A small RV branch was noted at the site of the thrombotic lesion, along with another tubular lesion in the posterior descending artery just distal to the bifurcation.

After wiring the posterior descending artery and the posterior LV branches with 0.014-inch workhorse guidewires, the posterior descending artery was predilated

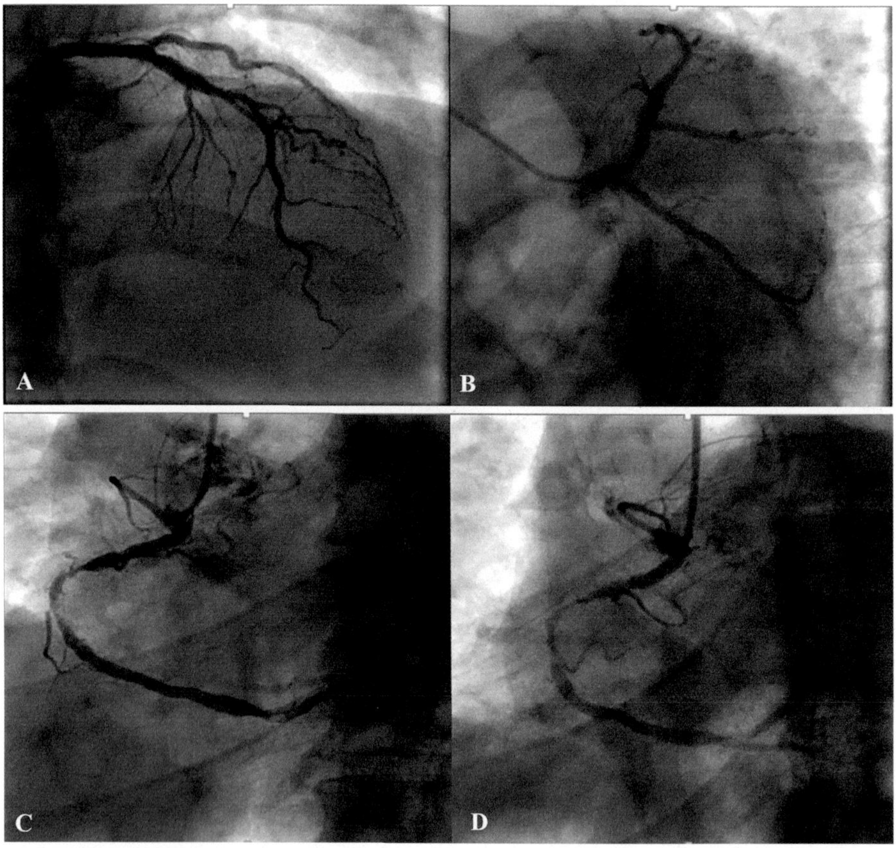

Fig. 1.1 (**a–d**) Selective coronary angiography performed via the right radial artery shows a normal LM, mild in-stent restenosis in the proximal segment of the LAD, and mild stenosis in the ostial and mid-segments of the LCX (**a**, **b**). The RCA is dominant and has a high take-off origin. Additionally, a diffuse thrombotic lesion is observed in the mid-segment of the RCA, with thrombus embolization to the posterior descending artery and the posterior LV branches. A significant tubular lesion is also noted in the posterior descending artery just distal to the crux (**c**, **d**)

and stented using a 2.75 × 28 mm Artemis balloon and an XIENCE stent, achieving good results (Fig. 1.2a). Subsequently, the culprit lesion was directly stented with a 3.5 × 38 mm XIENCE stent and postdilated using a 3.5 mm Apollo balloon (Fig. 1.2b).

Following the postdilation of the mid-RCA, a no-reflow phenomenon occurred. Vasodilator agents, including adenosine, nipride, verapamil, and epinephrine, were administered as distally as possible using a manual thrombosuction catheter. Fortunately, a TIMI flow grade III was achieved after several minutes. However, the patient suddenly developed refractory ventricular fibrillation 5 min later. Despite receiving more than 20 electrical shocks, the situation remained unresolved. Upon reviewing the PCI procedure, it was discovered that a small RV branch had been overlooked. Consequently, the RV branch was wired, and plain old balloon angioplasty was performed during cardiopulmonary resuscitation using an auto-pulse device. Unfortunately, no flow was achieved in the RV branch. Despite 3 h of continuous cardiopulmonary resuscitation, the ventricular tachycardia/ventricular fibrillation rhythm persisted. As a result, a venoarterial extracorporeal membrane oxygenation device was implanted, and the patient was transferred to the ICU. Cardiopulmonary resuscitation was continuously performed for 12 h, but no blood pressure was detected through the arterial line despite massive-volume transfusions and high-dose inotropic agents. The ECG revealed an RV infarction (Fig. 1.3). Transthoracic echocardiography indicated an LV ejection fraction of 40% with severe RV enlargement and dysfunction. The patient was in cardiogenic shock due to an RV MI. Within 24 h, he developed generalized anasarca, icteric appearance (Glasgow Coma Scale III), severe acidosis, and acute liver and renal failure (anuria). On the second day, volume therapy was adjusted under the guidance of central venous and pulmonary capillary wedge pressure (via Swan-Ganz catheter insertion). The patient's blood pressure was 80/60 mm Hg with high-dose inotropes and inodilators, and diuresis commenced. The platelet count dropped to 30,000, and

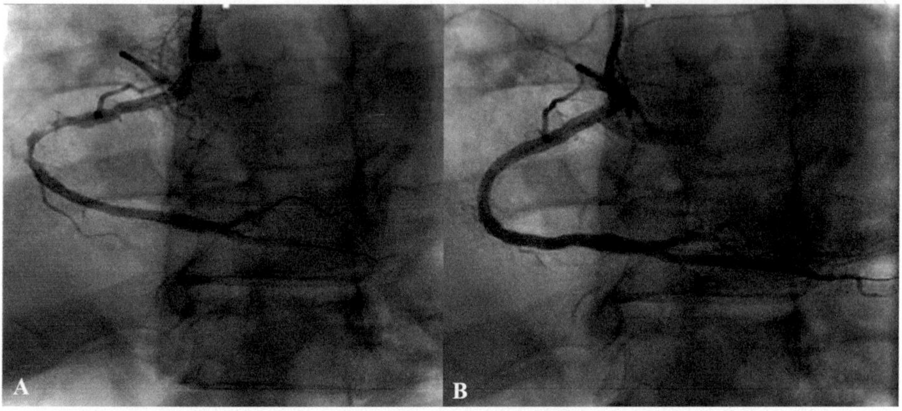

Fig. 1.2 (**a**, **b**). After the posterior descending artery and the posterior LV branches were wired with BMW guidewires, the posterior descending artery was predilated and stented (**a**). Subsequently, the mid-segment lesion was directly stented and postdilated, achieving acceptable final results (**b**)

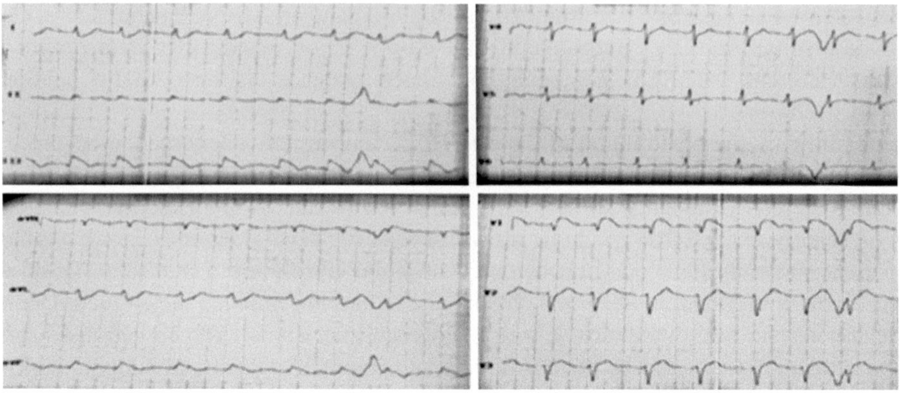

Fig. 1.3 The images showcase the patient's post-PCI ECG

the activated clotting time was 300 s, despite no anticoagulant administration. On the third day, the patient developed acute respiratory distress syndrome or alveolar hemorrhage, accompanied by signs of compartment syndrome in the right lower extremity. Through the transposition of an arterial cannula and conservative management, the mottling resolved, and arterial flow returned to normal. On the fourth day, the patient regained consciousness, and hemodynamic stability was achieved with a low dose of inotropic agents. The ECG appeared normal, and transthoracic echocardiography illustrated findings consistent with previous results. On the fifth day, the venoarterial extracorporeal membrane oxygenation device was removed, and an IABP was implanted. Within 1 week, liver and renal function improved, and the platelet count increased to 150,000. During this period, the patient experienced several episodes of coffee ground vomiting, which were managed conservatively. On the 10th day, he was successfully extubated, and by the 15th day, he regained the ability to walk. The final transthoracic echocardiography showed normal left and right ventricular sizes, mild LV dysfunction, and mild RV dysfunction. After 18 days, the patient was discharged in very good condition.

Discussion

One of the most severe complications during primary PCI is the no-reflow phenomenon, which can occur even in the presence of an epicardial vessel TIMI flow grade III, indicating microvascular obstruction. In other words, a TIMI flow grade II or lower or a normal TIMI flow grade with a TIMI myocardial perfusion grade (TMPG) of 0 to I is diagnosed as no-reflow [4]. The no-reflow phenomenon is more common after angioplasty in degenerated SVGs and thrombus-containing lesions, as well as following coronary atherectomy. Although the exact mechanism of the no-reflow phenomenon remains unclear, clinical and laboratory findings suggest that it may be related to capillary bed embolism, ischemic injury, vascular endothelial dysfunction, oxygen-free radical production, inflammatory reactions, stress responses, calcium overload, vasospasm, and other factors [5]. Despite ongoing advancements in reperfusion techniques for STEMI, the no-reflow phenomenon can still result in a poor prognosis. Routine PCI no-reflow rates, as measured by the TIMI flow grade, range from 1% to 5%, while the incidence rate of this phenomenon in patients with acute MI is between 2.3% and 41% [6]. Nonetheless, even with good TIMI flow grades, myocardial perfusion remains suboptimal in 15%–40% of cases, exhibiting grades 0 to I TMPG [7]. Achieving grade III TMPG is necessary for adequate and effective myocardial perfusion. Reperfusion no-reflow is typically preceded by ischemic cell injury and is often localized to the irreversibly damaged necrotic zone, with potential exacerbation at the time of reperfusion. Reperfusion no-reflow serves as an independent predictor of adverse clinical outcomes following acute MI, irrespective of infarct size, and is linked to heart failure and heightened mortality rates [8]. While RV function recovery is generally faster than LV function recovery, mortality rates following cardiogenic shock are comparable between the two. Preoperative evaluation of no-reflow risk in patients enables early intervention and proactive implementation of a treatment strategy to prevent its occurrence. Various risk factors have been identified to predict the risk of reperfusion no-reflow, including age >65 years, a minimum thrombus burden of 4 points, thrombosuction use, a target lesion diameter ≥3.5 mm, collateral circulation below grade II, elevated blood glucose levels, prolonged total ischemic time, and non-use of angiotensin-converting enzyme inhibitors/angiotensin receptor blockers. Additionally, increased manipulation during PCI, either via predilation or postdilation, has been associated with a higher risk of the no-reflow phenomenon [9]. The clinical manifestations of the no-reflow phenomenon can range from asymptomatic to persistent chest pain, vomiting, unresolved ST elevations, arrhythmia, cardiogenic shock, and cardiac arrest. Currently, no specific intervention effectively prevents this catastrophic event. Nevertheless, the upstream use of glycoprotein IIb/IIIa inhibitors, intracoronary adenosine, nitroprusside, and calcium-channel blockers has shown potential in reducing the likelihood of no-reflow occurrence [10]. In our experience, minimizing manipulation during PCI (e.g., avoiding predilation, employing slow stent deployment, and bypassing postdilation) or opting for deferred stenting in cases with high thrombus grades can help reduce the risk of the no-reflow phenomenon. Although no definitive treatment exists, intracoronary injection of vasodilators, adenosine, nicorandil, calcium-channel blockers, glycoprotein IIb/IIIa inhibitors, intracoronary thrombolytic agents, and epinephrine (particularly when administered through a microcatheter as distally as possible) has been reported to improve microvascular perfusion in refractory cases [11]. Acute MI contributes to 81% of cardiogenic shock cases [12]. Cardiogenic shock complicates 5%–10% of MI cases, most frequently resulting from LV dysfunction, mechanical complications, and, in rare instances, RV MI. The established criteria for cardiogenic shock include frank hypotension (systolic blood pressure <90 mm Hg or a reduction in mean blood pressure of 30 mm Hg), cardiac index <1.8 L/min/m^2 without support or <2.2 L/min/m^2 with support, elevated end-diastolic pressures on the right (>10–15 mm Hg) and/or left (>18 mm Hg) sides of the heart, and evidence of end-organ hypoperfusion [13].

The RV exhibits less adaptability to pressure afterload and higher tolerance to volume overload than the LV, explaining its inability to tolerate significant elevations in pulmonary artery pressure. RV failure leads to RV dilation, displacing the interventricular septum into the LV space, thus compromising LV diastolic filling and exacerbating systemic hypoperfusion [14]. For patients experiencing cardiogenic shock following RV MI, fluid resuscitation under hemodynamic monitoring via a pulmonary artery catheter is crucial, as additional volume infusion may result in pulmonary congestion. Vasopressors should be titrated to achieve a mean arterial pressure goal of >65 mm Hg. Vasopressin, which induces less pulmonary vasoconstriction than norepinephrine, may be considered a first-line vasopressor for patients with cardiogenic shock and acute RV failure (Fig. 1.4) [15]. RV afterload can be

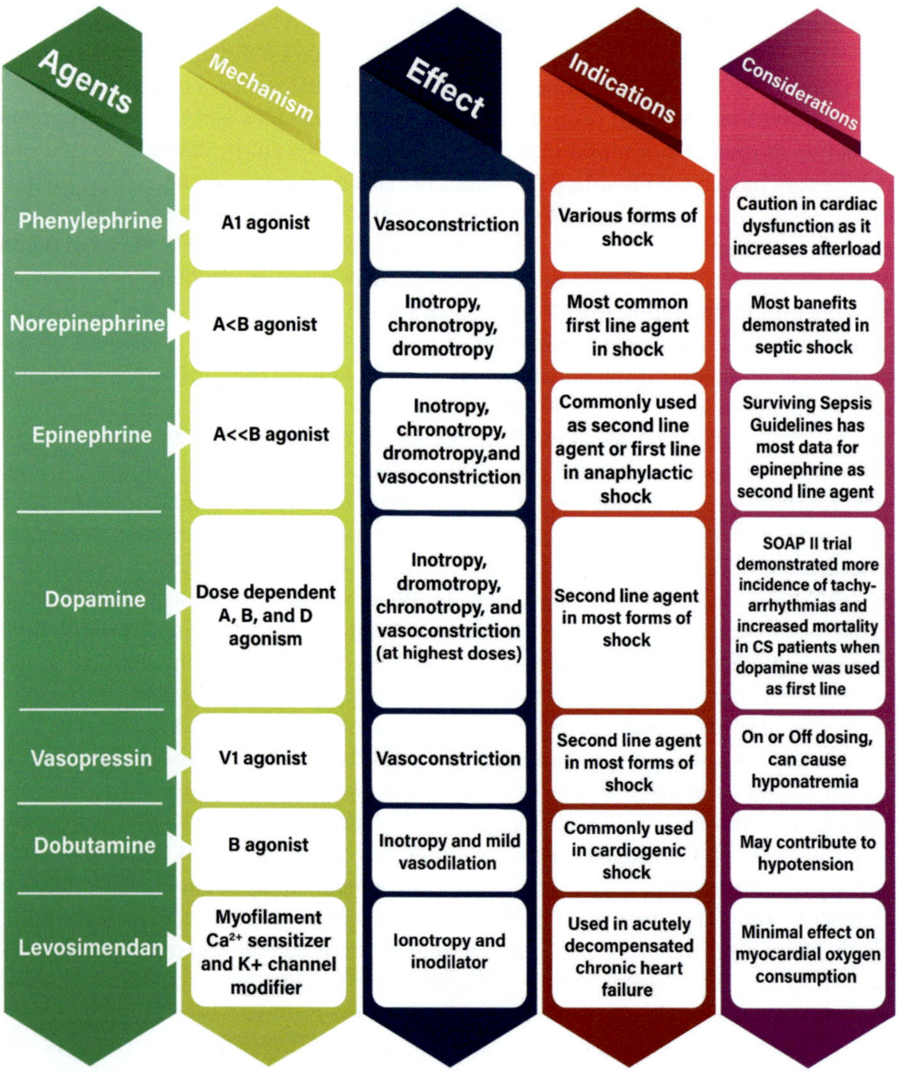

Fig. 1.4 The image presents a summary of systemic vasopressors

reduced using inodilators, phosphodiesterase III inhibitors, or nitric oxide under careful hemodynamic monitoring (Fig. 1.5). In patients who undergo invasive ventilation, the recommended tidal volume is 5–7 mL/kg.

Considering the potential for increased mortality associated with the use of vasopressors and inodilators during the acute MI phase, it is important to maintain a low threshold for implementing MCS, even though these modalities appear to reduce in-hospital mortality but have not yet shown definitive evidence for improved long-term survival. Further information about MCS devices can be found in section "Case 6".

Conclusions

Although slow-flow and no-reflow phenomena are potential complications during primary PCI in STEMI, identifying high-risk patients and employing preventive strategies can significantly improve outcomes in most cases. However, cardiogenic shock may still occur, leading to an in-hospital mortality rate exceeding 50%. Despite these challenges, meticulous monitoring and intensive care can sometimes result in successful management and improved patient outcomes, as demonstrated in our case.

Agent	Mechanism	Route	Side Effect
Nitric Oxide	↑cGMP	Inhaled	Blurred vision, confusion,sweating, malaise, headache, bleeding
Milrinone	Phosphodiesterase 3 inhibitor	Intravenous	Bleeding, hypotension, chest pain, tremors, bronchospasm, hypokalemia
Prostacyclin	↑cAMP, ↑K, ↓ET-1, and ↑K+	Inhaled or intravenous	Bleeding, arrhythmias, diarrhea, edema, fevers, chills
Dobutamine	B agonist	Intravenous	Hypotension, tachyarrhythmia, headache, thrombocytopenia

Fig. 1.5 The image presents a summary of vasoactive agents within the pulmonary circuit

Primary PCI on Unprotected LM Lesions

Case 2

A 65-year-old man with a medical history of uncontrolled hypertension, poorly controlled diabetes mellitus, and chronic coronary syndromes (previous PCI on the LAD) presented to the emergency department with severe retrosternal chest pain, diaphoresis, and vomiting lasting 1 h. Upon examination, the patient's hemodynamic status was stable, with a systolic blood pressure of 100 mm Hg, a heart rate of 100 beats per minute, a respiratory rate of 18 breaths per minute, and an oxygen saturation level of 90%. Cardiac examination revealed a mid-systolic murmur in the aortic zone. Pulmonary examination showed fine crackles midway up the chest, and all pulses were normal and symmetric. ECG demonstrated diffuse ST-segment depression in the inferior leads and leads V_2–V_6 and ST-segment elevation in leads AVR and V_1.

Bedside echocardiography revealed normal LV size, severe LV systolic dysfunction (ejection fraction = 30%), and global hypokinesia. No significant valvular heart disease was observed. Given the patient's symptoms and ECG findings, primary PCI was scheduled.

Selective coronary angiography performed through the right radial artery revealed significant thrombotic distal LM stenosis (Fig. 1.6a). Immediately following the initial contrast injection, the patient experienced a cardiac arrest with ventricular fibrillation rhythm. After prompt cardioversion, a full dose of heparin was administered. Subsequently, the LAD and the LCX were wired with 0.014-inch workhorse guidewires.

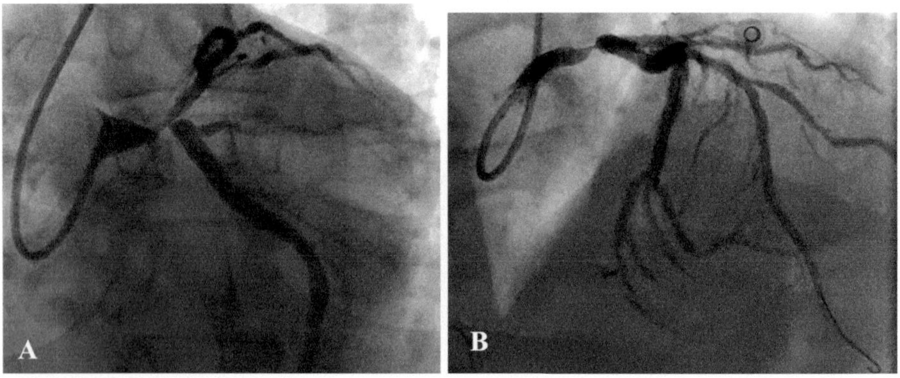

Fig. 1.6 (**a**, **b**) The images illustrate a significant distal LM lesion (sub-total) and significant stenosis of the ostioproximal segment of the LAD, with a patent previous stent in the mid-segment (**a**, **b**). The dominant LCX appears normal

Given the dominance of the LCX, direct stenting of the LM-LCX was performed using a 4.5 × 28 XIENCE stent, followed by the proximal optimization technique with a 5 × 8 Apollo balloon. After rewiring the LAD, the proximal strut was dilated using a 1.5 × 15 Emerge balloon. Subsequently, the LAD was stented with a 3.5 × 28 Ultimaster stent, and the final kissing balloon technique was executed using a stent balloon (the LAD) and a 3.5 × 15 Emerge-NC balloon, resulting in satisfactory final outcomes (the TAP technique) (Fig. 1.7).

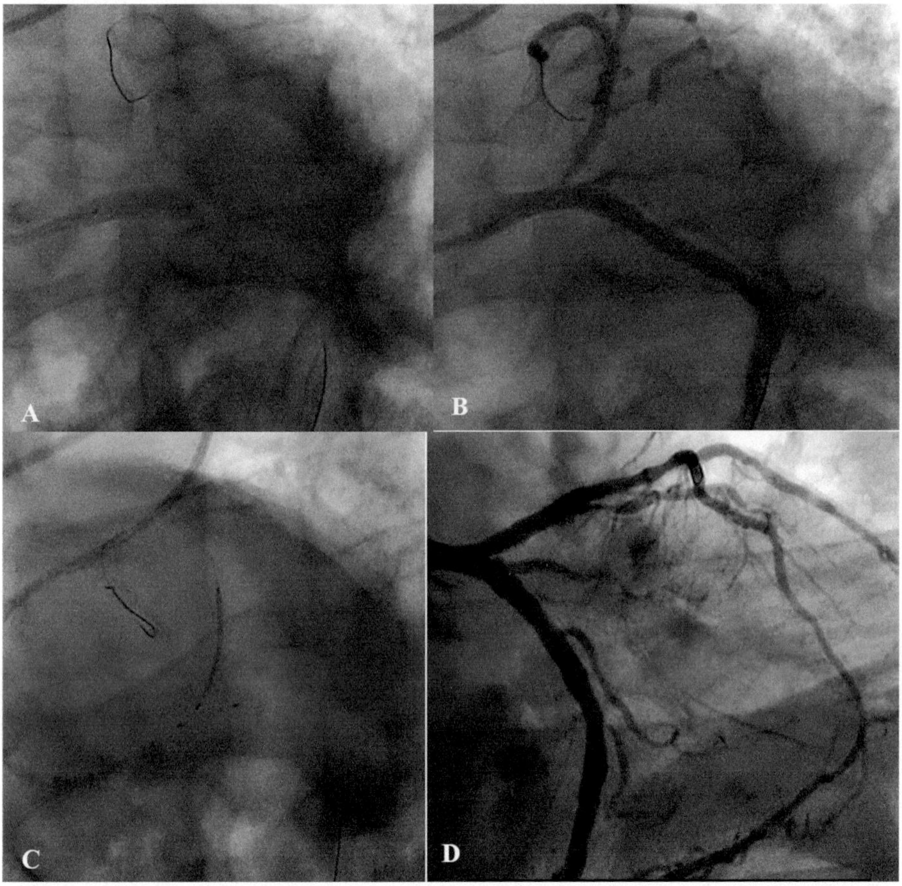

Fig. 1.7 (**a**–**d**). The images demonstrate the stenting of the LM–LCX using a 4.5 × 28 XIENCE stent (**a**). The proximal optimization technique was then performed with a 5 × 8 Apollo balloon (**b**). After rewiring the LAD, the proximal stent cell was dilated using a 1.5 × 15 Emerge balloon. The TAP technique was employed to stent the LAD with a 3.5 × 28 Ultimaster stent (**c**). Finally, the kissing balloon technique was carried out using a stent balloon and a 3.5 × 15 Emerge-NC balloon, achieving satisfactory final results (**d**)

Case 3

A 53-year-old man was referred to the catheterization laboratory from the exercise stress test department while undergoing cardiopulmonary resuscitation. In the days preceding the event, the patient's primary complaint was exertional dyspnea (New York Heart Association functional class II–III) without chest discomfort. Consequently, a cardiologist ordered a noninvasive test (the exercise stress test) for further evaluation. During the first stage of the test, the patient experienced dyspnea and dizziness, with monitoring ECG revealing diffuse down-sloping ST-segment depression and sudden-onset polymorphic ventricular tachycardia. He was transferred to the catheterization laboratory during ongoing cardiopulmonary resuscitation, and angiography was initiated (Fig. 1.8).

At the end of the procedure, the patient developed refractory ventricular fibrillation and ventricular tachycardia despite multiple administrations of antiarrhythmic agents. Additionally, he experienced cardiogenic shock despite the use of high-dose inotropic and vasopressor agents. After 30 min, due to persistent hemodynamic instability, a venoarterial extracorporeal membrane oxygenation (VA-ECMO) device was implanted in the catheterization laboratory. The patient was subsequently admitted to the ICU. Seven days later, he was discharged in good clinical condition.

Fig. 1.8 (**a**–**e**) During cardiopulmonary resuscitation, the patient was transferred to the catheterization laboratory. His baseline rhythm was polymorphic ventricular tachycardia. While the anesthesiology team focused on stabilizing the patient's hemodynamic status, the first operator obtained arterial access via the common femoral artery. Heparin was administered immediately, and the first angiographic injection revealed a total occlusion of the LM at the ostial segment (**a**). With the aid of a 0.014-inch guidewire, the LAD and the LCX were wired using Pilot-50 and SION blue guidewires, respectively (**b**). After balloon dottering of the LAD, direct stenting of the LM–LAD was performed using a 4 × 32 mm Ultimaster stent. The final proximal optimization technique was then carried out using a 5 × 8 mm AccuForce–NC balloon (**c**), achieving acceptable final results (a final TIMI flow grade III) (**d**). The final angiographic image demonstrated that the RCA was completely occluded and was retrogradely filled via the left coronary system (**e**)

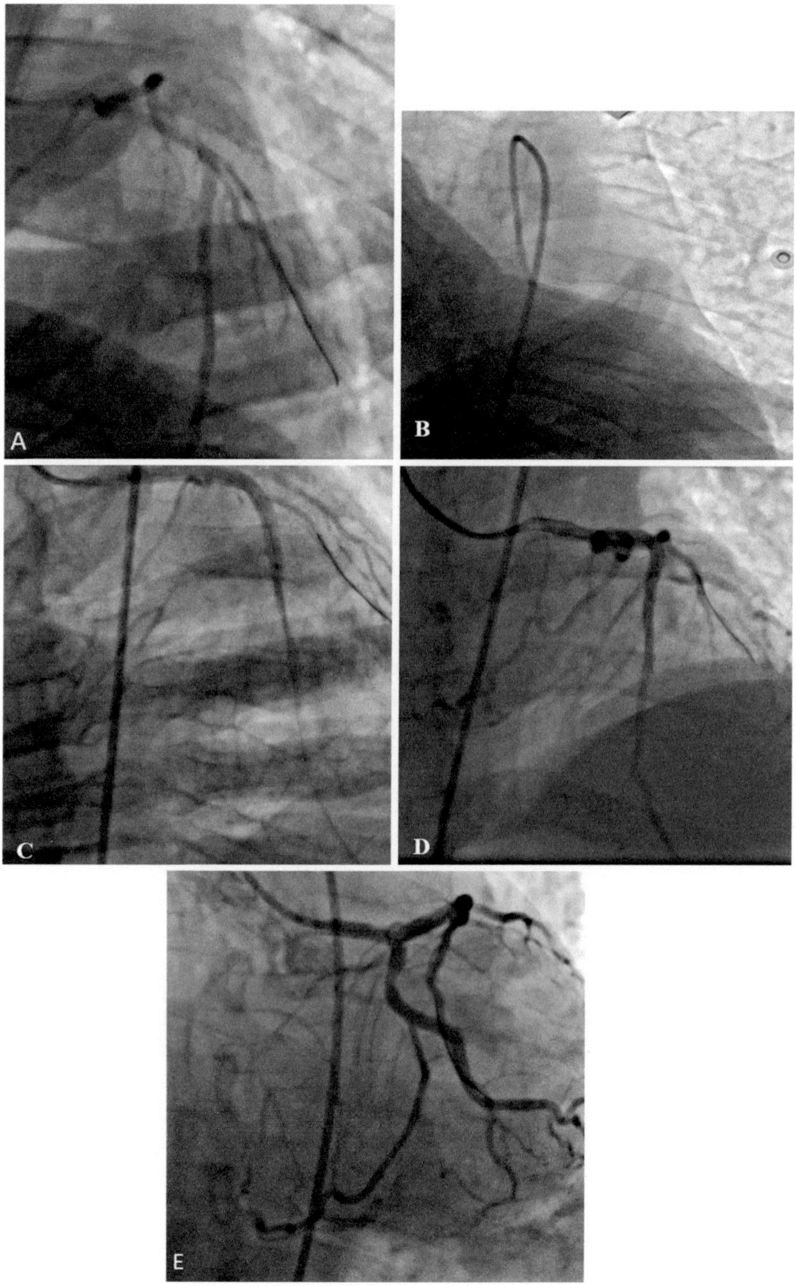
A
B
C
D
E

Case 4

A 70-year-old woman with a history of insulin-dependent diabetes mellitus, hypertension, and cerebrovascular accident presented to a PCI-capable center with severe retrosternal chest pain. ECG revealed ST-segment elevation in leads V_1–V_4, aVR, and aVL, with reciprocal ST-segment depression in the inferior leads. On physical examination, she was irritable, but her hemodynamic status was stable. The patient was scheduled for primary PCI.

Selective coronary angiography performed via the right common femoral artery showed a normal and dominant RCA, which retrogradely filled the LAD up to the ostial segment. Selective LM injection revealed a normal LM and a total thrombotic occlusion of the LAD at the ostial segment, with some thrombus extension into the LM. The LCX was normal (Fig. 1.9a, b).

Given the time-sensitive nature of the case, the LAD and the LCX were immediately wired using 0.014-inch workhorse guidewires. The operator decided to perform direct stenting of the LAD precisely at the ostioproximal segment (Fig. 1.9c). However, the patient suddenly developed ventricular fibrillation and suffered cardiac arrest. After several minutes of cardioversion, angiography revealed that the thrombus had protruded into the LM and the LCX, compromising the flow of the latter (Fig. 1.9d).

The LM–LAD was directly stented using a 4 × 18 mm XIENCE stent. The final angiographic injection showed good results, with a TIMI flow grade III in both the LAD and the LCX. Due to the patient's critical condition, the final proximal optimization technique was postponed to a later session.

After 18 h of eptifibatide infusion, the patient was discharged on day 3 without complications.

Case 5

A 35-year-old man presented to a PCI-capable center with a 1-h history of severe retrosternal chest pain. His hemodynamic status was stable, and no significant findings were noted on physical examination. The ECG demonstrated sinus tachycardia with diffuse ST-segment elevation in leads V_1–V_3 and reciprocal ST-segment depression in the inferior leads. Bedside echocardiography revealed normal LV size, moderate LV systolic dysfunction (ejection fraction = 35%), and severe hypokinesia of the anterior, anteroseptal, and anteroapical segments. The RV was of normal size and function. Mild-to-moderate mitral regurgitation was also observed.

Based on these findings, the patient was scheduled for primary PCI.

Selective coronary angiography via the right radial artery revealed a normal LM, a massive thrombus at the distal LM extending to the LAD, and some thrombus embolization in the first septal and diagonal branches and the distal LAD (Fig. 1.10b).

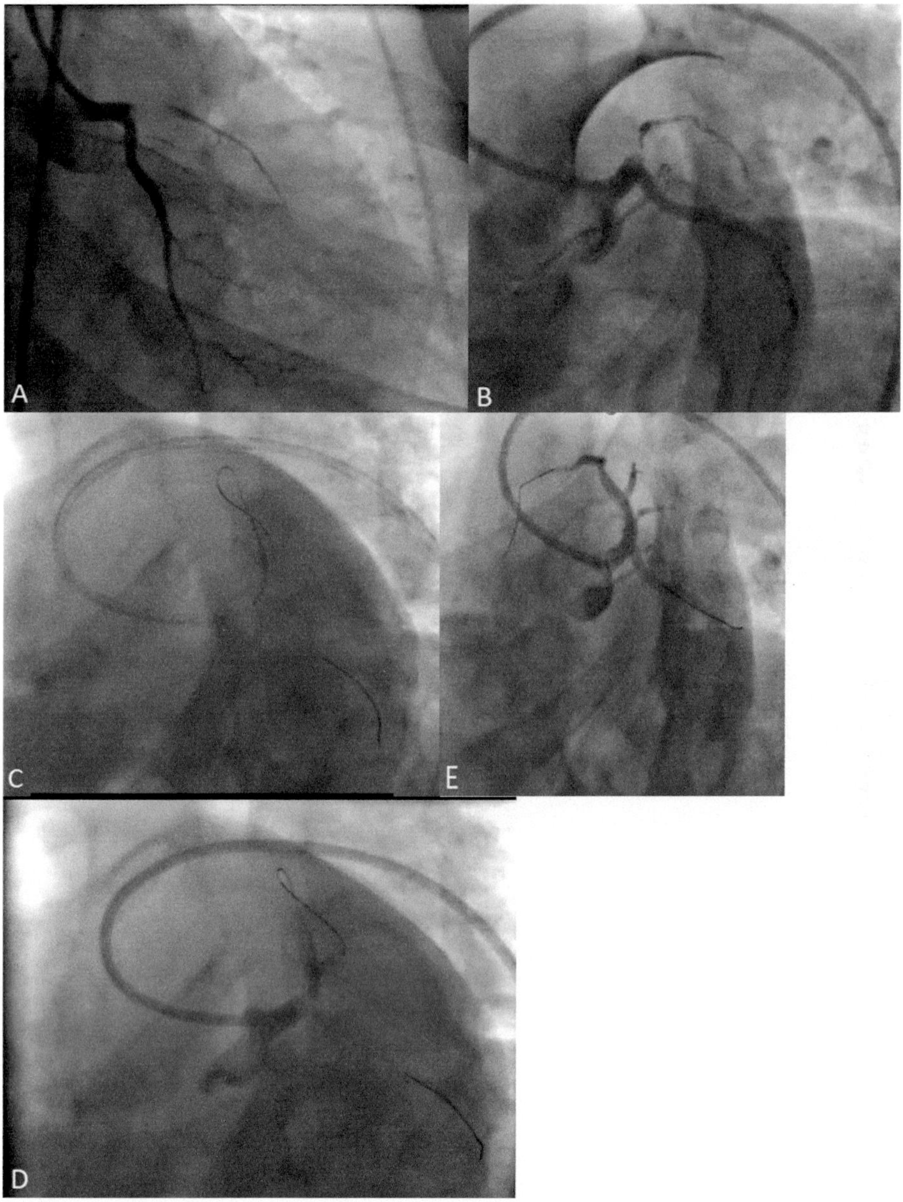

Fig. 1.9 (**a–e**) The LAD was totally thrombotic with some thrombus protrusion into the LM (**a**, **b**). After wiring the LAD and the LCX using 0.014-inch workhorse wires, direct precise stenting of the ostium part of the LAD was done using a 3.5 × 23 XIENCE stent (**c**). At this moment, the patient's hemodynamic status deteriorated. Following cardioversion, angiography showed the protrusion of the thrombus into the LM and the LCX. (The TIMI flow grade of the LCX was 0) (**d**). Immediate LM–LAD stenting using a 4 × 18 XIENCE stent demonstrated a good final result

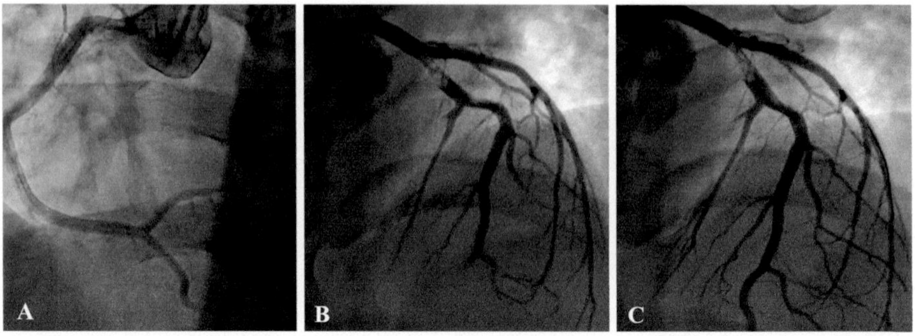

Fig. 1.10 (**a**–**c**) The RCA was normal and dominant (**a**). In the distal part of the LM, there was a large bulk of a thrombus extending to the LAD with clot embolization to the first septal and diagonal branches and the distal part of the LAD (**b**). After wiring the LAD with a 0.014-inch workhorse wire, the final TIMI flow grade of the LAD was III (**c**)

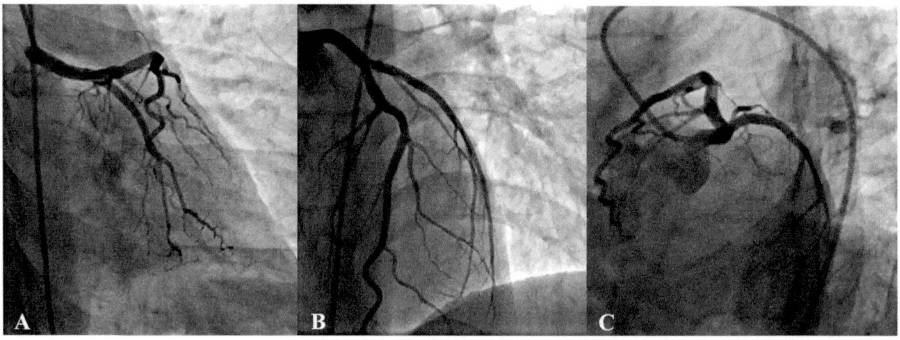

Fig. 1.11 (**a**–**c**) The LM was elongated and normal (**a**). No evidence of a thrombus was visualized, but the distal part of the LM and the ostial part of the LAD had eccentric lesions (**b**, **c**). Intravascular ultrasound confirmed fibrofatty plaque erosion at the distal part of the LM and the ostial part of the LAD without evidence of complete resolution of the LM-LAD thrombus (**a**, **b**). Intravascular ultrasound imaging exhibited a mild, eccentric fibrofatty plaque at the distal part of the LM and the ostial part of the LAD with plaque eruption and a minimal intraluminal thrombus (the LM's minimal lumen area = 15 mm^2 and the LAD's minimal lumen area = 9 mm^2) without visible dissection flaps or calcification

As the LAD was cut off at the distal part, it was wired using a 0.014-inch workhorse wire, and intracoronary integrilin was injected. The final angiography showed reperfusion of the LAD and the diagonal branch (Fig. 1.10c). Following the normalization of the ST segments, the procedure was terminated. Highly intensive antiplatelet therapy, including ASA, clopidogrel, and glycoprotein IIb/IIIa inhibitors (integrilin), was started for the patient for 48 h, and then enoxaparin (80 SQ BID) was prescribed for another 2 days.

The results of follow-up coronary angiography and intravascular ultrasound imaging 4 days later are presented in Fig. 1.11.

As the patient lacked traditional cardiovascular risk factors, a complete evaluation for thrombophilia factors, lipid profile, lipoprotein(a), homocysteine tests, and vasculitis markers was ordered. The patient was discharged with low-dose aspirin (81 mg) and ticagrelor (90 mg BID).

Discussion

We herein reported four cases of STEMI with the LM as the culprit lesion. Given that the LM supplies 80% of the myocardium, stenosis at this location can cause devastating consequences, including malignant arrhythmias and sudden cardiac death. While some of the patients had a relatively stable hemodynamic status, one patient was referred to the catheterization laboratory during cardiopulmonary resuscitation. STEMI with unprotected LM culprit lesions accounts for approximately 1%–4% of primary PCI cases. Cardiogenic shock is the most common presenting symptom, which is associated with higher risks of major adverse cardiac events and mortality rates, even when treated with prompt reperfusion therapy. Acute occlusion of the LM stem, resulting in STEMI, is a highly dangerous condition associated with severe adverse effects. Patients can be managed with primary PCI or CABG, but in patients who are too high-risk to tolerate delay, primary PCI is often favored. Despite the gravity of this issue, the literature contains limited data [16–18].

Patients with LM STEMI are usually older and frequently have multiple cardiovascular and non-cardiovascular comorbidities [5]. The in-hospital and 30-day mortality rates for these patients hover around 50%. However, those who survive have a significantly reduced risk of cardiovascular events in the following months. Additionally, mortality rates have decreased dramatically in the past decade, compared with previous reports, owing to advanced medications, instruments, assist devices, and ICU care [19].

Multiple factors contribute to mortality, including the pattern of presentation (cardiogenic shock is the strongest predictor of mortality, not pre-PCI cardiac arrest), timely interventions, collateral circulation, dominance of the right or left coronary system, baseline and final TIMI flow grades, final blush grades, operator experience, and center volume. A higher likelihood of favorable outcomes is associated with the presence of collateral circulation, better initial and final TIMI flow grades, RCA dominance, and expert interventionists [20–22].

A registry study, ATOLMA (Primary Angioplasty in a Catastrophic Presentation: Acute Left Main Coronary Total Occlusion—The ATOLMA Registry), compared patients with totally occluded LM stems (zero antegrade flow) with those having subtotal occlusion of the LM. Patients with totally occluded LM stems were found to be in a far more critical condition than those with subtotal LM occlusions, with a higher prevalence of cardiogenic shock (41%) and utilization of assist devices (41% IABPs and 6.5% venoarterial extracorporeal membrane oxygenation). Still, risk factors for mortality and morbidity in patients with totally occluded LM stems were similar to those of other LM STEMI patients, comprising the presence of collateral flow, post-PCI TIMI flow grades, and RCA dominance. Notably, initial presentation with cardiac arrest was found to have limited influence on long-term outcomes [23].

The Asia-Pacific Left Main ST-Elevation Registry (ASTER) found that patients with acute STEMI due to an unprotected LM culprit lesion had a heightened risk, morbidity, and mortality than other acute STEMI patients. While primary PCI is a viable treatment option for this high-risk group, the in-hospital mortality rate remains elevated, particularly for patients with comorbidities such as advanced age, diabetes mellitus, chronic kidney disease, and failure to restore epicardial flow post-PCI. Nevertheless, long-term outcomes appear promising, provided patients successfully navigate the early PCI phase and in-hospital period [24].

Conclusions

In this report, we describe four patients with STEMI and culprit lesions in the unprotected LM, which is considered extremely high risk, potentially resulting in cardiogenic shock or life-threatening arrhythmias. Similar to other STEMI cases, total ischemic time is a critical prognostic factor for patients with this condition. Additional significant prognostic factors include cardiogenic shock, absence of collateral vessels, and low final TIMI flow grades. The final TIMI flow grade appears to be the most important prognostic factor, underscoring the importance of direct stenting to minimize no-reflow and microvascular obstruction. Appropriate case selection for LM primary PCI (immediate or deferred stenting) is essential, with consideration of intracoronary imaging, the underlying thrombosis cause (section "Case 4"), and the operator's experience (section "Case 3"). Furthermore, teamwork is critical for favorable outcomes. An effective team should comprise a cardiologist, interventional cardiologist, cardiac surgeon, and anesthesiologist.

Primary PCI on Vein Graft and Staged PCI on the Native Vessel

Case 6

A 65-year-old woman, a known case of advanced diabetes mellitus and post-CABG status (10 years previously), presented with complaints of retrosternal chest pain and a burning sensation in the right arm of 2 weeks' duration, despite full guideline-directed medical therapy. When the patient arrived at the emergency department at 1:00 AM, she developed sudden onset cardiac arrest (asystole rhythm), for which she received cardiopulmonary resuscitation for about 2 h. An ECG demonstrated diffuse ST-segment depression in the precordial leads. Transthoracic echocardiography showed moderate LV enlargement, severe LV systolic dysfunction (ejection fraction = 10%), hypokinesia in the anterior, anterolateral, and anteroseptal regions, normal RV size, mild-to-moderate RV systolic dysfunction, severe mitral regurgitation with a central jet, and high pulmonary artery pressure (65 mm Hg). As the patient was intubated and her hemodynamic status remained unstable (blood pressure = 60/40 mm Hg, sinus tachycardia, decreased consciousness, and no urine output), she was scheduled for an immediate invasive approach after 1 h, despite receiving high-dose norepinephrine and dobutamine infusions. Notably, the surgeon refused to perform redo-CABG. After a comprehensive discussion with the patient's family regarding the risks and benefits of the procedure, they agreed to the invasive strategy. Consequently, the patient was considered for an immediate invasive approach.

Within 2 h of her presentation, the patient was transferred to the catheterization laboratory, where a cardiac surgeon, an anesthesiologist, and a heart failure team were present. Short sheaths (7 Fr) were inserted through the right and left common femoral arteries. While the second operator inserted the IABP via the left access, selective coronary angiography from the right access demonstrated the following: the LM was highly calcified and significantly stenotic in the distal part; the LAD was highly calcified and completely occluded in the mid-portion; the LCX was a small branch with significant stenosis in the mid-portion; the RCA was dominant and completely occluded from the proximal portion; the SVG on the LCX was patent; the SVG on the RCA was occluded; the left internal mammary artery on the LAD exhibited "the string sign" (nonfunctional); and the SVG on the LAD was totally thrombotic and occluded from the ostial part.

Immediately, the wiring of the vein graft on the LAD was performed using a 0.014-inch workhorse wire, followed by direct stenting of the ostioproximal part of the graft, which was highly calcified. This procedure resulted in 20% final residual stenosis (Fig. 1.12). The last injection revealed that the LAD filled the RCA retrogradely.

The patient experienced a very complex hospital course. Eight hours after the IABP implantation, she developed acute limb ischemia. Consequently, the balloon pump was removed, and she was managed conservatively. She suffered from multi-organ failure, massive upper gastrointestinal bleeding, and pneumosepsis, for which she received intensive care for 1 month. Ultimately, the patient was discharged in good condition. The final echocardiography showed an LV ejection fraction of 40%–45% with moderate mitral regurgitation.

As the bypass graft on the LAD was of the venous type, the LAD filled the RCA, and the final PCI result was suboptimal, the patient was scheduled for elective PCI on the LM to the LAD. After 2 months, the patient was admitted for PCI on the LM to the LAD. Since both the LM and the LAD were highly calcified, the decision was made to perform a rotational atherectomy. After wiring the LAD with a 0.014-inch workhorse wire, the wire was replaced with a floppy Rota wire using a microcatheter. Subsequently, direct rotablation (burr = 1.5) of the LM and the LAD was performed. Next, predilation with non-compliant (NC) balloons was conducted, followed by stenting from the LM to the LAD using a 3.5 × 38 mm XIENCE stent and a 3 × 38 mm XIENCE stent. The final proximal optimization technique was

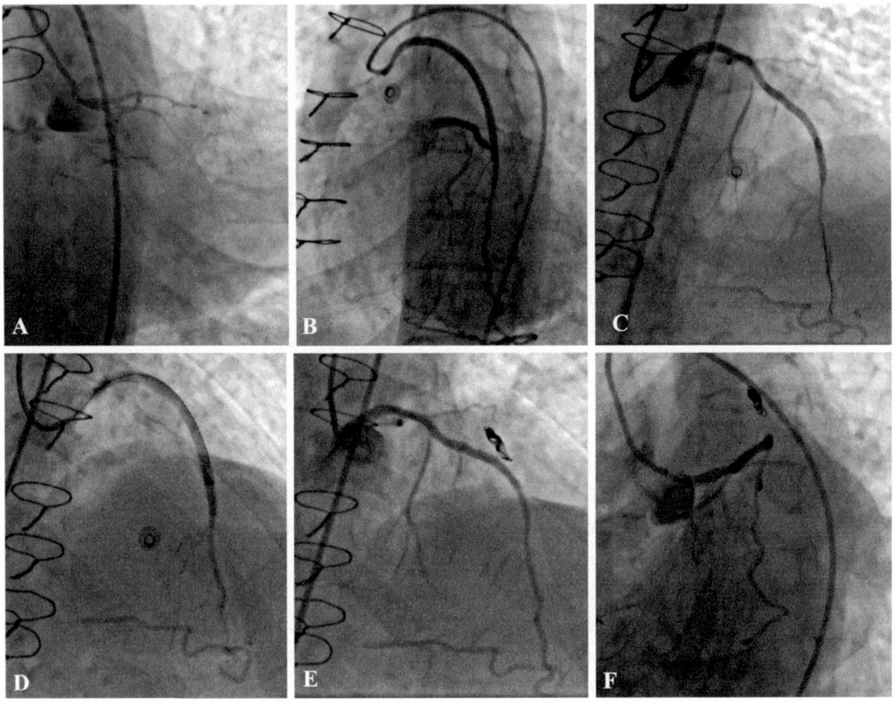

Fig. 1.12 LM injection showed a significant lesion in the ostial part. The LAD was occluded after the septal branch (**a**). Selective SVG injection revealed a patent ostioproximal stent, and the LAD was filled with good run-off (**b**). After the rotablation and preparation of the LM and the LAD, stenting was performed. Nonetheless, the competitive flow between the SVG and the native vessel resulted in poor filling of the LAD (**c**). Selective SVG injection demonstrated that the LAD was filled much better via the SVG (**d**), leading to the decision to coil the SVG. After coiling, selective LM injection showed significantly improved filling of the LAD and the RCA (**e**, **f**)

performed with a 4.5 × 8 mm Apollo balloon. The final injection demonstrated competitive flow between the LAD and the previously stented vein graft, which prevented adequate filling of the distal portion of the LAD. The patient experienced severe chest pain accompanied by ECG changes, including diffuse ST-segment depression in the precordial leads.

Consequently, a decision was made to coil the vein graft. Following the wiring of the vein graft and the advancement of a microcatheter over it, the proximal part of the graft was coiled using two detachable EV3 coils (6 × 30 mm). After 15 min, LM injection showed improved filling of the LAD and RCA (Fig. 1.12).

Discussion

For further discussion regarding the approach to post-CABG lesion revascularization, please refer to Chap. 6 Part I.

Eptifibatide (Integrilin)-Induced Thrombocytopenia and Thrombosis

Case 7

A 72-year-old man, with a history of coronary artery bypass graft surgery, presented at a PCI-capable hospital with severe retrosternal chest pain of 2 h duration. Physical examination revealed a stable hemodynamic state without abnormal findings, and ECG confirmed the diagnosis of posterolateral STEMI. As a result, the patient was promptly scheduled for primary PCI (Fig. 1.13).

The patient immediately received 325 mg of ASA and 600 mg of clopidogrel.

Selective coronary angiography revealed the following: a normal LM; an LAD that was completely occluded at the ostium but was filled via the left internal mammary artery with good run-off; a significant lesion in the mid-portion of the LCX with good run-off; an obtuse marginal artery that was completely occluded at the ostium; a dominant RCA that was completely occluded at the ostium but was filled via the left internal mammary artery; an occluded SVG to the RCA; a completely thrombotic-occluded SVG to the obtuse marginal artery, which was occluded in the proximal portion; and a patent left internal mammary artery graft to the LAD. PCI utilizing two stents was performed on the SVG on the obtuse marginal. During the procedure, 50 U/kg of unfractionated heparin and 180 μg/kg of two STAT boluses of integrilin followed by a 2 μg/kg/min infusion were administered. The patient was

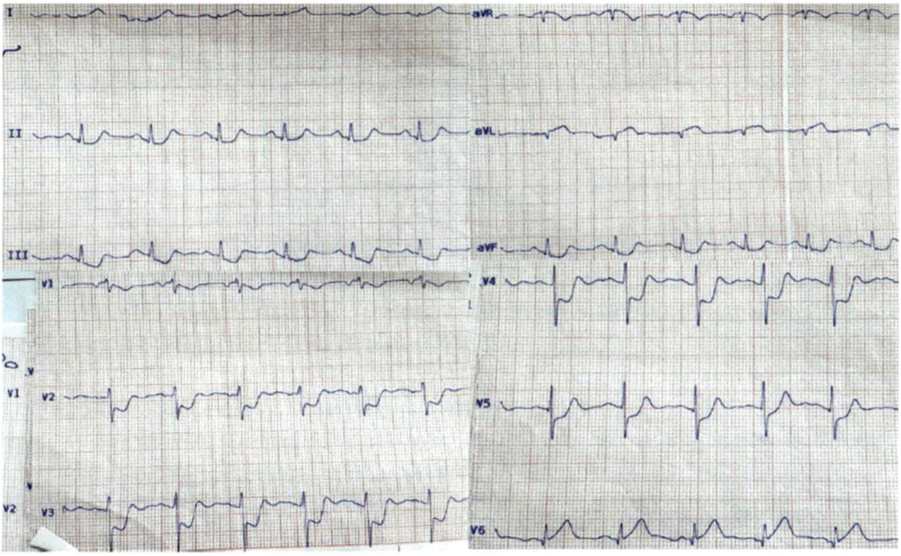

Fig. 1.13 The patient's initial ECG shows posterolateral STEMI

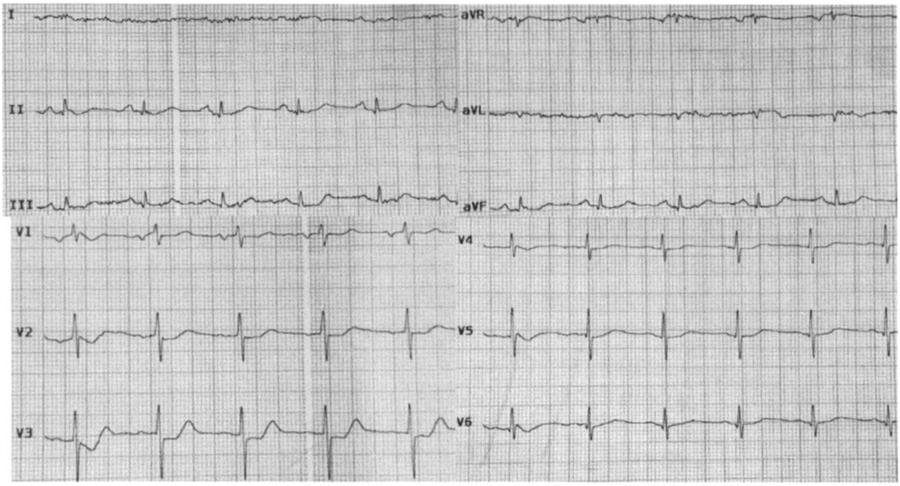

Fig. 1.14 The patient's ECG following PCI shows the reversal of the ST segments in the precordial, I, and aVL leads

then transferred to the CCU in a stable condition. Post-procedure, the patient received a daily regimen of 75 mg of clopidogrel, 80 mg of ASA, and a 12-h infusion of 2 μg/kg/min of integrilin. The post-PCI ECG is shown in Fig. 1.14.

Initial laboratory results showed a white blood cell count of 12,000, a blood urea nitrogen level of 18, a hemoglobin level of 14, a platelet count of 226,000, and a serum creatinine level of 1.2. Three hours after PCI, the platelet count was rechecked, and profound thrombocytopenia (31,000) was reported. The platelet count was rechecked in an EDTA-free tube, and a peripheral blood smear revealed a count of 24,000. Immediately, integrilin was discontinued, but ASA and clopidogrel were continued as there was no obvious bleeding. Additionally, dexamethasone (4 mg BID) was prescribed. Eight hours later, the platelet count increased to 40,000, and 24 h later, a peripheral blood smear showed no evidence of hemolytic anemia. The anti-PF4 antibody test was negative. Forty-eight hours after PCI, the patient complained of severe chest pain, and ECG revealed a posterior wall STEMI (Fig. 1.15). Given the patient's high bleeding risk, medical therapy was initiated, and heparin (12 U/kg/h) was started (platelet count = 120,000).

On the third day, the platelet count was 145,000, and the patient was pain-free. ECG changes had also resolved (Fig. 1.16).

On the fourth day of hospitalization, despite ongoing heparin infusion, the patient developed left leg pain and swelling. Color Doppler sonography revealed acute thrombosis in the popliteal vein. At that time, the patient's platelet count was 220,000. Transthoracic echocardiography revealed an RV of normal size with mild

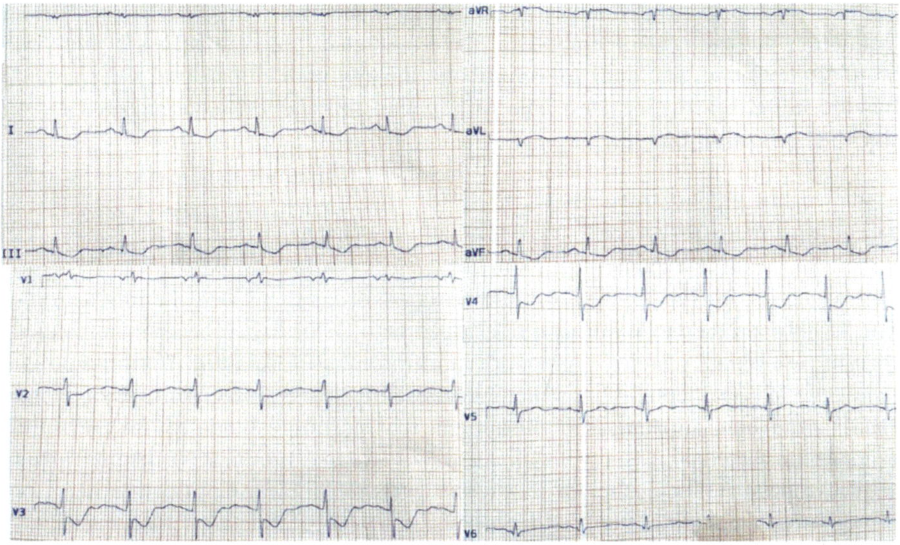

Fig. 1.15 The patient's ECG showcases a diffuse ST-segment depression in the precordial leads, in favor of a posterior wall STEMI

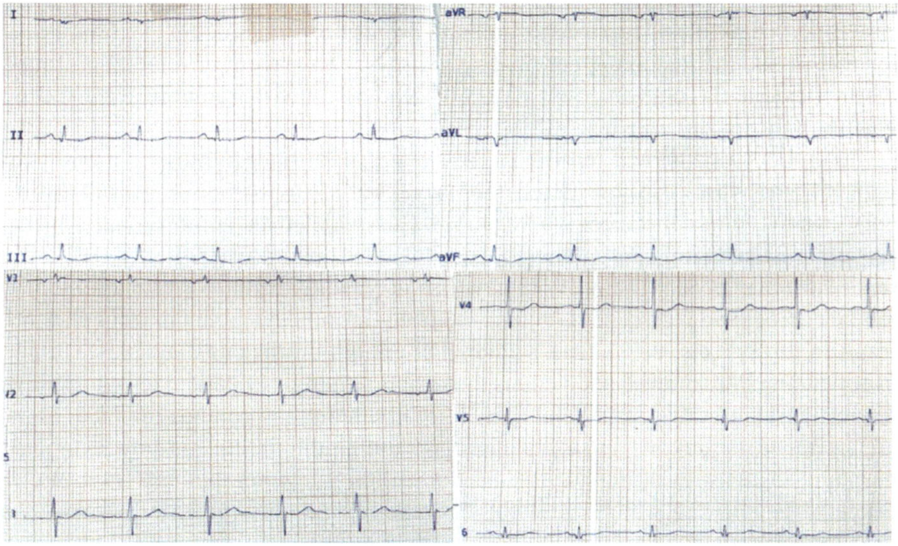

Fig. 1.16 The ECG shows the reversal of the ST segment in the precordial leads

systolic dysfunction. Given the patient's stable hemodynamic status, no further evaluation for pulmonary embolism was conducted. Heparin infusion and ASA were discontinued, and dual antithrombotic therapy with clopidogrel and rivaroxaban (15 mg BID) was initiated. The patient was ultimately discharged on the sixth day in good condition.

Discussion

Glycoprotein IIb/IIIa (GPIIb/IIIa) inhibitors are drugs that block the interaction of von Willebrand factor (VWF) and fibrinogen with GPIIb/IIIa receptors on platelets, thereby inhibiting platelet aggregation and thrombus formation. Eptifibatide is a small cyclic heptapeptide that structurally resembles arginine-glycine-aspartic acid (RGD), the binding domain of fibrinogen that normally binds to the GPIIb/IIIa receptor on platelets. It is a well-known GPIIb/IIIa inhibitor that has been shown to reduce mortality and ischemia in patients with ACS [25]. Eptifibatide has renal elimination, a half-life of approximately 2.5 h, and its antiplatelet effects persist for approximately 4–6 h [26]. One rare adverse side effect of eptifibatide is thrombocytopenia, ranging from mild (2.8%–4.9%) to severe (0.1%) [27, 28]. A handful of case reports have documented profound thrombocytopenia (platelet count <20,000) as early as 4 h after the initiation of eptifibatide, with platelet counts returning to normal within hours to 2–4 days after discontinuation of the drug [29–31]. Rare case reports have also described thrombotic events occurring alongside thrombocytopenia in patients receiving eptifibatide [32–34]. Our patient developed profound thrombocytopenia (platelet count <20,000), leading to two thrombotic events: deep vein thrombosis and transient STEMI.

Thrombocytopenia and thrombosis associated with Integrilin use is a rare phenomenon that should be considered after ruling out other potential causes of thrombocytopenia. In rare instances where multiple antiplatelet and antithrombotic agents are administered, mild thrombocytopenia (platelet count <70,000 or a 10%–30% reduction) may occur within 24–48 h, which is typically benign. While thrombocytopenia associated with type II heparin-induced thrombocytopenia usually does not result in thrombosis, our patient's platelet count did not fit this diagnosis. Despite experiencing both arterial and venous thrombotic events, our patient tested negative for anti-PF4 antibodies, ruling out type II heparin-induced thrombocytopenia. There are isolated case reports linking clopidogrel use to thrombotic thrombocytopenic purpura [35]. However, our patient did not exhibit a low hemoglobin level or evidence of hemolysis, which ruled out thrombotic thrombocytopenic purpura as a probable diagnosis.

Integrilin-induced thrombocytopenia and thrombosis is a rare adverse effect of integrilin with an onset between 4 and 8 h after initiation of the drug. Fewer than 0.3% of patients develop profound thrombocytopenia, which typically resolves within hours to a few days after discontinuation of the drug. The pathophysiology of integrilin-induced thrombocytopenia and thrombosis remains poorly understood, and it is unclear if the previous use of integrilin is a risk factor.

Gao et al. [36] suggested that integrilin-induced thrombocytopenia and thrombosis could be an immunologic process where antibodies against eptifibatide bind to platelet antigens, leading to platelet aggregation and thrombus formation. Another theory proposes that activation of Fcy FII signaling by platelet IIb/IIIa receptors could increase platelet granule secretion and aggregation in an IIb/IIIa-independent manner, contributing to the development of this adverse effect.

Conclusions

Despite the rarity of integrilin-induced thrombocytopenia and thrombosis, routine platelet count monitoring at arrival and 2–6 h after drug initiation may be prudent. If a significant decrease in platelet count is observed, discontinuation of the drug and evaluation for potential thrombotic events should be pursued.

Pharmacoinvasive PCI on a Patient with the Anomalous Origin of the RCA Through the Retrograde Approach

Introduction

Recent advances in PCI technologies, including MCS devices, have improved the treatment of high-risk patients with low LV systolic function as well as patients with acute MI complicated by cardiogenic shock. These high-risk patients are generally poor candidates for CABG due to the high risks of surgical morbidity and mortality. The ability to provide adequate hemodynamic support using MCS devices could potentially enable safer PCI and improve outcomes compared with an unprotected PCI strategy, surgical revascularization, or medical therapy alone [37]. Nonetheless, despite improvements in hemodynamics with the use of MCS devices, randomized trials using IABP or Impella 2.5 devices have not demonstrated a significant reduction in mortality during high-risk PCI compared with unprotected PCI [38, 39]. A clinical benefit from the use of MCS in the setting of cardiogenic shock complicating acute MI has also not yet been conclusively demonstrated. In the past decade, with the development of new-generation MCS devices offering greater hemodynamic support and lower device-associated complications—combined with appropriate case selection and optimal timing of device utilization—a growing body of data suggests improvements in procedural success and clinical outcomes [40].

Case 8

A 42-year-old woman, with known insulin-dependent diabetes mellitus and ischemic cardiomyopathy, and a history of primary PCI on the LAD, was referred to a non-PCI-capable hospital with chest pain and dyspnea. ECG revealed inferior wall STEMI, leading to the administration of intravenous fibrinolytic (reteplase) therapy within 30 min. Ninety minutes post-thrombolytic infusion, the patient reported pain relief, and ECG showed over 50% ST-segment resolution from baseline. Consequently, she was referred to our PCI-capable center for pharmacoinvasive treatment, arriving 3 h later with pulmonary edema and low cardiac output. Bedside transthoracic echocardiography demonstrated moderate LV enlargement, severe LV systolic dysfunction (ejection fraction = 5%–10%), akinesia and loss of tissue in the LAD territory, and hypokinesia in the inferior and inferolateral segments. RV size was mildly enlarged, with moderate dysfunction. Severe anteriorly directed functional mitral regurgitation without evidence of flailing was noticed. Moderate tricuspid regurgitation (gradient = 65 mm Hg) and moderate pulmonary hypertension (mean pulmonary arterial pressure = 30 mm Hg) were also noted. No ventricular septal rupture, pseudoaneurysms, or free wall rupture were detected. In light of the patient's condition, characterized by acute liver failure, acute kidney injury, metabolic acidosis, high lactate levels, and severe cardiogenic shock, intravenous diuretic, vasodilator, and inotropic therapy were initiated. Additionally, a bedside IABP was implanted.

Over the next 3 days, intravenous medical therapy and stabilization were continued, culminating in the removal of the IABP. Following a comprehensive assessment by the heart team, coronary computed tomography angiography was scheduled to provide more comprehensive information about the patient's coronary anatomy. The coronary computed tomography angiography revealed a normal LM, a patent LAD at the site of the previous stent with good run-off, and a normal LCX. Notably, the RCA was dominant and had an anomalous origin from the left aortic cusp. The RCA was totally thrombotic, with occlusion extending from the proximal to the mid-portion, presenting as a 2 cm-long proximal cap with a blunted origin. The RV branch originated at the proximal cap and was retrogradely perfused by septal branches. Cardiac magnetic resonance imaging with gadolinium injection demonstrated transmural infarction in the LAD territory, while LCX and RCA territories were fully viable.

Based on this information, the patient was scheduled for MCS-assisted PCI on the RCA. The PulseCath iVAC 2L catheter was selected as the appropriate MCS device for this procedure.

In such a high-risk PCI procedure, a multidisciplinary team is essential for optimal patient care. The presence of a cardiologist, interventional cardiologist, cardiac surgeon, heart failure team, anesthesiologist, and echocardiographer ensures that the patient's hemodynamic stability is closely monitored throughout the procedure.

After the insertion of right and left common femoral artery short sheaths (6 Fr) and the deployment of two ProGlide (suture-based/Abbott Vascular) vascular closure devices in the left common femoral artery, the sheath was replaced with a 10 Fr short sheath. The right radial artery access was prepared. Subsequently, a pigtail catheter was advanced through a J-tip 0.035-inch wire and positioned in the LV apex. The catheter was then replaced with a 260 cm super-stiff wire. Next, an 18 Fr

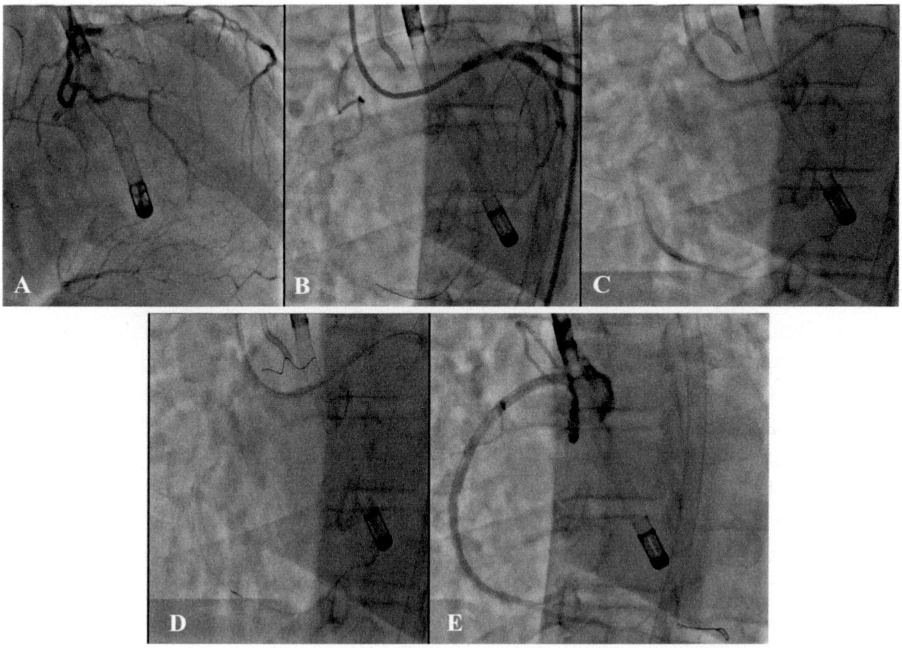

Fig. 1.17 (**a–e**) The tip and bidirectional valve of the iVAC catheter are positioned in the LV outflow tract and the aortic root, respectively (**a**). Due to the difficulty in selectively engaging the RCA (anomalous origin), a retrograde approach (RWE) was considered (**b**). After advancing the microcatheter over the SUOH03 wire (**c**), the wire was escalated to a tapered-tip CTO wire (Gaia second). Following penetration of the distal cap and exteriorization of the wire and microcatheter (**d**), the wire was exchanged for an RG3 wire. Next, the wire was snared, and the AL1 guiding catheter was used to engage the RCA. Ultimately, the entire procedure was performed antegradely without complications

long sheath was inserted, and heparin was administered to achieve an activated clotting time >250 s. After removal of the stylet, the iVAC catheter (a Nitinol braided catheter) was carefully de-aired and advanced over the wire, ensuring that the distal radio-opaque marker tip was positioned in the LV outflow tract and the other radio-opaque marker (bidirectional valve, 73 mm apart) was placed in the aortic root (Fig. 1.17a). At this stage, while the second operator de-aired the membrane pump with heparinized saline to rule out any bubbles in the system carefully, the first operator ligated the proximal hub with a dedicated clamp. At this stage, the end of the pump and the catheter were connected, while the other free side of the pump was connected to the console of the IABP. The catheter was now ready for pulsatile blood ejection (2 L/min). The iVAC2L system (PulseCath BV, Amsterdam, the Netherlands) consists of a 100 cm long single-lumen catheter positioned in the LV outflow tract through the aortic valve. During systole, blood is aspirated from the LV, and during diastole, it is directed back into the aortic arch. The pulsatile function is controlled by ECG or aortic pressure. In contrast to the Impella 2.5 system, the iVAC2L generates pulsatile flow and is compatible with conventional IABP consoles, which remain widely available in many hospitals. Selective LM injection revealed a normal LM, a patent previous LAD stent, and a normal LCX. Additionally, the second septal branch filled the RCA retrogradely up to its mid-portion. The origin of the RCA was anomalous, arising from the left cusp; consequently, several attempts using multiple guiding catheters to selectively engage the RCA were unsuccessful (Fig. 1.17b). Therefore, a decision was made to approach the RCA lesion via a retrograde approach. With the support of a Caravel microcatheter (150 cm), a SION blue wire was advanced to reach the second septal branch, and the microcatheter was then advanced over the wire. Injection of a minimal amount of dye through the microcatheter confirmed the correct target septal branch. Subsequently, with the aid of a SUOH03 wire, the wire was advanced into the posterior descending artery branch through fine and gentle rotational movements. The microcatheter was then advanced over the wire using clockwise and counterclockwise rotations (Fig. 1.17c). At this stage, the wire was exchanged for a Gaia second wire, and with minimal rotation, the wire was advanced past the distal cap and directly positioned into the aortic cusp (Fig. 1.17d). After confirming the position of the microcatheter in the aortic root, the wire was exchanged for an RG3 wire (Asahi Intecc Medical, 0.01 inch, 330 cm). The wire was exteriorized using loop snaring through the right common femoral artery, and an AL1 guide catheter was advanced over the wire to engage the RCA. Next, the entire procedure was completed antegradely with acceptable final results (Fig. 1.17e).

During the procedure, the patient's hemodynamic status remained stable, without the need for inotropic agent administration. Fifteen minutes later, after assessing her mental status, urine output, and mean blood pressure, the device and sheath were removed.

Discussion

In cardiogenic shock complicating acute MI, the LV is unable to effectively unload. This condition leads to elevated LV end-diastolic pressure, reduced cardiac output, increased wall stress and stroke work, decreased myocardial perfusion, impaired end-organ perfusion, and a myocardial demand-supply mismatch. These factors create a vicious cycle of hemodynamic disturbance, systemic inflammatory response, and worsening myocardial ischemia. In patients with a large ischemic burden and poor LV function undergoing PCI, further ischemia results in an additional reduction in the contractility of an already depressed ventricle, predisposing them to circulatory collapse [41]. Therefore, MCS has the potential to exert hemodynamic effects by unloading the LV, reducing LV end-diastolic pressure and end-diastolic volume, decreasing afterload, increasing cardiac output, and, consequently, improving systemic and coronary perfusion. The ideal MCS device can provide hemodynamic support to maintain adequate systemic perfusion, improve mean arterial pressure, and reduce myocardial oxygen demand through a concomitant reduction in LV preload and afterload while maintaining or enhancing coronary perfusion. The available MCS devices and their characteristics are summarized in Fig. 1.18 [42].

Features	IABP	Impella 2.5	Impella CP	iVAC21	tandemHeart	VA-ECMO
Inflow / Outflow	Aora	LV aora	LV aora	LV aora	LV aora	RA aora
Mechanism of action	Pneumatic	Axial flow	Axial flow	Pulsatile flow	Centrifugal flow	Centrifugal flow
Inserition approach	Pc (FA)	Pc(FA)	Pc(FA)	Pc(FA)	PC (FA/FV)	Pc(FA/FV)
Sheath size	7-8 F	13F	14F	17F	venous: 21F Arterial:12-19F	venous: 17-21F Arterial:16-19F
Flow (L/min)	0.5-0.3	Max 2.5	3.7- 4.0	Max 2.8	Max 4.0	Max 7.0
Pump speed (RPM)	N/A	Max 51.000	Max 51.000	40m / beat	Max 7500	Max 5000
Duration of support	2-5 days	6h-10 days	6h-10 days	6h- 10 days	UP to 14 days	7-10 days
LV function dependency	+	-	-	-	-	-
Synchrony with the cardiac cycle	+	-	-	-	-	-
LV unloading	+	++	+++	+	+++	-
Afterload	↓	↓	↓	↓	↑	↑↑
MAP	↑	↑↑	↑↑	↑↑	↑↑	↑↑
Cardiac index	↑	↑↑	↑↑↑	↑↑	↑↑↑	↑↑↑
PCWP	↓	↓	↓↓	↓	↓↓	▣
LVEDP	↓	↓↓	↓↓	↓↓	↓↓↓	▣
Coronary Perfusion	↑	↑	↑	↑	↑	▣
Myocardial Oxygen	↓	↓↓	↓↓	↓↓	↓↔	▣
Anticoagulation	+	+	+	+	+	+
Implant complexity	+	++	++	++	+++	++
Management complexity	+	++	++	++	+++	+++
Complications	Limb ischemia, bleeding	Hemolysis limb ischemia, bleeding	Hemolysis limb ischemia, bleeding	Hemolysis limb ischemia, bleeding	Limb ischemia, bleeding, hemolysis	Bleeding, limb ischemia, hemolysis
Contraindications	Moderate-to-severe AR-severe PAD	Severe AS/AR mechanical Aov , LV thrombus CI to AC	Severe AS/AR mechanical Aov , LV thrombus CI to AC	Severe AS/AR mechanical Aov , LV thrombus CI to AC	Moderate-to-severe AR, severe PAD, CI to AC.LA thrombus	Moderate-to-severe AR severe PAD, CI to AC
CE-certification	+	+	+	+	+	+
FDA approval	+	+	+	-	+	+

Fig. 1.18 The image provides comparisons of different mechanical support devices. AC: anticoagulation, AoV: aortic valve, AR: aortic regurgitation, AS: aortic stenosis, CI: contraindication, FA: femoral artery, FDA: US Food and Drug Administration, FV: femoral vein, LV: left ventricle, LVEDP: left ventricular end-diastolic pressure, IABP: intra-aortic balloon pump, MAP: mean arterial pressure, Max: maximum, PAD: peripheral arterial disease, Pc: percutaneous, PCWP: pulmonary capillary wedge pressure, PS: peripheral surgical, RA: right atrium, RPM: rotations per minute, VA-ECMO: venoarterial extracorporeal membrane oxygenation

Indications for Percutaneous MCS-Assisted PCI (Fig. 1.19)

1. Cardiogenic shock;
2. Post-MI mechanical complications; and
3. High-risk PCI without cardiogenic shock, including:
   - High thrombus burden with a high risk of the no-reflow phenomenon;
   - Degenerated vein grafts;
   - The last remaining coronary conduit;
   - Unprotected LM PCI with LCX dominance;

Fig. 1.19 The figure suggests an algorithmic approach toward MCS in patients with cardiogenic shock complicating AMI and PCI. MCS: mechanical circulatory support, AMI: acute myocardial infarction, PCI: percutaneous coronary intervention, IABP: intra-aortic balloon pump, VA-ECMO: veno-arterial extracorporeal membrane oxygenation

- Highly anticipated ischemic events;
- Need for extensive revascularization;
- Need for a retrograde approach;
- Need for atherectomy; and
- Severe LV dysfunction (ejection fraction <20%–30%).

MCS in Cardiogenic Shock

Cardiogenic shock is defined by a combination of persistent hypotension (systolic blood pressure <80–90 mm Hg or mean arterial pressure 30 mm Hg lower than baseline), a cardiac index <1.8 L/min/m^2 without support or <2.2 L/min/m^2 with support, elevated end-diastolic pressures on the right (>10–15 mm Hg) and/or left (>18 mm Hg) side of the heart, and evidence of end-organ hypoperfusion (e.g., cool extremities, decreased urine output, elevated lactate levels, or altered mental status). Cardiogenic shock complicates 6%–8% of patients with acute MI. Despite advances in PCI techniques, the availability of MCS, and improved supportive care, the mortality rate of cardiogenic shock has remained unchanged over the past 20 years. While early and appropriate revascularization of the culprit vessel in this setting improves survival, the evidence for mechanical support is much more limited, and no study to date has demonstrated that MCS use leads to an improvement in clinical outcomes.

IABP-SHOCK II [43] is the only large-scale randomized trial examining the use of MCS in acute MI. There was no difference in mortality between the two groups at 30 days, and this finding was sustained at 5 years of follow-up. The results from the IABP-SHOCK II trial have significantly impacted clinical guideline recommendations. The routine use of the IABP in cardiogenic shock has been downgraded from Class I to Class III in the European guidelines [44]. In the IMPRESS trial [45], mechanically ventilated patients with cardiogenic shock were randomized 1:1 to receive either the IABP or the Impella device. The 30-day and 6-month mortality rates were identical between the two groups (50% Impella vs. 46% IABP). Schrage et al. [46] matched 237 patients from the IABP-SHOCK II trial with 237 patients treated with the Impella LP 2.5/CP for cardiogenic shock complicating acute MI from the EUROSHOCK registry and found no reduction in 30-day mortality with Impella use compared with the IABP or standard medical therapy. Thiele et al. [47] compared the use of the IABP with percutaneous LV assist devices (TandemHeart) in patients with cardiogenic shock complicating acute MI. Forty-one patients, all of whom were mechanically ventilated, were randomly assigned (1:1). The ventricular assist device (TandemHeart) significantly improved the cardiac power index and effectively lowered serum lactate levels compared with the IABP; however, it was associated with a significantly higher rate of vascular and bleeding complications, with no difference in 30-day mortality between the groups (45% vs. 43%).

MCS During High-Risk PCI

There is currently no uniform definition of high-risk PCI. Nevertheless, various factors potentially increase procedural complications, and they can be broadly grouped into the following categories: patient comorbidities/presentation; LV function/valvular disease; and complexity of the coronary anatomy. In patients with a large ischemic burden and poor LV systolic function, sustained hypotension with reduced coronary perfusion during PCI can lead to profound myocardial ischemia and further insult to an already impaired LV. Thus, it is critical that MCS device insertion in this setting be performed before PCI, as this minimizes the risk of preceding ischemia and circulatory collapse, which may subsequently require emergent bail-out MCS implementation. The suggested algorithmic approach for MCS in patients with cardiogenic shock complicating acute MI and high-risk PCI is depicted in Fig. 1.19 [42].

The Balloon Pump–Assisted Coronary Intervention Study (BCIS-1) [48] and the Intra-Aortic Balloon Counterpulsation and Infarct Size in Patients with Acute Anterior Myocardial Infarction Without Shock (CRISP AMI) [49] both demonstrated no significant benefit of MCS in protecting during PCI among high-risk patients. The PROTECT II trial [39], the largest prospective randomized study examining Impella hemodynamic support in high-risk PCI, enrolled 452 patients with complex three-vessel disease or unprotected LM disease accompanied by severe LV dysfunction. Participants undergoing non-emergent high-risk PCI were randomized to receive either IABP or Impella 2.5 support. The primary endpoint of 30-day major adverse events showed no significant difference between the groups, with rates of 35.1% for the Impella and 40.1% for the IABP ($P = 0.227$) in the intention-to-treat analysis, and 34.3% for the Impella and 42.2% for the IABP ($P = 0.092$) in the per-protocol analysis. At 90-day follow-up, a trend toward reduced adverse events emerged in the intention-to-treat population (40.6% vs. 49.3%; $P = 0.066$), reaching statistical significance in the per-protocol analysis (40.0% vs. 51.0%; $P = 0.023$). While 30-day outcomes were comparable, the Impella demonstrated superior hemodynamic support and clinically relevant reductions in late adverse events, particularly evident in protocol-adherent patients.

Conclusions

While hemodynamic support demonstrates theoretical advantages, randomized clinical trials have not yet translated these benefits into improved clinical outcomes for patients with cardiogenic shock or those undergoing high-risk PCI. Vascular complication-related morbidity remains the primary limitation of all MCS devices. Clinical decisions regarding their use should, therefore, be individualized through heart team evaluation, carefully balancing potential hemodynamic improvements against risks of vascular injury and bleeding complications.

Primary PCI on the LCX and Cardiogenic Shock Following Acute Mitral Regurgitation

Case 9

A 57-year-old woman with a history of insulin-dependent diabetes mellitus and systemic hypertension presented with a 24-h history of retrosternal chest pain radiating to the left shoulder and dyspnea. Her baseline hemodynamic status was stable, with a blood pressure of 140/80 mm Hg, a heart rate of 85 beats per minute, and an oxygen saturation level of 98%. Physical examination was unremarkable except for fine bilateral basilar crackles. ECG (Fig. 1.20) indicated an inferoposterior STEMI. Bedside echocardiography revealed normal LV size, moderate LV systolic dysfunction (ejection fraction = 40% and posterior wall akinesia), normal RV measurements, mild mitral regurgitation, mild aortic insufficiency, and no evidence of mechanical complications.

The patient was promptly transferred to the catheterization laboratory for primary PCI. The culprit lesion was identified as a dominant, totally thrombotic LCX

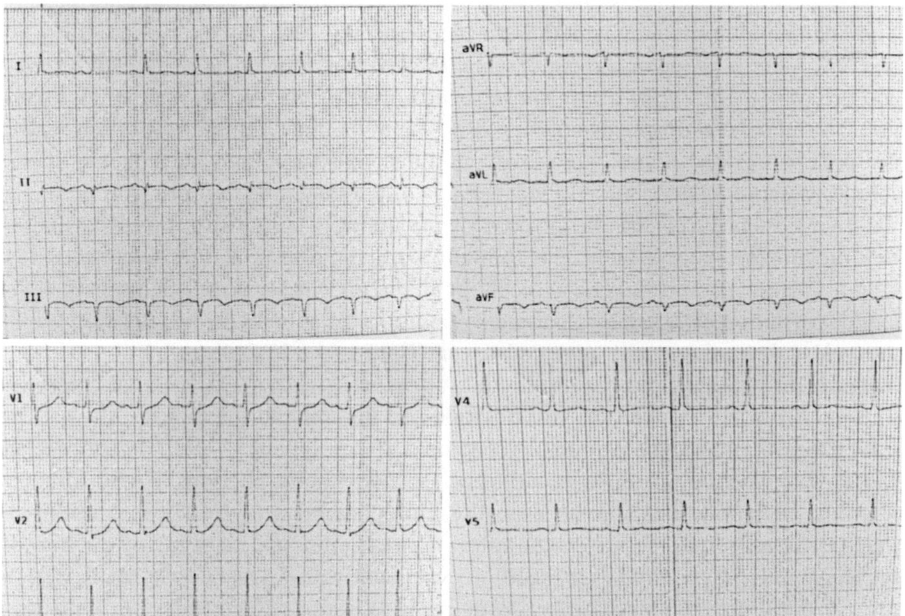

Fig. 1.20 The patient's surface ECG shows sinus rhythm, a Q wave in the inferior leads with a T-wave inversion, and a tall R wave in leads V_1–V_3 with a mild ST depression, suggesting an inferoposterior MI

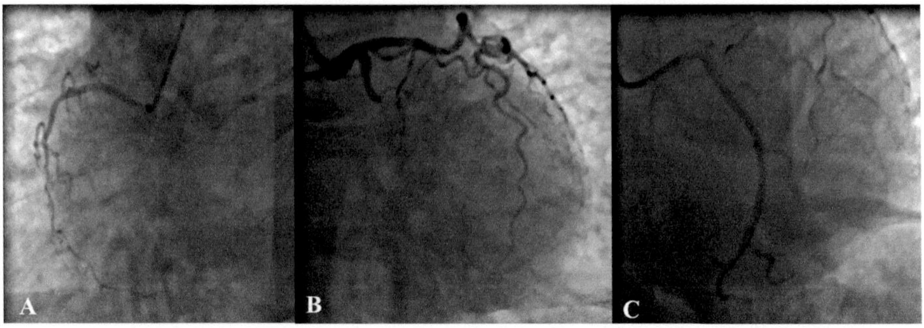

Fig. 1.21 (**a–c**) Selective coronary angiography via right radial access demonstrated a diminutive RCA (**a**), a normal LM with moderate stenosis in the proximal LAD, and a dominant LCX with total thrombotic occlusion in the proximal segment (**b**). After wiring the LCX and performing balloon dottering, direct stenting of the LCX was performed using a 3.5 × 38 mm XIENCE stent (the final TIMI flow grade = II) (**c**)

with a proximal occlusion (Fig. 1.21). Consequently, primary PCI on the LCX was planned. A 0.014-inch workhorse wire was used to access the LCX, after which balloon dottering and direct stenting with a 3.5 × 38 XIENCE stent were performed. The final TIMI flow grade was II, indicating partial reperfusion.

During the procedure, the patient experienced a sudden onset of severe dyspnea and orthopnea with pink frothy sputum, indicative of pulmonary edema. Her oxygen saturation level dropped to 60%, prompting immediate intubation. Subsequently, the patient became hemodynamically unstable and went into cardiogenic shock (blood pressure = 40/20 mm Hg, heart rate = 120 bpm, sinus rhythm). At this stage, the no-reflow phenomenon occurred. Despite multiple administrations of vasodilators, adenosine, and epinephrine through a microcatheter at the distal part, her hemodynamic instability persisted. Although the final TIMI flow grade was II, indicating partial reperfusion, the patient's condition remained critical. Bedside transthoracic echocardiography revealed severe ischemic mitral regurgitation without evidence of papillary muscle or chordal rupture (Fig. 1.22a). Despite the administration of high-dose diuretics, vasodilators, and norepinephrine therapy, the patient continued to experience pulmonary edema and cardiogenic shock. Therefore, an IABP was inserted to provide additional hemodynamic support. Thirty minutes after the insertion, a follow-up transthoracic echocardiography showed improvement, with only mild-to-moderate mitral regurgitation and no signs of chordal or papillary muscle rupture (Fig. 1.22b). Consequently, the patient was transferred to the CCU for further monitoring. After 24 h of stability, the balloon pump was removed, and she was discharged in good condition.

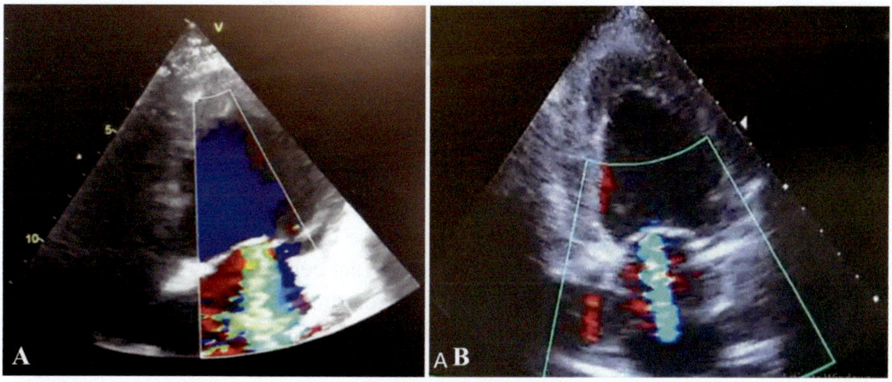

Fig. 1.22 (**a**, **b**) Initial transthoracic echocardiography during the patient's cardiogenic shock demonstrated severe acute mitral regurgitation with a central jet (systolic pulmonary artery pressure = 80 mm Hg) (**a**). Thirty minutes after the implantation of an IABP and initiation of hemodynamic support, echocardiography revealed mild-to-moderate mitral regurgitation (systolic pulmonary artery pressure = 25 mm Hg) (**b**)

Discussion

Despite advancements in revascularization strategies for STEMI, the prevalence of mechanical complications associated with MI has not significantly decreased in recent years. Delayed medical contact, untimely revascularization, and inadequate restoration of blood flow in the coronary bed (no-reflow or slow-flow phenomena) are among the most common contributing factors to mechanical complications. These complications include acute mitral regurgitation, ventricular septal rupture, free wall rupture, and pump failure [50]. Given the high mortality and morbidity rates associated with mechanical complications of MI and the intricate nature of their treatment, a multidisciplinary team approach involving an interventional cardiologist, a cardiac surgeon, and a heart failure specialist is essential [51, 52]. Patients who develop mechanical complications following MI often share similar characteristics. They tend to be older, female, and have a history of heart failure and chronic kidney disease. Additionally, these patients frequently present with their first acute MI [53].

Revascularization serves as the foundation for preventing mechanical complications in patients with MI. Recent studies have demonstrated no significant differences between PCI and fibrinolytic therapy concerning the incidence of mechanical complications [54].

Mitral Regurgitation

Acute mitral regurgitation, like other mechanical complications of MI, has a reduced incidence in the reperfusion era, although its mortality rate remains around 40%. The most common cause of acute mitral regurgitation is papillary muscle or chordal rupture. Notably, mitral regurgitation occurs more frequently in posteromedial papillary muscles because anterolateral papillary muscles receive dual perfusion [55].

For accurate evaluation of the underlying cause of mitral regurgitation (acute or chronic, mechanical or functional), transesophageal echocardiography has demonstrated higher sensitivity than transthoracic echocardiography. In patients with acute mitral regurgitation, coronary angiography typically reveals single- or double-vessel disease [56].

Initial treatment options for acute mitral regurgitation include various pharmacological and mechanical approaches. In normotensive patients, diuretics and vasodilators are often used to manage pulmonary congestion and reduce regurgitant volume. Respiratory support may be necessary for those experiencing respiratory distress. For patients in cardiogenic shock, LV assist devices can provide crucial hemodynamic support by increasing cardiac output, decreasing afterload, and reducing regurgitant volume [57, 58].

The definitive treatment for acute mitral regurgitation resulting from papillary muscle or chordal rupture is mitral valve replacement. This surgical intervention aims to correct the underlying mechanical issue and improve hemodynamic stability. While less common, mitral valve repair may be considered for less critically ill patients or those with partially ruptured papillary muscles.

The choice between mechanical and biological valves for mitral valve replacement in acute mitral regurgitation cases remains a topic of debate. Nonetheless, studies have suggested survival benefits associated with mechanical valves [59]. In patients experiencing acute mitral regurgitation following STEMI, mitral valve replacement is typically the preferred course of action.

Aside from papillary muscle or chordal rupture, other mechanisms contributing to mitral regurgitation in acute MI include papillary muscle dysfunction and tethering of the posterior leaflet. As seen in our patient, who showed no signs of papillary muscle or chordal rupture, these alternative pathophysiologies can play a significant role in the development of mitral regurgitation. Similar to papillary muscle rupture, papillary muscle dysfunction more commonly affects posterior papillary muscles [60–62].

In patients with STEMI and acute severe ischemic mitral regurgitation, timely reperfusion can lead to a significant reduction in mitral regurgitation and a favorable long-term prognosis. Conversely, a delay in reperfusion can result in a very poor prognosis, highlighting the importance of prompt intervention.

Ischemic mitral regurgitation is generally recognized as a chronic issue and is seldom documented in the acute phase of MI.

In the presented case, immediate revascularization of the culprit artery was performed to address the patient's acute condition. Although the procedure initially led to an increase in mitral regurgitation severity due to the no-reflow phenomenon, prompt and appropriate cardiopulmonary resuscitation contributed to her successful recovery from frank pulmonary edema and cardiogenic shock.

This case serves as a compelling example of the importance of identifying the underlying cause of cardiogenic shock in the context of STEMI and implementing an effective treatment strategy.

Ventricular Septal Rupture

Ventricular septal rupture is an infrequent cause of acquired ventricular septal defects that typically develop within 3–5 days following a transmural acute MI. Patients often present with pulmonary edema, acute illness, and hemodynamic instability. A characteristic physical examination finding includes a harsh holosystolic murmur with trills auscultated in the left parasternal border, and the presence of S3, which is considered pathognomonic.

Echocardiography is vital in assessing the location and characteristics of the defect, which can vary depending on the infarction territory. In cases of anterior MI, ventricular septal rupture is usually simple and slit-like in shape, involving the apical portion of the interventricular septum.

In cases of inferior or posterior MI, ventricular septal rupture tends to be more complex, involving the base of the interventricular septum. Moreover, ventricular septal rupture is frequently associated with ischemic mitral regurgitation [63]. Echocardiography plays a pivotal role in evaluating prognostic predictors, such as the function of both ventricles, the size of the defect, the magnitude of the left-to-right shunt, and pulmonary arterial pressure.

Urgent surgical repair of post-MI ventricular septal defects in patients experiencing cardiogenic shock and respiratory failure is associated with a 40% mortality rate. For individuals in whom surgery poses excessive risks, percutaneous septal repair may serve as a suitable alternative. (Please refer to the Chap. 9 for further details.) Some experts advocate for postponing surgical intervention in hemodynamically stable patients without end-organ damage, whenever possible, to achieve better outcomes.

Free Wall Rupture

Free wall rupture, a catastrophic complication of acute MI, often leads to sudden cardiac death. Early studies indicated a higher incidence of free wall rupture in patients receiving fibrinolytic therapy than in those undergoing PCI. High clinical suspicion, rapid diagnosis through echocardiography, and immediate surgical intervention are critical for managing this life-threatening event. Moreover, afterload reduction and the use of assist devices, particularly the extracorporeal membrane oxygenation device, are essential for stabilizing patients and facilitating their transition to surgery.

LV Pseudoaneurysms

Post-MI LV pseudoaneurysms, although rare (<2%), represent a severe clinic-pathologic entity. These outpouchings form when cardiac rupture is confined by adherent pericardium or scar tissue in the absence of myocardial tissue. Unlike LV pseudoaneurysms, true ventricular aneurysms feature an outer layer consisting of the endocardium and the myocardium, with an intact heart wall. Given that both pathologies involve weak heart-wall protrusions and share nonspecific clinical findings, differentiating between them poses a diagnostic challenge. Echocardiographic identification of a narrow neck and wide apex serves as characteristic features of a pseudoaneurysm, aiding in diagnostic accuracy.

Cardiac magnetic resonance imaging is a valuable tool in distinguishing LV pseudoaneurysms from true aneurysms, demonstrating a reported sensitivity of 100%. The most frequently observed location for these outpouchings is the inferior or posterolateral wall. Clinical presentation can often be nonspecific, encompassing symptoms such as congestive heart failure, chest pain, dyspnea, or arrhythmia, which may lead to diagnostic delays. Therefore, maintaining a high index of clinical suspicion is crucial for prompt diagnosis and effective management of LV pseudoaneurysms.

Urgent surgery is generally advised for all patients with LV pseudoaneurysms, given the high risk of perforation (48%) and the unpredictable nature of this complication. However, the high surgical mortality rate (23%–28%) warrants careful consideration of the benefits and risks associated with the surgical approach.

For smaller pseudoaneurysms with a low risk of massive rupture, thromboembolism, and cardiac failure, percutaneous approaches may be suitable alternatives. Septal occluder devices have been employed to close pseudoaneurysms via the percutaneous approach. In high-risk surgical patients and those requiring redo cardiovascular surgery, percutaneous device closure of LV pseudoaneurysms is preferable. Data regarding the medical management of these cases is scarce and controversial, contributing to ongoing debate within the medical community.

Conclusions

Although mechanical complications of acute MI are rare, prompt diagnosis is crucial when they do occur. In all patients with STEMI who present with symptoms of heart failure or hemodynamic instability, it is essential to rule out mechanical complications before attributing these symptoms to pump failure.

Treatment options for these complications include medical and mechanical afterload reduction strategies. However, definitive treatment typically involves surgical intervention, with percutaneous transcatheter approaches being appropriate in certain situations.

Spontaneous LM Dissection

Case 10

A 45-year-old man presented to a PCI-capable center with a complaint of acute-onset, severe retrosternal chest pain of 2 h duration. On physical examination, the patient's hemodynamic status was stable, without significant findings. ECG demonstrated diffuse ST-segment elevation in leads V_1–V_3 and the aVR lead, with reciprocal ST-segment depression in the inferior leads. Bedside transthoracic echocardiography showed normal LV size, severe LV systolic dysfunction (ejection fraction = 30%), severe hypokinesia of the anterior, anteroseptal, and apical segments, normal RV size and function, mild-to-moderate mitral regurgitation, and no evidence of mechanical complications or aortic root dissection. The patient was referred to the catheterization laboratory for primary PCI.

Selective coronary angiography through the right common femoral artery showed a suspicious dissected flap in the distal part of the LM and a totally occluded LAD from the ostium. The RCA was normal, and it retrogradely filled the LAD and the LM (Fig. 1.23).

The patient was a young man with no traditional cardiovascular risk factors. Additionally, he had not only coronary arteries free of any atherosclerotic plaques upon visual inspection but also a visible dissection flap in the LM. Accordingly, the SCAD of the LM was highly suspected. Given the patient's stable hemodynamic status, he was scheduled for primary PCI rather than primary CABG.

First, the LAD and the LCX were wired with 0.014-inch SION blue and BMW wires. Next, the diagnosis of SCAD was confirmed using an intravascular

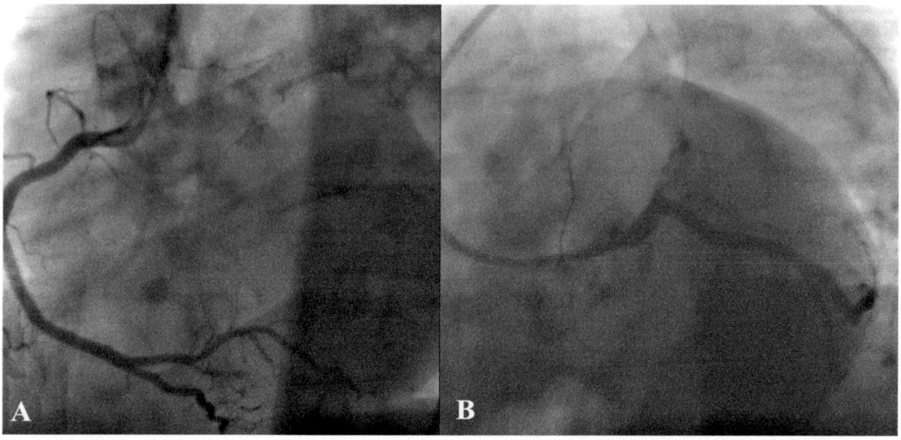

Fig. 1.23 (**a**, **b**) The patient's RCA was dominant and normal, and it retrogradely filled the LAD and the LM (**a**). LM injection showed a suspicious dissected flap at the distal part. The LAD was cut off at the ostial part, while the LCX was normal (**b**)

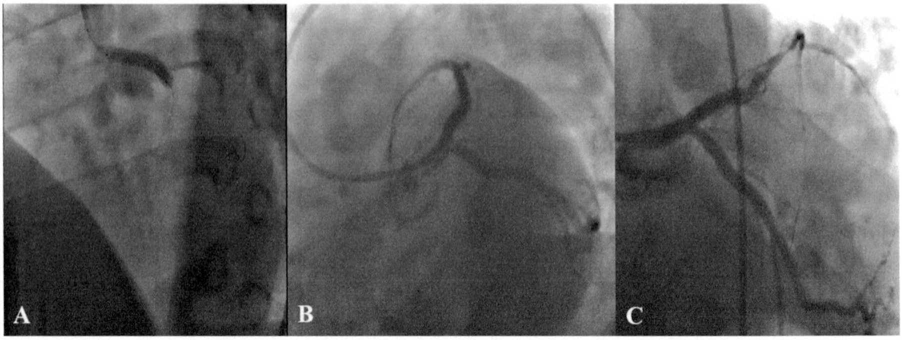

Fig. 1.24 (**a–c**) The patient's LAD and LCX were wired. Next, the correct positioning of the LAD wire in the true lumen was confirmed, and the artery was predilated. Then, the LM–LAD was directly stented using a 4 × 24 XIENCE stent. Afterward, the LM was postdilated with a 5 × 12 Apollo balloon, achieving an acceptable final result

ultrasound catheter. The position of the LAD wire in the true lumen was then reassessed. Afterward, the LAD was predilated with a 2.5 × 15 mm Artemis balloon, and the LM-LAD was directly stented without additional contrast injection using a 4 × 24 mm Ultimaster stent. Subsequently, the LM was postdilated using a 5 × 12 mm Apollo balloon, achieving a final TIMI flow grade III (Fig. 1.24).

Spontaneous LM Dissection

Case 11

The patient was a 34-year-old single woman who was referred to a PCI-capable center with complaints of severe constrictive retrosternal chest pain and 3 h of diaphoresis. Her hemodynamic status was stable. In physical examination, no significant findings were denoted. ECG showed a significant diffuse ST-segment depression in leads V_I–V_6 and an ST-segment elevation in lead V_I. Bedside transthoracic echocardiography illustrated normal LV size, mild LV systolic dysfunction (global hypokinesia), normal RV size and function, no significant valvular heart disease, and no evidence of mechanical complications.

Consequently, the patient was referred to the catheterization laboratory for primary PCI.

Selective coronary angiography demonstrated a dissected flap from the ostial part of the LM with extension to the LAD and the LCX with a TIMI flow grade II (Fig. 1.25).

After the final injection, the patient developed ventricular fibrillation/ventricular tachycardia and experienced cardiac arrest. While receiving resuscitation, she was transferred to the operating room, where CABG was successfully performed. She was discharged 7 days later without complications.

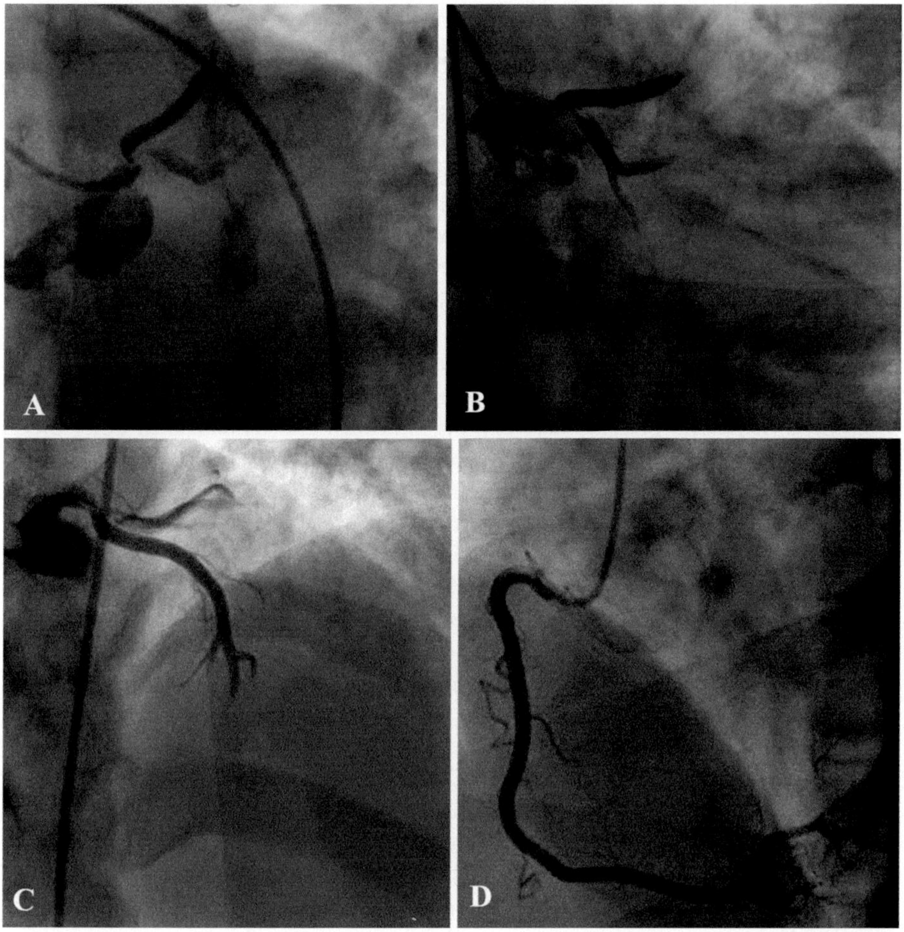

Fig. 1.25 (**a**–**d**) Selective coronary angiography shows a dissected flap from the ostial part of the LM that extends to the LAD and the LCX (TIMI flow grade I) (**a**–**c**). The RCA is dominant and normal

Discussion

Contrary to earlier beliefs, SCAD is not an uncommon presentation in patients presenting to the emergency department with acute MI. Until recently, SCAD was frequently underdiagnosed and misidentified as normal coronary arteries, coronary vasospasm, thromboembolism, or Takotsubo cardiomyopathy, leading to potential mismanagement and increased mortality. Increased awareness among physicians and cardiologists has led to a better understanding of SCAD's incidence, pathophysiology, risk factors, and management. Previous case reports and studies revealed that nearly one-third of women under 50 years old presenting to the emergency department with chest pain were diagnosed with SCAD [64]. The majority of patients affected by SCAD are middle-aged women (approximately 90% of cases), often lacking traditional cardiovascular risk factors [65]. SCAD appears to involve intimal tearing (inside-out) or hemorrhagic events in the vasa vasorum (outside-in) within a non-atherosclerotic segment, leading to compression of the intramural hematoma on the true lumen. The expansion of the hematoma, driven by pressure, results in the propagation of the dissection plane and the formation of a true lumen and a thrombus-containing false lumen [66]. Around 50% of SCAD cases present without detectable underlying triggers. Nevertheless, potential predisposing factors include emotional or physical stressors, connective tissue disorders, inflammatory disorders, pregnancy and the postpartum phase, sympathomimetic drugs, or hormonal influences [67]. Recent studies have confirmed the hypothesis that SCAD is a form of fibromuscular dysplasia affecting coronary arteries. In fact, a significant proportion of SCAD patients also exhibit fibromuscular dysplasia in other blood vessels, including renal, vertebral, iliac, and brachial arteries [68].

Nearly all patients with SCAD present to the emergency department with chest pain (90%) [69], dyspnea, or sudden cardiac arrest. Studies indicate that over 95% of these patients exhibit elevated troponin levels [70]. On average, 29.7%, 59.9%, and 8% of SCAD patients present with STEMI, non-STEMI, and sudden cardiac arrest (ventricular tachycardia/ventricular fibrillation), respectively [71]. Accurate diagnosis of SCAD relies on a combination of awareness of clinical conditions and appropriate utilization of imaging modalities.

Selective coronary angiography, with or without dedicated intracoronary imaging tools such as intravascular ultrasonography or optical coherence tomography, is the preferred modality for confirming SCAD [72]. Relying solely on angiography, which visualizes the lumen but not the arterial wall, and basing the diagnosis on multiple lumens or contrast staining of the arterial wall can lead to a missed diagnosis in over 70% of SCAD cases [73]. The most common angiographic feature of SCAD is abrupt vessel narrowing, which may extend to the distal segments of the vessel [74]. In young female patients with otherwise smooth coronary arteries, the presence of diffuse long coronary lesions should always prompt consideration of SCAD and intramural hematoma as part of the differential diagnosis [12].

In the Saw classification, patients with SCAD are categorized as follows:

- **Type I:** Radiolucent flaps and linear double lumens with contrast hold-ups (29%–48%);
- **Type II:** The most common type (52%–67%), which is further divided into:
  - **IIa:** Long, diffuse, and smooth stenosis in the mid-to-distal portion, with a return to normal vessel caliber distally; and
  - **IIb:** Stenosis extending to the end of the vessel;
- **Type III:** Angiographically indistinguishable from focal atherosclerotic stenosis (0.3%–9%); and
- **Type IV:** Total occlusion of the vessel.

According to numerous trials and case reports, the LAD, particularly its mid to distal segments, is the most frequently affected vessel in SCAD. The involvement of these LAD segments is associated with a better prognosis than proximal LAD involvement [75]. However, the pattern differs in cases of SCAD occurring during pregnancy and the postpartum period, where the proximal part of the LAD or the LM is more commonly affected. These patients typically present with STEMI, resulting in a worse prognosis and higher maternal and fetal mortality rates [76].

Multiple trials have found that 40%–86% of SCAD patients exhibit fibromuscular dysplasia of noncoronary vessels, dissections, and aneurysms [68]. Given these findings, the latest guidelines recommend the use of whole-body computed tomography angiography or magnetic resonance angiography to detect these abnormalities [71]. Still, computed tomography angiography is not an ideal modality for detecting coronary SCAD, as normal results in coronary computed tomography angiography do not exclude SCAD. Nonetheless, it is a good modality for follow-up in SCAD patients, particularly in cases involving large vessels or proximal segment dissection [77].

Observational data suggest that a significant proportion (70%–97%) of SCAD lesions demonstrate angiographic healing in patients who were followed up with repeat imaging weeks to months after an initial conservative management approach. According to the American Heart Association (AHA) statement, patients with SCAD should be closely monitored for 3–5 days due to the high risk (5%–10%) of propagation of the index dissection flap and involvement of other coronary vessels during the same hospitalization. In stable patients without high-risk anatomy (the LM or the proximal segments of two vessels), a conservative management approach is recommended, whereas urgent revascularization (either PCI or CABG) is advised for those with ongoing ischemia or hemodynamic instability. The AHA statement also recommends repeat coronary angiography at 6–8 weeks [71].

Observational studies have revealed that during follow-ups, *de novo* dissections are more likely to occur in different coronary segments, while in the acute phase, extension of the initial dissection is more common [78].

Coronary tortuosity, characterized by a corkscrew appearance, has been identified as the most frequent predictor of recurrent dissections in observational trials [79].

Managing high-risk SCAD patients who are at risk of antegrade or retrograde propagation of the dissection flap, or involvement of other coronary vessels during the initial hospitalization or after discharge, remains a subject of debate. The risks associated with PCI failure (around 30%) or iatrogenic dissection of other vessels (3.4% vs. <0.2%) are considerably higher in these patients. Additionally, the resorption of intramural hematoma over time increases the likelihood of late stent malapposition and stent thrombosis, further complicating management decisions. Considering these challenges, it is crucial to carefully weigh the risks and benefits of various treatment strategies for high-risk SCAD patients.

The optimal medical management of SCAD remains a subject of debate, although aggressive hypertension control and the use of beta-blockers are generally recommended for managing the acute phase and preventing recurrence. Antithrombotic therapy poses a challenge, as physicians must carefully balance the risk of dissection propagation against the risk of intraluminal obstruction. The use of statins in SCAD patients is also controversial, and their prescription is generally limited to high-risk patients for primary prevention purposes [79].

Prolonged angina is a common issue in SCAD patients, often leading to readmission despite normal ischemia testing and lesion healing. This persistent angina may result from endothelial dysfunction, vasospasm, microvascular disease, or noncardiac causes [80]. In such cases, calcium channel blockers, nitrates, and ranolazine are recommended to help alleviate symptoms [71].

Despite a low mortality rate of approximately 1%–2% within 3 years, SCAD has a significant recurrence rate of around 17%–18% within 4–5 years, with a higher frequency in patients with hypertension. Consequently, beta-blocker therapy may prove beneficial for preventing recurrence in these patients [81].

Given that pregnancy is a common risk factor for SCAD, contraception is recommended to mitigate this risk. In cases where a patient is already pregnant, a multidisciplinary team approach is advised to provide comprehensive care and manage potential complications [71].

In section "Case 10", the patient presented with LM–LAD SCAD and stable hemodynamics. The decision was made to perform primary PCI on the LM-LAD, which yielded positive outcomes. Whole-body computed tomography angiography revealed no signs of fibromuscular dysplasia in the extra-coronary arteries. Over a 2-year follow-up period, no recurrence of SCAD was observed.

In section "Case 11", the patient presented with LM, LAD, and LCX SCAD, along with unstable hemodynamics. Due to the extent of the lesions and the patient's clinical status, emergent CABG was performed, using SVGs on the LAD and the LCX. In-hospital whole-body magnetic resonance angiography revealed a diagnosis of Takayasu arteritis. Consequently, immunosuppressive therapy was initiated promptly. Over the subsequent 5 years following the operation, the patient has been regularly monitored by a rheumatologist through clinical, laboratory, and imaging assessments. No flare-up episodes have been documented during this period. The most recent coronary computed tomography angiography demonstrated no signs of coronary artery dissection and confirmed patent vein grafts.

Conclusions

SCAD is an increasingly identified cause of ACS, particularly in women. Although the complete pathophysiology of this non-atherosclerotic event is not yet fully understood, it is often associated with an underlying arteriopathy. While an invasive approach is typically recommended for atherosclerotic causes of ACS, conservative management is generally preferred for SCAD patients due to the high rate of spontaneous healing observed within 6–8 weeks in most cases. However, invasive interventions, such as revascularization, may be necessary in specific circumstances.

There is a risk of approximately 5%–10% for propagation of dissected segments antegradely or retrogradely or involvement of another coronary artery during the initial hospitalization. Consequently, monitoring patients for at least 3–5 days is crucial to ensure prompt detection and management of such complications. Moreover, intermediate and long-term recurrence rates of SCAD are significant, highlighting the importance of long-term follow-up to assess patients' progress and identify any potential recurrent events.

Concomitant fibromuscular dysplasia of extra-coronary arteries should also be taken into account when managing SCAD patients. Retrospective trials and case series have revealed low long-term mortality rates but high morbidity rates, primarily due to recurrence. Prospective, multicenter, randomized trials are necessary to establish optimal management strategies for this patient population.

Current guidelines from the European Society of Cardiology (ESC) and American guidelines recommend a conservative approach to SCAD management. Nonetheless, there is consensus that noninvasive management should be initiated unless patients present with hemodynamic instability and ongoing ischemia. In high-risk cases, such as the two patients discussed, revascularization may be a beneficial therapeutic option.

We herein described two SCAD patients, highlighting the importance of individualized treatment approaches based on each patient's condition and coronary anatomy. The first patient underwent successful PCI, while the second patient required emergent surgery due to her specific clinical presentation.

In light of these cases, we emphasize the significance of managing SCAD patients at high-quality centers with extensive expertise. This approach ensures that patients have access to a multidisciplinary team of specialists, state-of-the-art diagnostic and therapeutic tools, and comprehensive care tailored to their unique needs.

References

1. Virani SS, Alonso A, Benjamin EJ, et al. Heart disease and stroke statistics-2020 update: a report from the American Heart Association. Circulation. 2020;141(9):e139–596.
2. Thygesen K, Alpert JS, Jaffe AS, et al. Fourth universal definition of myocardial infarction. J Am Coll Cardiol. 2018;72(18):2231–64.
3. Ibanez B, James S, Agewall S, Antunes MJ, Bucciarelli-Ducci C, Bueno H, Caforio ALP, Crea F, Goudevenos JA, Halvorsen S, Hindricks G, Kastrati A, Lenzen MJ, Prescott E, Roffi M, Valgimigli M, Varenhorst C, Vranckx P, Widimsky P, ESC Scientific Document Group. ESC Guidelines for the management of acute myocardial infarction in patients presenting with ST-segment elevation: the Task Force for the management of acute myocardial infarction in patients presenting with ST-segment elevation of the European Society of Cardiology (ESC). Eur Heart J. 2017;2018(39):119–77.
4. Van't Hof AW, Liem A, Suryapranata H, et al. Angiographic assessment of myocardial perfusion in patients treated with primary angioplasty for acute myocardial infarction: myocardial blush grade. Circulation. 1998;97:2302–6.
5. Hori M, Inoue M, Kitakaze Y, et al. Role of adenosine in hyperemic response of coronary blood flow in microembolization. Am J Phys. 1986;250:H509–18.

6. Harrison RW, Aggarwal A, Ou FS, et al. Incidence and outcomes of no-reflow phenomenon during percutaneous coronary intervention among patients with acute myocardial infarction. Am J Cardiol. 2013;111:178–84.
7. De Luca G, van't Hof AW, Ottervanger JP, et al. Unsuccessful reperfusion in patients with ST-segment elevation myocardial infarction treated with primary angioplasty. Am Heart J. 2005;150:557–62.
8. Resnic FS, Wainstein M, Lee MKY, et al. No-reflow is an independent predictor of death and myocardial infarction after percutaneous coronary intervention. Am Heart J. 2003;145:42–6.
9. Yang L, Cong H, Yali L, et al. Prediction of no-reflow phenomenon in patients treated with primary percutaneous coronary intervention for ST-segment elevation myocardial infarction. Medicine. 2020;99:26.
10. Mahaffey KW, Puma JA, Barbagaleta NA, et al. Adenosine as an adjunct to thrombolytic therapy for acute myocardial infarction: results of a multicenter, randomized, placebo-controlled trial: the Acute Myocardial Infarction STudy of ADenosine (AMISTAD) trial. J Am Coll Cardiol. 1999;34:1711–20.
11. Skelding KA, Goldstein JA, Mehta L, et al. Resolution of refractory no-reflow with intracoronary epinephrine. Cathet Cardiovasc Interv. 2002;57:305–9.
12. Harjola V-P, Lassus J, Sionis A, Køber L, Tarvasmäki T, Spinar J, Parissis J, Banaszewski M, Silva-Cardoso J, Carubelli V, Di Somma S, Tolppanen H, Zeymer U, Thiele H, Nieminen MS, Mebazaa A, for the CardShock study investigators and the GREAT network. Clinical picture and risk prediction of short-term mortality in cardiogenic shock: clinical picture and outcome of cardiogenic shock. Eur J Heart Fail. 2015;17:501–9.
13. Van Diepen S, Kats JN, Albert NM, et al. Contemporary management of cardiogenic shock: a scientific statement from the American Heart Association. Circulation. 2017;136:e232–68.
14. De Asua I, Rosenberg A. On the right side of the heart: medical and mechanical support of the failing right ventricle. J Intensive Care Soc. 2017;18:113–20.
15. Gordon AC, Wang N, Walley KR, Ashby D, Russell JA. The cardiopulmonary effects of vasopressin compared with norepinephrine in septic shock. Chest. 2012;142:593–605.
16. Lee MS, Bokhoor P, Park S-J, et al. Unprotected left main coronary disease and ST-segment elevation myocardial infarction. J Am Coll Cardiol Intv. 2010;3(8):791–5.
17. Stone GW, Kappetein AP, Sabik JF. Five-year outcomes after PCI or CABG for left main coronary disease. N Engl J Med. 2019;381(19):1820–30.
18. SerruysPocock NR, Makikallio T, Lindsay MM. Percutaneous coronary angioplasty versus coronary artery bypass grafting in the treatment of unprotected left main stenosis: updated 5-year outcomes from the randomised, noninferiority NOBLE trial. Lancet. 2020;395(10219):191–9.
19. Patel N, De Maria GL, Kassimis G, et al. Outcomes after emergency percutaneous coronary intervention in patients with unprotected left main stem occlusion: the BCIS national audit of percutaneous coronary intervention 6-year experience. JACC Cardiovasc Interv. 2014;7:969–80.
20. Xu M, Virtanen K, Hekali P, Frick MH. Survival with total occlusion of the left main coronary artery. Significance of the collateral circulation. Catheter Cardiovasc Diagn. 1979;5(3):269–75.
21. Abu-Assi E, Castiñeira-Busto M, González-Salvado V, et al. Coronary artery dominance and long-term prognosis in patients with ST-segment elevation myocardial infarction treated with primary angioplasty. Rev Esp Cardiol. 2016;69(1):19–27.
22. Patel N, De Maria GL, Kassimis G, et al. Outcomes after emergency percutaneous coronary intervention in patients with unprotected left main stem occlusion. J Am Coll Cardiol Intv. 2014;7(9):969–80.
23. Gutiérrez-Barrios A, Gheorghe L, Camacho-Freire S, et al. Primary angioplasty in a catastrophic presentation: acute left main coronary total occlusion—the ATOLMA Registry. Hindawi J Interv Cardiol. 2020;2020:Article ID 5246504, 8 pages. https://doi.org/10.1155/2020/5246504.
24. Yap J, Singh GD, Kim J-S, et al. Outcomes of primary percutaneous coronary intervention in acute myocardial infarction due to unprotected left main thrombosis: the Asia-Pacific Left

Main ST-Elevation Registry (ASTER). J Interv Cardiol. 2018;31:129–35. Wiley Interventional Cardiology, accepted October 2017.
25. Boersma E, Harrigton RA, Moliterno DJ, et al. Platelet glycoprotein IIb/IIIa inhibitors in acute coronary syndromes: a meta-analysis of all major randomized clinical trials. Lancet. 2002;359:189–98.
26. Harrington RA, Kleiman NS, Kottke-Marchant K, et al. Immediate and reversible platelet inhibition after intravenous administration of a peptide glycoprotein IIb/IIIa inhibitor during percutaneous coronary intervention. Am J Cardiol. 1995;76:1222–7.
27. Randomized placebo-controlled trial of effect of eptifibatide on complications of pervcutaneous coronary intervention: IMPACT II. Integrilin to minimize Platelet aggregation and Coronary Thrombosis-II. Lancet. 1997;349:1422–8.
28. Inhibition of platelet glycoprotein IIb/IIIa with eptifibatide in patients with acute coronary syndromes. The PURSUIT trial investigators. Platelet glycoprotein IIb/IIIa inhibitors in unstable angina; receptor suppression using integrilin therapy. N Engl J Med. 1998;339:436–43.
29. Hongo RH, Brent BN. Association of eptifibatide and acute profound thrombocytopenia. Am J Cardiol. 2001;88:428–31.
30. Rezkalla SH, Hayes JJ, Curtis BR, et al. Eptifibatide-induced acute profound thrombocytopenia presenting as refractory hypotension. Catheter Cardiovasc Interv. 2003;58:76–9.
31. Salengro E, Mulvihil NT, Farah B. Acute profound thrombocytopenia after use of eptifibatide for coronary stenting. Catheter Cardiovasc Interv. 2003;23:374–9.
32. Michael W, Tempelhof KH, et al. Eptifibatide – induced thrombocytopenia with thrombosis and disseminated intravascular coagulation immediately after left main coronary artery percutaneous coronary angioplasty. Texas Heart Inst J. 2012;39(1):86–91.
33. Christopher G, Stephania BE, Hani K, et al. Eptifibatide – induced thrombocytopenia and subsequent thrombosis. Pharmacy Tech J. 2012;8:146–50.
34. Epelman S, Nair D, Downey R, et al. Eptifibatide-induced thrombocytopenia and thrombosis. J Thromb Thrombolysis. 2006;22:151–4.
35. Bennett CL, Kim B, Zakarija A, et al. Two mechanistic pathways for theinopyridine-associated thrombotic thrombocytopenia purpura: a report from the SERF-TTP Research Group and the RADAR Project. J Am Coll Cardiol. 2007;50(12):1138–43.
36. Gao C, Boylan B, Bougie D, et al. Eptifibatide–induced thrombocytopenia and thrombosis in humans require FcgammaRIIa and the integrin beta3 cytoplasmic domain. J Clin Invest. 2009;119:504–11.
37. Waldo SW, Secemsky EA, O'Brien C, Kennedy KF, Pomerantsev E, Sundt TM 3rd, McNulty EJ, Scirica BM, Yeh RW. Surgical ineligibility and mortality among patients with unprotected left main or multivessel coronary artery disease undergoing percutaneous coronary intervention. Circulation. 2014;130:2295–301. [CrossRef] [PubMed].
38. Perera D, Stables R, Thomas M, Booth J, Pitt M, Blackman D, de Belder A, Redwood S, BCSI-1 Investigators. Elective intra-aortic balloon counterpulsation during high-risk percutaneous coronary intervention: a randomized controlled trial. JAMA. 2010;304:867–74. [CrossRef] [PubMed].
39. O'Neill WW, Kleiman NS, Moses J, Henriques JP, Dixon S, Massaro J, Palacios I, Maini B, Mulukutla S, Dzavík V. A prospective, randomized clinical trial of hemodynamic support with Impella 2.5 versus intra-aortic balloon pump in patients undergoing high-risk percutaneous coronary intervention: the PROTECT II study. Circulation. 2012;126:1717–27. [CrossRef] [PubMed].
40. Ameloot K, Bastos MB, Daemen J, Schreuder J, Boersma E, Zijlstra F, Van Mieghem NM. New generation mechanical circulatory support during high-risk PCI: a cross sectional analysis. EuroIntervention. 2019;15(5):427–33. [CrossRef].
41. Shreenivas SS, Wilensky RL. Percutaneous circulatory support during percutaneous coronary intervention. Interv Cardiol. 2012;4(4):449–60.
42. Asleh R, Resar JR. Utilization of percutaneous mechanical circulatory support devices in cardiogenic shock complicating acute myocardial infarction and high-risk percutaneous coro-

nary interventions. J Clin Med. 2019;8(8):1209. https://doi.org/10.3390/jcm8081209. PMID: 31412669; PMCID: PMC6724052.
43. Herrera-Perez D, Haslam A, Crain T, Gill J, Livingston C, Kaestner V, Hayes M, Morgan D, Cifu AS, Prasad V. A comprehensive review of randomized clinical trials in three medical journals reveals 396 medical reversals. elife. 2019;8:e45183. [CrossRef] [PubMed].
44. Ibanez B, James S, Agewall S, Antunes MJ, Bucciarelli-Ducci C, Bueno H, Caforio ALP, Crea F, Goudevenos JA, Halvorsen S, et al. 2017 ESC Guidelines for the management of acute myocardial infarction in patients presenting with ST-segment elevation. Eur Heart J. 2018;39:119–77. [CrossRef] [PubMed].
45. Ouweneel DM, Eriksen E, Sjauw KD, van Dongen IM, Hirsch A, Packer EJ, Vis MM, Wykrzykowska JJ, Koch KT, Baan J, et al. Percutaneous mechanical circulatory support versus intra-aortic balloon pump in cardiogenic shock after acute myocardial infarction. J Am Coll Cardiol. 2017;69:278–87. [CrossRef] [PubMed].
46. Schrage B, Ibrahim K, Loehn T, et al. Impella support for acute myocardial infarction complicated by cardiogeic shock. Circulation. 2019;139:1249–58.
47. Thiele H, et al. Randomized comparison of intra-aortic balloon support with a percutaneous left ventricular assist device in patients with revascularized acute myocardial infarction complicated by cardiogenic shock. Eur Heart J. 2005;26(13):1276–83.
48. Perera D, Stables R, Clayton T, de Belder M, Ludman P, Katritsis D, et al. "Long-term mortality data from the Balloon Pump-Assisted Coronary Intervention Study (BCIS-1): a randomized, controlled trial of elective balloon counterpulsation during high-risk percutaneous coronary intervention." Circulation. 2013;127(2):207–212. https://doi.org/10.1161/CIRCULATIONAHA.112.134593.
49. Patel MR, et al. Intra-aortic balloon counterpulsation and infarct size in patients with acute anterior myocardial infarction without shock: the CRISP AMI randomized trial. JAMA. 2011;306(12):1329–37.
50. Peterson ED, Shah BR, Parsons L, Pollack CV Jr, French WJ, Canto JG, Gibson CM, Rogers WJ. Trends in quality of care for patients with acute myocardial infarction in the National Registry of Myocardial Infarction from 1990 to 2006. Am Heart J. 2008;156:1045–55. https://doi.org/10.1016/j.ahj.2008.07.028.
51. O'Gara PT, Kushner FG, Ascheim DD, Casey DE Jr, Chung MK, de Lemos JA, Ettinger SM, Fang JC, Fesmire FM, Franklin BA, et al. 2013 ACCF/AHA guideline for the management of ST-elevation myocardial infarction: a report of the American College of Cardiology Foundation/American Heart Association Task Force on Practice Guidelines. Circulation. 2013;127:e362–425. https://doi.org/10.1161/CIR.0b013e3182742cf6.
52. Ibánez B, James S, Agewall S, Antunes MJ, Bucciarelli-Ducci C, Bueno H, Caforio ALP, Crea F, Goudevenos JA, Halvorsen S, et al. 2017 ESC Guidelines for the management of acute myocardial infarction in patients presenting with ST-segment elevation. Rev Esp Cardiol (Engl Ed). 2017;70:1082. https://doi.org/10.1016/j.rec.2017.11.010.
53. Rogers WJ, Frederick PD, Stoehr E, Canto JG, Ornato JP, Gibson CM, Pollack CV Jr, Gore JM, Chandra-Strobos N, Peterson ED, et al. Trends in presenting characteristics and hospital mortality among patients with ST elevation and non-ST elevation myocardial infarction in the National Registry of Myocardial Infarction from 1990 to 2006. Am Heart J. 2008;156:1026–34. https://doi.org/10.1016/j.ahj.2008.07.030.
54. Armstrong PW, Gershlick AH, Goldstein P, Wilcox R, Danays T, Lambert Y, Sulimov V, Rosell Ortiz F, Ostojic M, Welsh RC, STREAM Investigative Team, et al. Fibrinolysis or primary PCI in ST-segment elevation myocardial infarction. N Engl J Med. 2013;368:1379–87. https://doi.org/10.1056/NEJMoa1301092.
55. French JK, Hellkamp AS, Armstrong PW, Cohen E, Kleiman NS, O'Connor CM, Holmes DR, Hochman JS, Granger CB, Mahaffey KW. Mechanical complications after percutaneous coronary intervention in ST-elevation myocardial infarction (from APEX-AMI). Am J Cardiol. 2010;105:59–63. https://doi.org/10.1016/j.amjcard.2009.08.653.

56. Puerto E, Viana-Tejedor A, Martínez-Sellés M, Domínguez-Pérez L, Moreno G, Martín-Asenjo R, Bueno H. Temporal trends in mechanical complications of acute myocardial infarction in the elderly. J Am Coll Cardiol. 2018;72:959–66. https://doi.org/10.1016/j.jacc.2018.06.031.
57. Tehrani BN, Truesdell AG, Psotka MA, Rosner C, Singh R, Sinha SS, Damluji AA, Batchelor WB. A standardized and comprehensive approach to the management of cardiogenic shock. JACC Heart Fail. 2020;8:879–91. https://doi.org/10.1016/j.jchf.2020.09.005.
58. van Diepen S, Katz JN, Albert NM, Henry TD, Jacobs AK, Kapur NK, Kilic A, Menon V, Ohman EM, on behalf of the American Heart Association Council on Clinical Cardiology, Council on Cardiovascular and Stroke Nursing, Council on Quality of Care and Outcomes Research, Mission: Lifeline, et al. Contemporary management of cardiogenic shock: a scientific statement from the American Heart Association. Circulation. 2017;136:e232–68. https://doi.org/10.1161/CIR.0000000000000525.
59. Thompson CR, Buller CE, Sleeper LA, Antonelli TA, Webb JG, Jaber WA, Abel JG, Hochman JS. Cardiogenic shock due to acute severe mitral regurgitation complicating acute myocardial infarction: a report from the SHOCK Trial Registry. SHould we use emergently revascularize Occluded Coronaries in cardiogenic shocK? J Am Coll Cardiol. 2000;36(3 suppl A):1104–9. https://doi.org/10.1016/s0735-1097(00)00846-9.
60. Stout KK, Verrier ED. Acute valvular regurgitation. Circulation. 2009;119:3232–41.
61. Tanimoto T, Imanishi T, Kitabata H, et al. Prevalence and clinical significance of papillary muscle infarction detected by late gadolinium-enhanced magnetic resonance imaging in patients with ST-segment elevation myocardial infarction. Circulation. 2010;122:2281–7.
62. Damluji AA, van Diepen S, Katz JN, Menon V, Tamis-Holland JE, Bakitas M, Cohen MG, Balsam LB, Chikwe J, American Heart Association Council on Clinical Cardiology, Council on Arteriosclerosis, Thrombosis and Vascular Biology, Council on Cardiovascular Surgery and Anesthesia, Council on Cardiovascular and Stroke Nursing. Mechanical complications of acute myocardial infarction a scientific statement from the American Heart Association. Circulation. 2021;144(2):e16–35.
63. Crenshaw BS, Granger CB, Birnbaum Y, Pieper KS, Morris DC, Kleiman NS, Vahanian A, Califf RM, Topol EJ. Risk factors, angiographic patterns, and outcomes in patients with ventricular septal defect complicating acute myocardial infarction. GUSTO-I (Global Utilization of Streptokinase and TPA for Occluded Coronary Arteries) Trial Investigators. Circulation. 2000;101:27–32. https://doi.org/10.1161/01.cir.101.1.27.
64. Rashid HN, Wong DT, Wijesekera H, et al. Incidence and characterization of spontaneous coronary artery dissection as a cause of acute coronary syndrome: a single-center Australian experience. Int J Cardiol. 2016;202:336–8.
65. Gupta A, Wang Y, Spertus JA, et al. Trends in acute myocardiak infarction in young patients and differences by sex and race, 2001 to 2010. J Am Coll Cardiol. 2014;64:337–45.
66. Adlam D, Alfonso F, Mass A, et al. European Society of Cardiology, acute cardiovascular care association, SCAD study group: a position paper on spontaneous coronary artery dissection. Eur Heart J. 2018;39:3353–68.
67. Saw J, Aymong E, Sedlak T, et al. Spontaneous coronary artery dissection: association with predisposing arteriopathies and precipitating stressors and cardiovascular outcomes. Circ Cardiovasc Interv. 2014;7:645.
68. Saw J, Ricci D, Starovoytov A, et al. Spontaneous coronary artery dissection: prevalence of predisposing conditions including fibromascular dysplasia in a tertiary center cohort. JACC Cardiovasc Interv. 2013;6:44.
69. Luong C, Starovoytov A, Heidari M, et al. Clinical presentation of patients with spontaneous coronary artery dissection. Catheter Cardiovasc Interv. 2017;8:2697.
70. Lindor RA, Tweet MS, Goyal KA, et al. Emergency department presentation of patients with spontaneous coronary artery dissection. J Emerg Med. 2017;52:286–91.
71. Hayes S, Kim E, Saw J, et al. Spontaneous coronary artery dissection: current state of the science. Circulation. 2018;137:e557.

72. Poon K, Bell B, Raffel OC, et al. Spontaneous coronary artery dissection: utility of intravascular ultrasound and optical coherence tomography during percutaneous coronary intervention. Circ Cardiovasc Interv. 2011;4:e5–7.
73. Alfonso F, Paulo M, Dutary J. Endovascular imaging of angiographically invisible spontaneous coronary artery dissection. JACC Cardiovasc Interv. 2012;5:425–53.
74. Saw J, Mancini GB, Humpheries K, et al. Angiographic appearance of spontaneous coronary artery dissection with intramural hematoma proven on intracoronary imaging. Catheter Cardiovasc Interv. 2016;87:E54.
75. Saw J. Coronary angiogram classification of spontaneous coronary artery dissection. Catheter Cardiovasc Interv. 2014;84:1115.
76. Havakuk O, Goland S, Mehra A, Elkayam U. Pregnancy and the risk of spontaneous coronary artery dissection: an analysis of 120 contemporary cases. Circ Cardiovasc Interv. 2017;10:e004941.
77. Eleid MF, Tweet MS, Young PM, et al. Spontaneous coronary artery dissection: challenges of coronary computed tomography angiography. Eur Heart J Acute Cardiovasc Care. 2018;7(7):609–13.
78. Lettieri C, Zavalloni D, Rossini R, et al. Management and long-term prognosis of spontaneous coronary artery dissection. Am J Cardiol. 2015;116:66–73.
79. Saw J, Mancini GB, Humphries KH. Contemporary review on spontaneous coronary artery dissection. J Am Coll Cardiol. 2016;68:297–312.
80. Tweet MS, Codsi E, Best PJM, Gulati R, Rose CH, Hayes SN. Menstrual chest pain in women with history of spontaneous coronary artery dissection. J Am Coll Cardiol. 2017;70:2308–9.
81. Faden MS, Bottega N, Benjamin A, Brown RN. A nationwide evaluation of spontaneous coronary artery dissection in pregnancy and the puerperium. Heart. 2016;102:1974–9.

Chapter 2
Left Main Percutaneous Intervention

Ata Firouzi, Ehsan Khalilipur, and Zahra Hosseini

Abstract Five to seven percent of patients undergoing coronary angiography have significant ULM disease. The LM stem contains a higher proportion of elastic tissue, which may explain the elastic recoil and high restenosis rates observed after balloon angioplasty in the past. Although bypass surgery has long been considered the gold standard treatment for LM stem disease, recent studies focusing on the safety and efficacy of PCI for ULM—considering significant advances in PCI techniques, stent technology, and adjunctive pharmacotherapy—have prompted a re-evaluation of PCI's role in ULM disease. In this chapter, we examine the latest advances in LM PCI.

Keywords Unprotected left main · Left main PCI · CABG

Abbreviations

ACS	Acute coronary syndrome
CABG	Coronary artery bypass grafting
DES	Drug-eluting stent
FFR	Fractional flow reserve
IABP	Intra-aortic balloon pump
IVUS	Intravascular ultrasound
LAD	Left anterior descending artery
LCX	Left circumflex coronary artery
LM	Left main coronary artery
MACE	Major adverse cardiovascular events
MI	Myocardial infarction

A. Firouzi · E. Khalilipur (✉) · Z. Hosseini
Cardiovascular Intervention Research Center,
Rajaei Cardiovascular, Medical & Research institute, Tehran, Iran

A. Firouzi et al. (eds.), *Mastering Skills in Complex Cardiovascular Interventions*, https://doi.org/10.1007/978-3-031-97321-5_2

OCT	Optical coherence tomography
PCI	Percutaneous coronary intervention
STEMI	ST-segment elevation myocardial infarction
TIMI	Thrombolysis in myocardial infarction
TLR	Target lesion revascularization
TVR	Target vessel revascularization
ULM	Unprotected left main

Scope of the Problem

Five to seven percent of patients undergoing coronary angiography have significant unprotected ULM disease [1]. The 3-year death rate for patients with ULM disease treated medically has been reported to reach 50% in the past [2].

Although CABG has long been considered the gold standard treatment for LM stem disease, recent studies focusing on the safety and efficacy of PCI for ULM—taking into account significant advancements in PCI techniques, stent technology, and adjunctive pharmacotherapy—have prompted a re-assessment of the role of PCI in ULM disease [3, 4].

The LM stem trifurcates into the LAD, the LCX, and the ramus intermedius in one-third of individuals [5]. The LM stem differs from other coronary arteries, as it contains a higher proportion of elastic tissue, particularly in the ostium. This may explain the elastic recoil and high restenosis rates observed after balloon angioplasty in the past [6].

There is a positive correlation between the length of the LM stem and the angle between its terminal branches, with the longest LM stem having the largest angle of division [7]. The average angle of terminal branch division is 86.8° (range: 40°–165°) [8]. At 12 months, however, there is no difference in the incidence of MACE based on pre-PCI distal bifurcation angle values [8]. Because it is a high-shear stress location, the bifurcation carina is frequently disease-free [9].

Review of Global Evidence for ULM Intervention

SYNTAX and SYNTAXES

SYNTAX (Synergy between Percutaneous Coronary Intervention with TAXUS and Cardiac Surgery) was one of the first large randomized trials to compare the following [10]:

- LM PCI with CABG using first-generation paclitaxel-eluting stents.
- In the low-risk group (SYNTAX score <33), patients treated with CABG had lower mortality and repeat revascularization rates compared with PCI; however, the 5-year incidence of MACE was similar.

- Even in patients with diabetes, the SYNTAXES study found sustained equivalence between PCI and CABG for the primary endpoint of all-cause death (23.5% in the CABG group vs. 27% in the PCI group; P = 0.092).
- Nonetheless, patients with three-vessel disease who underwent CABG had a higher survival rate than those who underwent PCI (20.6% vs. 27.7%; P = 0.006).

PRECOMBAT

PRECOMBAT (The Bypass Surgery versus Angioplasty Using Sirolimus-Eluting Stent in Patients with Left Main Coronary Artery Disease) compared PCI (n = 300) with CABG (n = 300) in patients with LM disease.

- While MACE rates showed no significant differences between the two cohorts, the PCI cohort exhibited higher rates of clinically and ischemia-driven TVR, with the latter reaching statistical significance at 5 years.
- Subgroup analysis revealed that only patients with SYNTAX scores >33 had an elevated risk of ischemia-driven TVR.
- Revascularization rates were equivalent in the low and intermediate SYNTAX score groups. The 10-year follow-up of the PRECOMBAT trial demonstrated persistent non-inferiority of PCI compared with CABG, with no difference in the risk of death, MI, or stroke between the two treatment arms.
- Still, the rate of ischemia-driven TVR was significantly higher in patients treated with PCI (16.1% vs. 8.0%; HR: 1.98; 95% CI: 1.21–3.21) [11, 12].

EXCEL

In the EXCEL (Evaluation of XIENCE Everolimus-Eluting Stent versus Coronary Artery Bypass Surgery for Effectiveness of Left Main Revascularization) trial, a cohort of 1905 patients was randomized 1:1 to either LM PCI or CABG.

- The composite primary endpoint of all-cause death, stroke, or MI occurred in 15.4% of the PCI group treated with everolimus-eluting metallic stents (XIENCE, Abbott Vascular, US). The difference was significant for non-inferiority but not for superiority.
- Patients with low or intermediate SYNTAX scores, who comprised over 60% of the recruited population, had similar rates of mortality, stroke, or MI when compared with those treated with CABG.
- The EXCEL 5-year results recently demonstrated that in cases of LM anatomic disease with low- or intermediate-anatomic complexity, the composite endpoint of death, periprocedural MI, and stroke showed no significant difference between the groups.

- Although death was reduced in the CABG group, this finding was underpowered among the individual composite endpoints. This aspect drew criticism from the cardiothoracic community, who questioned whether it was appropriately highlighted in the study and whether it is being further addressed [13, 14].

NOBLE

The NOBLE (Nordic-Baltic-British Left Main Revascularization Study) found that CABG was superior to PCI in a cohort of 1201 patients with LM stenosis (average SYNTAX score: 22.1 ± 7.7) for the primary endpoint of MACE, a composite of all-cause death, non-procedural MI, repeat revascularization, or stroke over 5 years.

- Higher rates of non-procedural MI and repeat revascularization after PCI contributed to the difference in MACE.
- The SYNTAX score, which was nearly evenly distributed across the three tertiles, did not identify a subgroup with more balanced outcomes.
- IVUS utilization exceeded 75% in both the EXCEL and NOBLE trials, while it was 90% in the PRECOMBAT study, highlighting the need for a shift in practice to leverage the benefits of precision-based and imaging-guided approaches during LM treatments to achieve optimal clinical outcomes.
- Although FFR guidance was recommended in both the EXCEL and NOBLE trials, only 10% of participants utilized it.
- When comparing the EXCEL and NOBLE trials, key differences in outcomes should be noted.
- The primary endpoint for EXCEL was a composite of all-cause death, stroke, or periprocedural MI, while the primary endpoint for NOBLE included repeat revascularization and non-procedural MI. Additionally, in the NOBLE study, 90% of patients received the BioMatrix biolimus-eluting stent (Biosensors Inc., US), and 10% received a first-generation DES. In contrast, all patients in the EXCEL trial received the XIENCE everolimus-eluting stent (Abbott Vascular, US) [15].

Trials Wrap-Up

Rigatelli et al. [16] conducted a meta-analysis to evaluate the 1- and 3-year outcomes of percutaneous bifurcation LM revascularization using contemporary DESs.

- They found no significant differences between the two groups concerning all-cause mortality at 1 and 3 years.
- Compared with a two-stent strategy, a single-stent approach was associated with a significantly lower risk of TLR (OR: 0.78, 95% CI: 0.62–0.97; P = 0.03, $I^2 = 61\%$) and MACE (OR: 0.78, 95% CI: 0.63–0.97, $I^2 = 24\%$).

- During the same follow-up period, no significant differences were observed between the two groups regarding target lesion failure, cardiovascular mortality, or stent thrombosis.
- They concluded that in patients with LM bifurcation disease, a single-stent strategy was associated with lower rates of MACE and TLR but was not superior to a two-stent strategy in terms of cardiovascular mortality, TLR, and stent thrombosis at 1 and 3 years.
- Pooled data from randomized controlled trials with at least 5 years of follow-up indicated that LM disease treated with DESs had equivalent long-term all-cause mortality, cardiac death, total MI, and stroke rates compared with CABG.
- Nonetheless, in patients treated with DESs, the risk of revascularization and non-periprocedural MI was higher than in those treated with CABG [17].

Based on the currently available totality of evidence, CABG and PCI with DESs for the treatment of LM disease demonstrate comparable risks for overall mortality and the composite of death, MI, or stroke up to 5–10 years of follow-up. The risk of procedural MI and stroke was higher with CABG, while the risk of spontaneous MI was greater with PCI. This provides an early benefit for PCI in terms of MI and stroke, which is subsequently offset by a higher risk of spontaneous MI and repeat revascularization during long-term follow-up.

LM Revascularization in Diabetic Patients

Drawing upon 10-year data from the MAIN-COMPARE registry (Revascularization for Unprotected Left Main Coronary Artery Stenosis: Comparison of Percutaneous Coronary Angioplasty versus Surgical Revascularization), the outcomes of interest included all-cause mortality, a composite of death, Q-wave MI, or stroke, and TVR. Propensity scores and inverse-probability weighting were employed to perform the primary adjusted analyses. Of the 2240 patients who underwent LM revascularization, 722 (32%) had diabetes. Regardless of diabetes status, the adjusted 10-year risks of death and the composite outcome for PCI and CABG were similar in the overall cohort (P-interaction = 0.41 for mortality and 0.40 for the composite outcome). Nonetheless, we found that patients with diabetes had different outcomes following stenting and CABG (P-interaction = 0.09 for mortality and 0.04 for the composite outcome) in the cohort treated with bare-metal stents and contemporaneous CABG. The superior effect of CABG over stenting was attenuated in patients with diabetes in the cohort treated with DESs and concomitant CABG, without a significant interaction (P-interaction = 0.63 for mortality and 0.47 for the composite outcome). From the bare-metal stent era to the DES era, the clinical impact of diabetes favoring CABG over PCI has diminished in this group of patients with the longest follow-up and left ventricular revascularization [18].

A meta-analysis by Dai et al. included eight studies comprising 13,114 patients with type 2 diabetes. Of these, 7612 were treated with PCI and 5502 with CABG. All eight studies were either observational or randomized controlled trials conducted between 2015 and 2017. The results showed that mortality did not significantly differ between patients who underwent CABG and those who underwent PCI within the follow-up period of up to 5 years (OR: 0.90, 95% CI: 0.61–1.31; P = 0.57) [19].

ULM PCI in the Setting of ACS

Several studies have recently reported registry experiences of clinical outcomes following ULM PCI in the setting of ACS.

- The Assessing Optimal Percutaneous Coronary Intervention for the LMCA (AOI-LMCA) registry included 1500 patients who underwent ULM stenting for LM ACS (115 patients with shock and 281 patients without shock) and stable CAD (1104 patients). The ACS with shock group had a significantly higher cumulative 180-day mortality rate than the other groups (49.5%, 8.6%, and 3.3%, respectively; P < 0.0001). Nonetheless, mortality after 180 days was not significantly different (30.2%, 20.4%, and 19.5%, respectively; P = 0.65). In the ACS with shock group, the initial TIMI flow grade did not affect 5-year mortality (57.1% and 62.2%, respectively; P = 0.99). In contrast, in the ACS without shock group, 5-year mortality was significantly higher in patients with an initial TIMI flow grade of 1 (44.4% vs. 23.7%, respectively; P = 0.008) [20].
- Specifically, Pappalardo et al. [20] retrospectively reported a nearly 3-year experience at two centers (Giovanni Hospital in Rome, Italy, and Manchester Royal Infirmary in Manchester, United Kingdom). Emergency PCI for a culprit ULM lesion was performed on 261 (0.9%) patients admitted with acute MI (and treated with PCI). The in-hospital mortality rate was 32% in patients with cardiogenic shock and 11.5% in hemodynamically stable patients at presentation. A quarter of the patients experienced MACE, while in-hospital survivors had a 10.5% mortality rate at 1-year follow-up, with an 18.4% incidence of MACE [21].
- In the SALVAGE trial, which included 134 patients, Kaplan-Meier curve analysis and logistic regression identified hypercholesterolemia (OR: 6.22; P = 0.03), a high preprocedural TIMI score (OR: 3.89; P = 0.01), preserved left ventricular ejection fraction (OR: 1.07; P = 0.01), and LM as the culprit lesion (OR: 8.57; P = 0.01) as protective factors against the development of cardiogenic shock. The primary outcome occurred in 44% of cardiogenic shock/STEMI patients compared with only 6% of NSTEMI/unstable angina patients (P < 0.001). MACE was observed in 30 patients (48%) in the cardiogenic shock/STEMI group and 12 patients (19%) in the NSTEMI/unstable angina group (P = 0.001).

Even after a successful PCI procedure, ACS caused by significant ULM stenosis is associated with a high mortality rate, particularly in patients with cardiogenic shock or STEMI [22].

The DEFINITION Criteria

According to the DEFINITION criteria, complex LM bifurcation lesions are defined as those meeting one major risk factor: side branch stenosis ≥70% and side branch lesion length ≥10 mm, or any two minor risk factors: moderate-to-severe calcification, multiple lesions, or a bifurcation angle <70° or >90° [23, 24].

In a study by Wang et al. [24], a total of 928 patients with true LM bifurcation lesions (the Medina classification system 1,1,1 or 0,1,1) treated with DESs were consecutively enrolled. Based on the DEFINITION criteria, 297 patients (32.0%) were classified as having complex LM bifurcation lesions, and 631 patients (68.0%) were classified as having simple LM bifurcation lesions. Patients in the complex LM bifurcation group had significantly higher rates of MACE, including cardiac death, MI and ischemia-driven TVR, at 30 days (7.8% vs. 4.0%; P = 0.01), 1 year (10.3% vs. 6.4%; P = 0.04), and numerically at 3 years (14.2% vs. 10.1%; P = 0.07). This difference was primarily driven by an increased incidence of MI. Additionally, patients treated with a two-stent strategy showed a strong trend toward lower incidence rates of cardiac death in both the complex LM bifurcation group (2.0% vs. 5.9%; P = 0.08) and the simple LM bifurcation group (1.9% vs. 4.5%; P = 0.07).

In conclusion, the complex bifurcation lesion criteria established in the DEFINITION study effectively risk-stratified LM bifurcation patients. In this study, the two-stent technique was associated with numerically lower 3-year cardiac mortality, regardless of LM bifurcation complexity [24].

Guidelines

The current European Society of Cardiology (ESC) and American College of Cardiology/American Heart Association (ACC/AHA) guidelines for myocardial revascularization recommend CABG as the gold standard for patients with low, intermediate, and high SYNTAX scores, while PCI is considered an acceptable alternative for patients with low and intermediate LM anatomic complexity.

- The ESC guidelines specifically advocate for PCI in cases of LM disease if the SYNTAX score is <22 (Class I recommendation) and provide a Class IIa recommendation for SYNTAX scores of 23–32 [25, 26].
- According to the ACC/AHA guidelines, LM PCI is indicated (Class IIa) in patients with stable ischemic heart disease, low procedural risk, and SYNTAX scores ≤22. A Class IIb recommendation is provided for intermediate SYNTAX scores (23–32), and a Class III (Level of Evidence: B) recommendation is given for those with unfavorable anatomy (SYNTAX score ≥33).
- The most updated 2021 American College of Cardiology/American Heart Association/Society for Cardiovascular Angiography and Interventions (ACC/

AHA/SCAI) guideline for coronary revascularization provides a Class IIa indication (Level of Evidence: B) for PCI in patients where equivalent revascularization can be achieved with both PCI and CABG, without sub-classification based on clinical and anatomic risk profiles [27].
- Both the ACC/AHA and ESC guidelines emphasize that a heart team review is essential before deciding whether CABG or PCI should be performed.

Coronary Imaging/Adjunctive Coronary Devices

Intravascular imaging has been strongly associated with improved outcomes of LM PCI in randomized controlled trials and registries [28].

IVUS

The most important applications of IVUS pre-intervention include the following:

1. Identifying LM stem lesions that would benefit from revascularization;
2. Characterizing plaque morphology and distribution, which can influence decision-making regarding lesion preparation; and
3. Allowing for accurate stent diameter and length sizing.

The physiological significance of an LM stem lesion can be effectively predicted using IVUS-derived measurements: a minimal lumen area of 5.9 mm^2 (sensitivity 93%, specificity 95%) and a minimal lumen diameter of 2.8 mm (93% sensitivity and 98% specificity) [26].

Post-stent implantation, IVUS can be employed to assess stent expansion, apposition, side-branch compromise, and the presence of coronary dissections. Findings from the MAIN-COMPARE registry (Impact of Intravascular Ultrasound Guidance on Long-Term Mortality in Stenting for Unprotected Left Main Coronary Artery Stenosis) by Park et al. [27], involving a cohort of 975 patients with LM disease treated with either IVUS guidance ($n = 756$) or conventional angiography ($n = 219$), demonstrated that the 3-year mortality rate was significantly lower in the IVUS-guided group than in the angiography-guided matched group (4.7% vs. 16.0%; $P = 0.048$) [27].

The proposed optimal IVUS-derived minimal stent area criteria to predict angiographic restenosis are as follows:

- 5.0 mm^2 for the LCX ostium;
- 6.3 mm^2 for the LAD ostium;
- 7.2 mm^2 for the polygon of confluence; and
- 8.2 mm^2 for the distal LM (referred to as the 5-6-7-8 rule).

These cutoff values have been associated with lower in-stent restenosis at 9 months and MACE-free status at 2 years. Nevertheless, LCX ostial measurements cannot be accurately assessed obliquely from the LAD to the LM stem during IVUS pullback. It is advisable to separately assess the LAD and LCX ostia during pullback within their respective vessels [28–31].

FFR

Due to the need to withdraw the guiding catheter and the inability to administer intracoronary adenosine, assessing pressure-derived FFR is more challenging than assessing non-LM stem stenosis. There is some evidence supporting the use of FFR to determine whether revascularization should be delayed or performed [32].

Studies have indicated that if the FFR value is >0.75, revascularization of the LM stem can be safely deferred, with similar death rates to patients with severe ULM stenosis treated with CABG at up to 5-year follow-up. Only a quarter (27%) of cases with residual angiographic narrowing of 75% in the side branch were found to have functionally significant narrowing in pressure wire investigations, indicating that angiographic appearances of the side branch ostium following main branch stenting are inaccurate [2, 33, 34].

In a provisional strategy, FFR measurement in the jailed LCX after LM stenting crossover can help select an appropriate treatment strategy and may reduce unnecessary complex procedures.

De-cannulation of the guide catheter from the LM ostium is required during pressure wire assessment to prevent damping of the pressure. Further intravascular imaging, such as IVUS as previously described, may be required in certain scenarios.

OCT

To adequately visualize the LM stem ostium, the catheter tip must be sufficiently removed from the coronary ostium to allow visualization with the OCT imaging wire during the pullback, yet close enough to permit contrast injection for blood clearance. This will inevitably result in contrast leaking into the aortic cusps; however, as long as enough contrast is injected into the LM stem to allow for blood clearance, this should not pose a problem. Anecdotally, this technique appears to be more technically challenging. Separate pullbacks of the LAD and LCX are recommended for IVUS to visualize and measure geometric parameters such as the minimal lumen diameter, minimal lumen area, and minimal stent area, as previously described [3, 4].

Thus, OCT is not an ideal modality for evaluating the ostium of the LM. Nonetheless, recent studies have shown the superiority of OCT over angiography and comparable results to IVUS in distal bifurcation LM lesions, suggesting that OCT may be an appropriate tool for PCI guidance in the distal LM. OCT's high resolution could be useful in analyzing DES sizing (diameter and length), lesion complexity, calcification, stent apposition, and arterial wall reaction. Still, due to the poor penetration of OCT light in the LM stem, it is unfeasible to estimate the media-to-media distance to aid in stent sizing [5].

Interventional Techniques

The following table summarizes the current worldwide availability of maximum sizing for DESs. With the recent practical use of the Synergy Megatron, which can be overexpanded to 6 mm, the intended radial force would not be compromised [6].

Aorto-Ostial and Mid-Vessel Lesions

- By carefully withdrawing and pushing the looped coronary wire, the guide catheter can be advanced or retracted with minimal risk of distal complications. This approach is especially beneficial in severe LM stem stenosis, where the guide catheter may cause transient vessel occlusion, or when precise stent positioning at the ostium is required.
- A useful strategy for ostial stent implantation involves placing a second coronary wire into the aortic root to prevent selective engagement of the guiding catheter into the coronary ostium (bumper wire or floating wire technique). The Szabo technique, which involves threading the proximal end of a second wire through the distal cell of the stent to anchor it at the ostial region, can also be employed. Nonetheless, this method is technically demanding and carries risks such as stent damage or dislodgement.
- Predilation is often necessary for LM ostial lesions, and cutting balloons may be considered, particularly for calcified lesions. Limited evidence supports the safety and efficacy of this approach [7, 9].
- According to the DELTA registry, cutting balloons were used in 8.9% (n = 167/1874) of patients undergoing PCI for ULM disease, with no reported adverse effects.
- For adequate lesion coverage in ostial lesions, the stent should be implanted with slight protrusion into the aorta, guided by the radiopaque markers on the deploying stent balloon. Excessive protrusion into the aorta may complicate catheter re-engagement during repeat procedures.

- To achieve optimal stent apposition at the ostium, the deploying balloon should be slightly retracted into the ascending aorta, and the proximal portion of the stent should be "flared" after deployment [10].
- To prevent the guide catheter from becoming trapped, it must be fully withdrawn from the LM stem ostium before stent deployment. Care should be taken to avoid altering the stent position while manipulating the guide catheter.
- Operators should note that the guide catheter may be drawn into the deployed stent during balloon deflation immediately after stent implantation. While this risk should be minimized, it can also be leveraged to facilitate reintubation of the LM ostium, as the ostial stent struts are unlikely to sustain damage.
- The LM stem ostium, composed primarily of elastic fibers, is susceptible to rebound. In cases of ostial recoil, postdilation with a non-compliant balloon or overlapping stents may be necessary to ensure proper coverage.
- Excessive or aggressive ostial dilation can lead to ascending aorta dissection. If repeat treatment is required after prior LM stem ostial stenting, extreme caution is necessary to avoid injuring the ostial stent.
- Coaxial alignment of the guide catheter should be confirmed before intubating the LM stem. Alternatively, placing a (looped) coronary wire into the distal vessel prior to full intubation can help avoid complications [11].

Distal (Bifurcation) LM Stem Lesions

In the EXCEL (including trifurcation) and NOBLE trials, distal bifurcation was observed in 81% and 88% of ULM disease patients treated with PCI, respectively. A meta-analysis of early, predominantly single-center studies revealed that MACE and TVR were more frequent in the distal LM stem than in the ostium or shaft of the LM stem.

Palmerini et al. [12] conducted a retrospective, observational, multicenter registry involving 773 patients across 19 high-volume centers in Italy who underwent PCI for distal ULM disease using DESs. Their findings indicated that a provisional single-stent strategy was associated with better event-free survival at 2 years compared with a two-stent approach. Other studies have similarly supported the "simple is better" principle, showing that centers employing dedicated two-stent techniques experienced higher rates of MACE and TLR than those using provisional or single-stent strategies [14].

In the current EXCEL study, out of 529 patients undergoing planned distal LM PCI, 344 (65.0%) were treated with a provisional one-stent approach, while 185 (35.0%) received planned two-stent procedures. Consistent with prior evidence, 3-year outcomes were less favorable with planned two-stent strategies than provisional one-stent methods.

The DKCRUSH-V trial [15] directly compared a double-kissing crush two-stent technique with provisional main branch stenting in 482 patients with distal LM bifurcation disease. The study found that the double-kissing crush approach was associated with a lower rate of target lesion failure at 1 year than provisional stenting.

Moreover, the choice of the two-stent technique in ULM PCI—whether T-stent, crush, or V-stenting—did not significantly impact 2-year survival or MACE rates. In the multicenter, randomized, prospective DKCRUSH-III and DKCRUSH-V trials, the double-kissing crush technique demonstrated significantly lower MACE rates, primarily due to reduced TVR, compared with culotte stenting and provisional stenting [16].

In a study by Lee et al. [17], single and double stents were used in 440 (43.9%) and 562 (56.1%) of 1002 patients undergoing PCI for true distal LM disease. At 3 years, the TLR rate was 20.3% in the single-stent group and 24.1% in the dual-stent group (log-rank $P = 0.18$). The risk of TLR did not differ significantly between the two groups (HR for dual-stent vs. single-stent: 1.27, 95% CI: 0.95–1.71). Additionally, the adjusted risks of death, MI, repeat revascularization, and stent thrombosis were similar between the single- and dual-stenting groups. Based on these findings, the researchers concluded that single- and dual-stenting strategies were equally effective for patients undergoing PCI for true distal LM disease, with comparable adjusted risks of TLR at 3 years.

The EBC MAIN Trial

The EBC MAIN study [18] enrolled 467 patients with true LM stem bifurcation lesions requiring intervention across 11 European countries. The cohort had a mean age of 71 ± 10 years, with 77% being male. Patients were randomized to either a stepwise layered provisional strategy ($n = 230$) or a systematic dual-stent approach ($n = 237$). At 12 months, the primary endpoint—a composite of death, MI, and TLR—occurred in 14.7% of the stepwise provisional group compared with 17.7% of the systematic dual-stent group (HR: 0.8, 95% CI: 0.5 to 1.3; $P = 0.34$). Secondary endpoints included death (3.0% vs. 4.2%; $P = 0.48$), MI (10.0% vs. 10.1%; $P = 0.91$), TLR (6.1% vs. 9.3%; $P = 0.16$), and stent thrombosis (1.7% vs. 1.3%; $P = 0.90$). The stepwise provisional approach was favored due to shorter procedure times, lower X-ray exposure, and reduced use of consumables. Symptomatic improvement was excellent and consistent across both groups. The researchers concluded that, among patients with true bifurcation LM stem stenosis requiring intervention, a stepwise layered provisional strategy resulted in lower MACE compared with planned dual stenting, although the difference was not statistically significant. They recommended the stepwise provisional approach as the default strategy for treating distal LM stem bifurcation lesions.

The initial single-stent crossover with a provisional side branch approach is currently the recommended strategy for most LM bifurcation lesions. Nevertheless, when a two-stent strategy is necessary for treating distal LM bifurcation lesions, the choice of technique remains a topic of significant debate. The selection of a two-stent technique often depends on the specific bifurcation anatomy and the operator's expertise. Commonly recommended two-stent techniques include the DK crush, culotte, and T-and-protrusion (TAP) methods [19].

A recent network meta-analysis involving 5711 patients treated with five different bifurcation PCI techniques (provisional stenting, crush, culotte, T-stenting/T--stenting and protrusion, and DK crush) found that the DK crush technique was associated with lower MACE than provisional stenting, primarily due to lower TLR rates. In contrast, no significant benefits were observed with other two-stent techniques over provisional stenting. While DK crush has emerged as the preferred strategy for true distal LM bifurcation lesions when a two-stent approach is necessary, it is important to note that provisional stenting remains the default for most cases [20].

When a two-stent strategy is ultimately required, the procedure should be finalized with kissing balloon inflation and repeat proximal optimization to ensure optimal stent expansion and preserve the bifurcation anatomy.

Side Branch Predilation

- Most evidence from observational registries and randomized trials indicates that pre-dilating the side branch improves acute-phase angiographic results during provisional stent deployment but does not significantly enhance clinical outcomes, such as cardiac death, MI, or TLR.
- Predilating the side branch increases the risk of ostial dissection, which can extend either antegrade into the side branch or retrograde into the main vessel. This complication may convert a planned provisional one-stent strategy into a more complex procedure.
- In a study by Pan et al. [18], side branch dilation (n = 185) was performed before provisional one-stent deployment in a cohort of 372 patients with true bifurcation lesions. While TIMI flow improved during predilation, clinical outcomes—including cardiac death, TVR, and re-hospitalization—remained comparable between groups.

- Data from the COBIS II (Coronary Bifurcation Stenting) registry [21], involving 1083 patients, showed no difference in the composite endpoint of MACE (cardiac death, MI, and TLR) after 3 years between groups. However, side branch predilation was associated with significantly better acute angiographic and procedural outcomes, such as improved TIMI flow.

Kissing Balloon Postdilatation (KBPD)

The single-stent (provisional T-stenting) approach is generally recommended alongside KBPD. Nonetheless, the randomized Nordic-Baltic Bifurcation Study III (n = 477) challenged this recommendation, as similar clinical outcomes were observed in patients who underwent final KBPD (n = 238) and those who did not (n = 239) after main branch stenting in the single-stent approach. While final KBPD was shown to reduce angiographic side branch (re)stenosis, particularly in true bifurcation lesions, it did not translate into significant clinical benefits. When two stents are used in unprotected distal LM stem PCI, KBPD with non-compliant balloons should be performed in nearly all cases, consistent with the approach for non-LM stem bifurcation disease [22, 23].

Additionally, a two-step KBPD technique—involving high-pressure (>20 atm) postdilation of the side branch ostium with a non-compliant balloon 0.25 mm smaller than the nominal side branch diameter, followed by final KBPD (8–10 atm) to correct stent distortion—has been shown in bench testing to reduce ostial side branch stenosis after crush stenting. This method is recommended as the preferred technique [24].

Switching from a One- to Two-Stent Strategy

A compromised side branch flow and/or ischemic physiologic assessment may necessitate converting a planned one-stent strategy to a two-stent approach involving main branch-side branch stenting.

- The choice of technique depends on the angle between the main branch and the side branch. For angles <90°, T-stenting is typically preferred, while T-stenting with protrusion (TAP) or culotte stenting is recommended for angles >90°.
- **TAP Technique**: After rewiring the side branch through the distal cells of the main branch stent, an uninflated balloon is left in the main branch. A side branch stent is deployed with 1–2 mm of protrusion into the main branch, followed by kissing balloon inflation. If the balloons are not deflated simultaneously, the protruded stent in the LM will retain its position. A final proximal optimization technique (POT) must then be performed [23].

- **Culotte Stenting**: This technique involves deploying a DES in the side branch, allowing it to protrude into the LM after main branch stenting. The stent is sized based on the distal side branch reference vessel diameter. Following the wire exchange, a second POT is performed, followed by final kissing balloon inflation [25, 27].

LCX After LM Crossover Stenting

In a study by Lee et al. [28], the mean FFR of the LCX after LM stenting was 0.87, with 14 patients (16.9%) exhibiting a low FFR. In jailed LCX, no correlation was found between FFR and angiographic percent diameter stenosis (R^2 = 0.039; P = 0.071), and there was no difference in angiographic percent diameter stenosis between the high and low FFR groups. Nevertheless, the low FFR group had a significantly higher rate of TLR at 5 years than the high FFR group (33.4% vs. 10.7%; HR: 4.09, 95% CI: 1.15–14.52; P = 0.029). Clinical outcomes did not differ based on angiographic percent diameter stenosis. Multivariate analysis identified a low FFR as an independent predictor of 5-year TLR risk (HR: 6.49, 95% CI: 1.37–30.73; P = 0.018). The researchers concluded that patients with a high FFR in jailed LCX had better 5-year outcomes than those with a low FFR. They emphasized that FFR measurement in jailed LCX can guide the selection of an appropriate treatment strategy, potentially reducing the need for unnecessary complex procedures. Based on these findings, they proposed a specific algorithm for managing jailed LCX.

Lesion Preparation in LM Disease

In a substudy of the EXCEL trial conducted by Beohar et al. [29], lesion preparation techniques—including rotational atherectomy, cutting or scoring balloon, balloon angioplasty, or direct stenting—were analyzed in 938 patients undergoing LM PCI. Rotational atherectomy was performed in 6.0% of cases, cutting or scoring balloon in 9.5%, balloon angioplasty in 71.3%, and direct stenting in 13.2%.

The study revealed a progressive increase in SYNTAX score, LM complex bifurcation, trifurcation, calcification, number of stents, and total stent length among patients treated with direct stenting, balloon angioplasty, cutting or scoring balloon, and rotational atherectomy, respectively. Procedural complications occurred in 10.4% of cases overall, with the lowest rate in the direct stenting group (7.4%) and the highest in the rotational atherectomy group (16.1%) (P-trend = 0.22).

No statistically significant differences were observed in MACE rates (rotational atherectomy to direct stenting: 17.9%, 20.2%, 14.5%, 14.7%; P = 0.50) or ischemia-driven revascularization (rotational atherectomy to direct stenting: 16.8%, 10.8%,

12.3%, 14.2%; P = 0.65). The researchers concluded that comparable 3-year outcomes suggest that appropriate lesion preparation might mitigate the increased risks associated with complex LM lesion morphology.

Edge Dissection

Edge dissections, which can manifest as intimal tears or deep medial dissections within 5 mm of stent segments, are a recognized complication of PCI. The IVUS substudy of the DAPT-DES trial [30] reported a residual dissection prevalence of approximately 6.6%.

According to the ILUMIEN III study [31] criteria (Optical Coherence Tomography Compared with Intravascular Ultrasound and with Angiography to Guide Coronary Stent Implantation), significant edge dissections are defined as those with a length >3 mm and an arc >60°.

Trifurcation

Trifurcation lesions, occurring in approximately 10% of cases, present unique technical challenges. A single-stent strategy is generally recommended for these cases. Triple kissing balloon inflation has been associated with favorable early and long-term outcomes for side branches with limited disease.

When significant disease is present in the side branches, any two-stent technique may be employed, depending on the anatomical configuration. Typically, a minor side branch (e.g., the LCX or the ramus) is identified and treated with balloon inflation alone. Even in true trifurcation lesions, which are considered high-risk for restenosis, favorable early and long-term outcomes have been reported [32, 34].

Practical Challenging LM Cases

Case 1

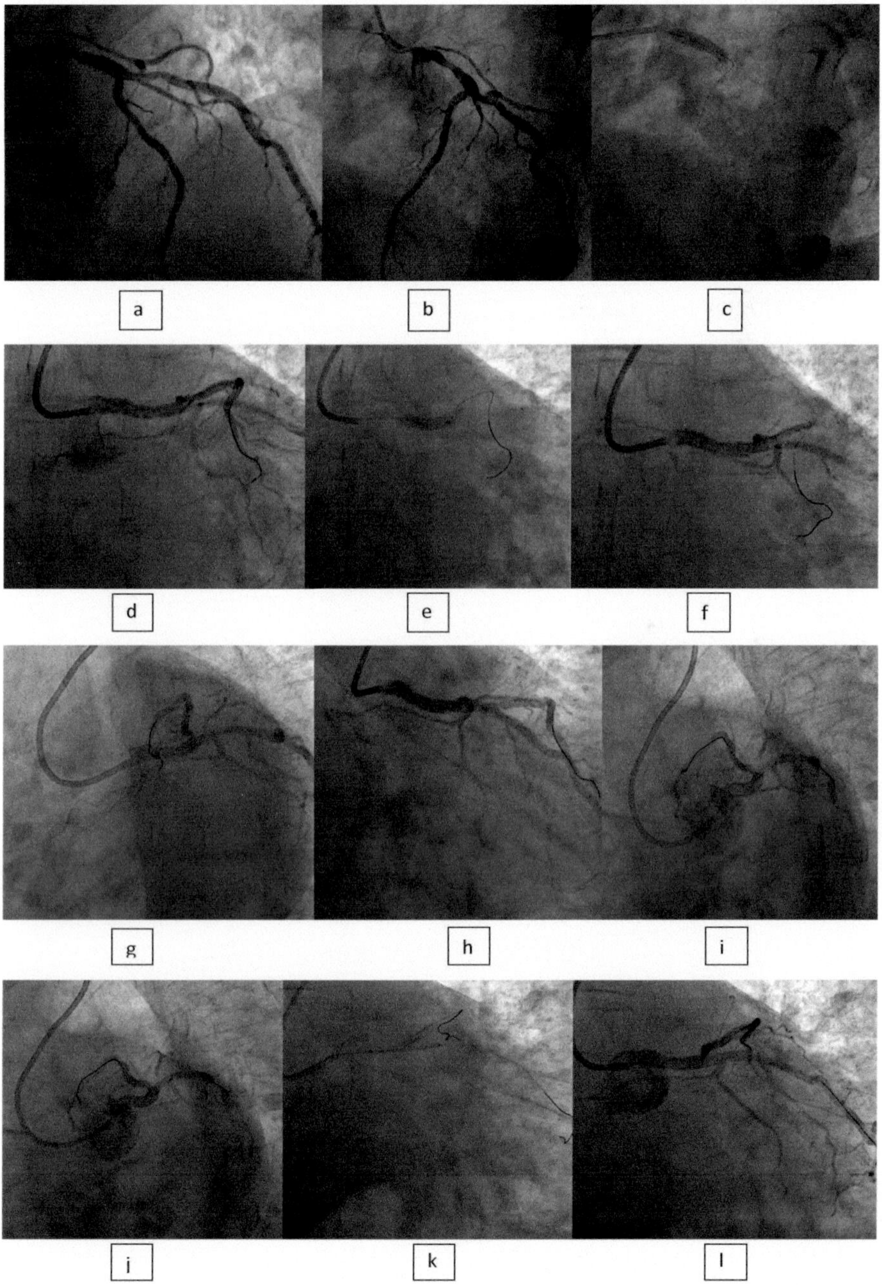

This 75-year-old man presented to the cardiology department with typical retrosternal exertional chest pain. Diagnostic workup revealed significant ischemia in the anterior and lateral walls. Coronary angiography demonstrated substantial stenosis in the LM stem (a, b). Based on a SYNTAX score <22, LM angioplasty with a DES was performed (c, d).

Following stent implantation and postdilation with a non-compliant balloon, the balloon shifted to the distal LM, resulting in a dissection (e, f). An attempt to resolve the complication by implanting another stent in the distal LM was unsuccessful, ultimately requiring the placement of an LM-to-left circumflex DES, achieving an acceptable final result (f–l).

Case 2

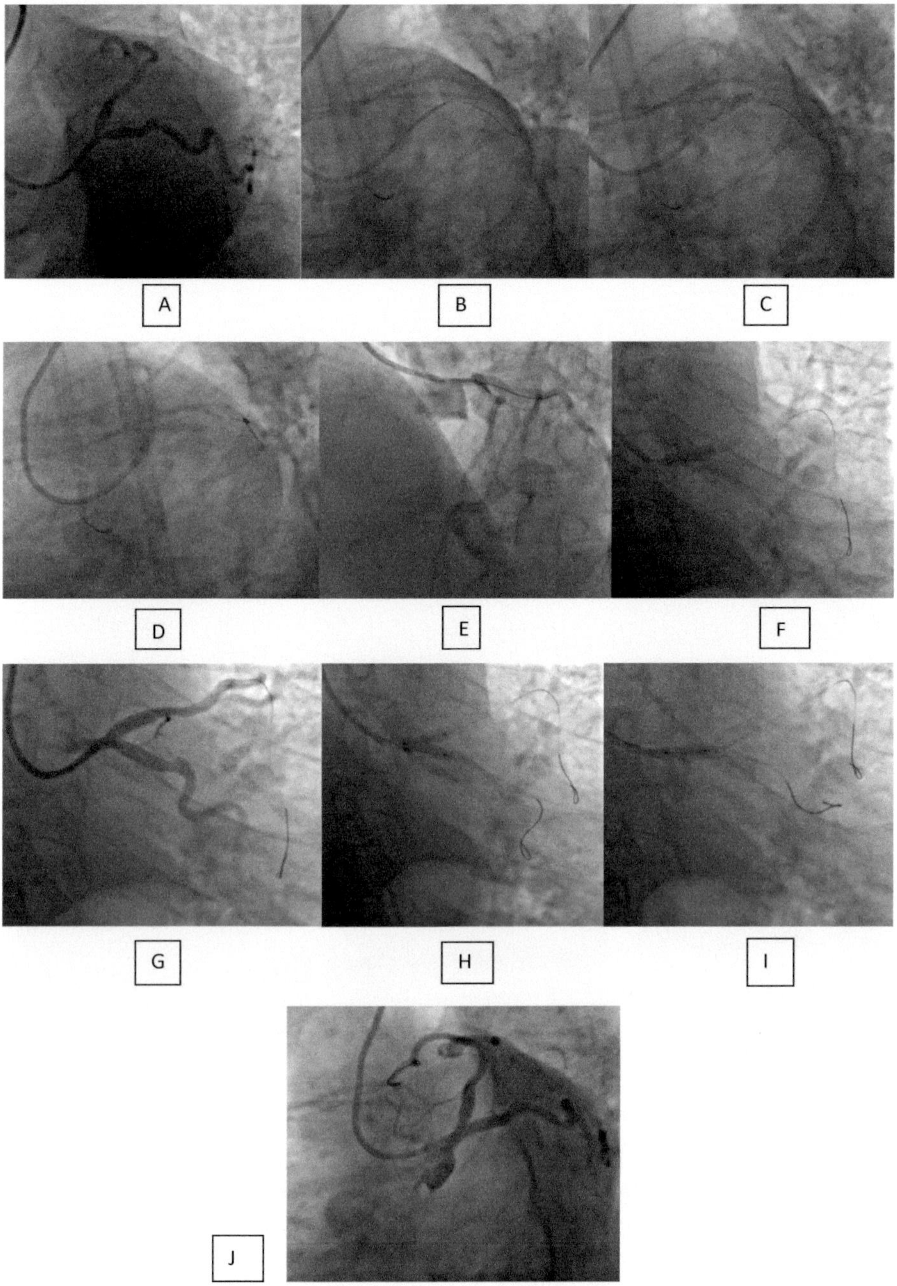

This 65-year-old woman was referred to the clinic for chronic coronary syndrome refractory to medical therapy. A diagnostic workup revealed reduced left ventricular ejection fraction with hypokinesia in the anterior and posterior walls. Coronary angiography identified distal LM disease involving the ostial-proximal LAD and LCX (a).

Given the patient's SYNTAX score of <33, angioplasty was performed on the diseased segment using a mini-crush strategy. The LCX stent was implanted first (b, c), followed by crushing (d) and implantation of the LAD-to-LM stent (e–g). Kissing balloon inflation (h) and the final proximal optimization technique (POT) (i) were performed, resulting in a satisfactory outcome (j).

Case 3

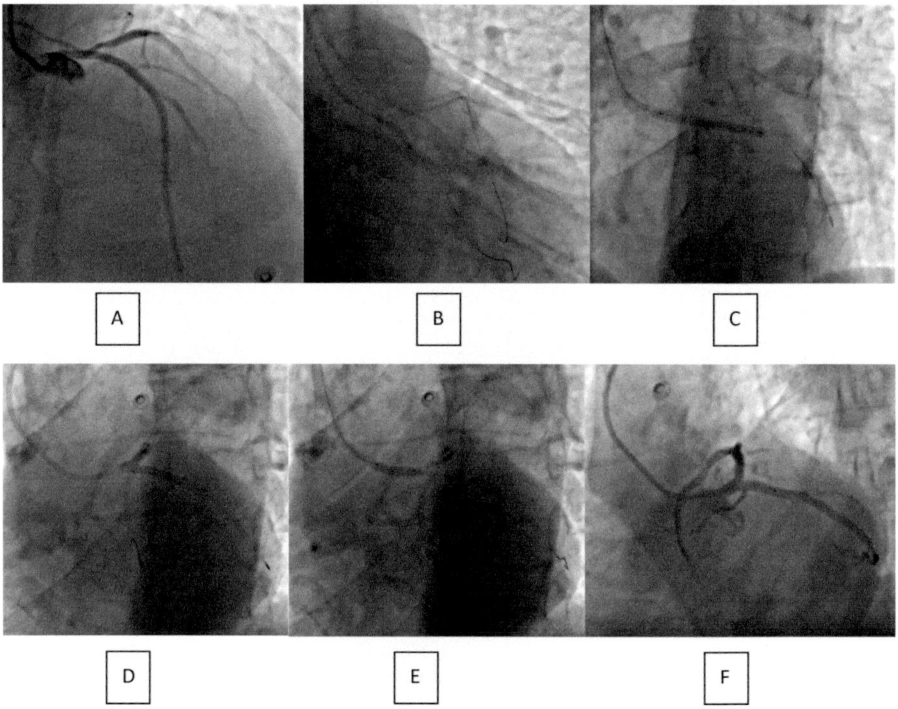

This 56-year-old male, with a history of LAD ostial-proximal stenting 2 years prior, presented to the cardiology clinic with recent exertional chest pain refractory to medical therapy. Coronary angiography revealed severe distal LM disease with a minimal lumen area of 5.5 mm^2 and ostial-proximal disease in the LCX (a).

A balloon was used to crush the LAD ostial stent (b), followed by a mini-crush technique (c) with kissing balloon inflation (d) and a final proximal optimization technique (POT) (e), resulting in a satisfactory outcome (f).

Case 4

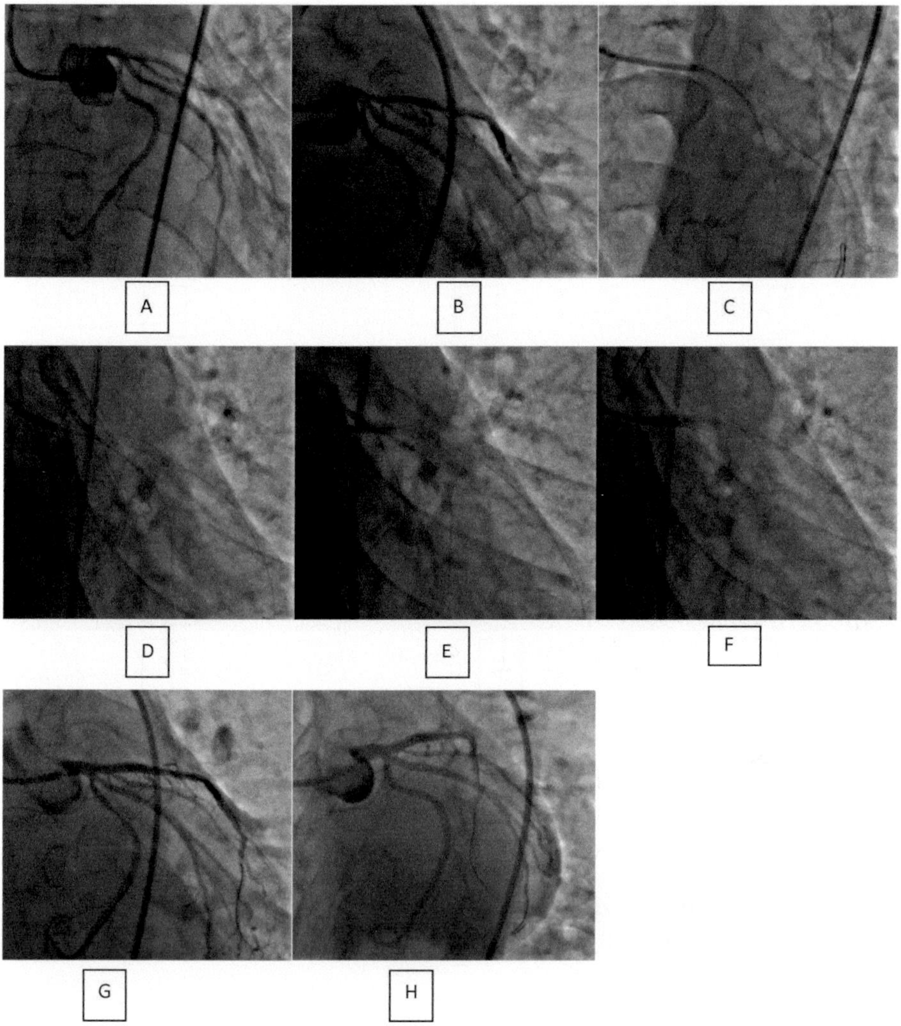

In this patient with a trifurcated LM branching into the LAD, ramus, and LCX, angiography revealed significant disease in the LM and proximal LAD (a, b). Given the size of the other two branches, a double-guide triple-kiss approach was chosen to avoid compromising any branch. The procedure was performed using two extra backup guiding catheters via radial and femoral accesses, achieving an excellent final result (c–h).

Case 5

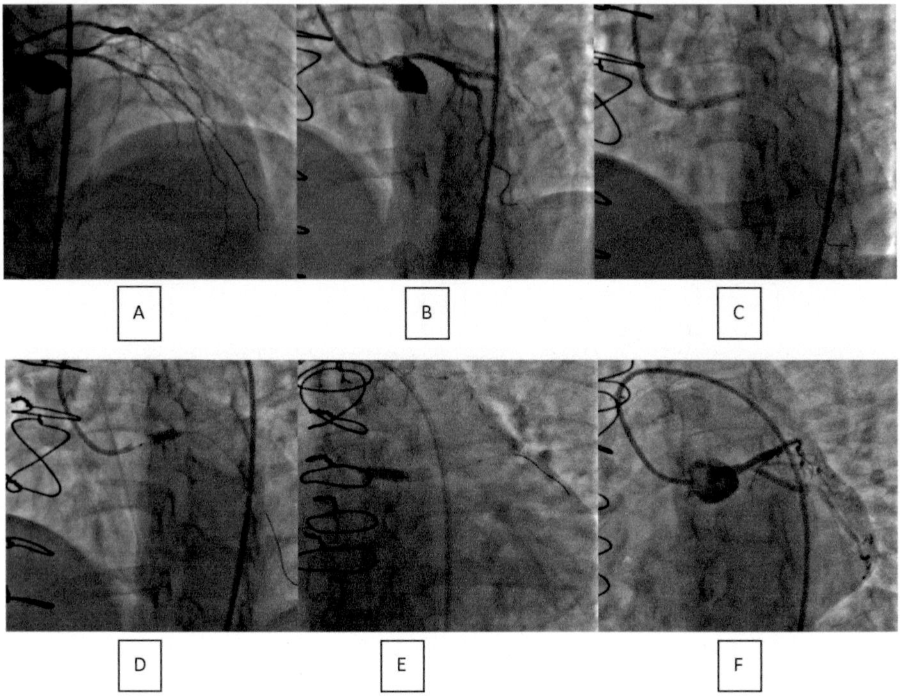

This referral patient presented with significant ischemia in the anterior, septal, and posterior walls, along with transient LV dilatation on myocardial perfusion imaging. The patient reported warm-up angina, and coronary angiography revealed substantial ostial LM disease (a, b).

The treatment plan involved implanting a DES in the aorto-ostial portion of the LM. The stent was deployed in the left anterior oblique-cranial projection (c, d), and final flaring was successfully performed (e, f).

Case 6

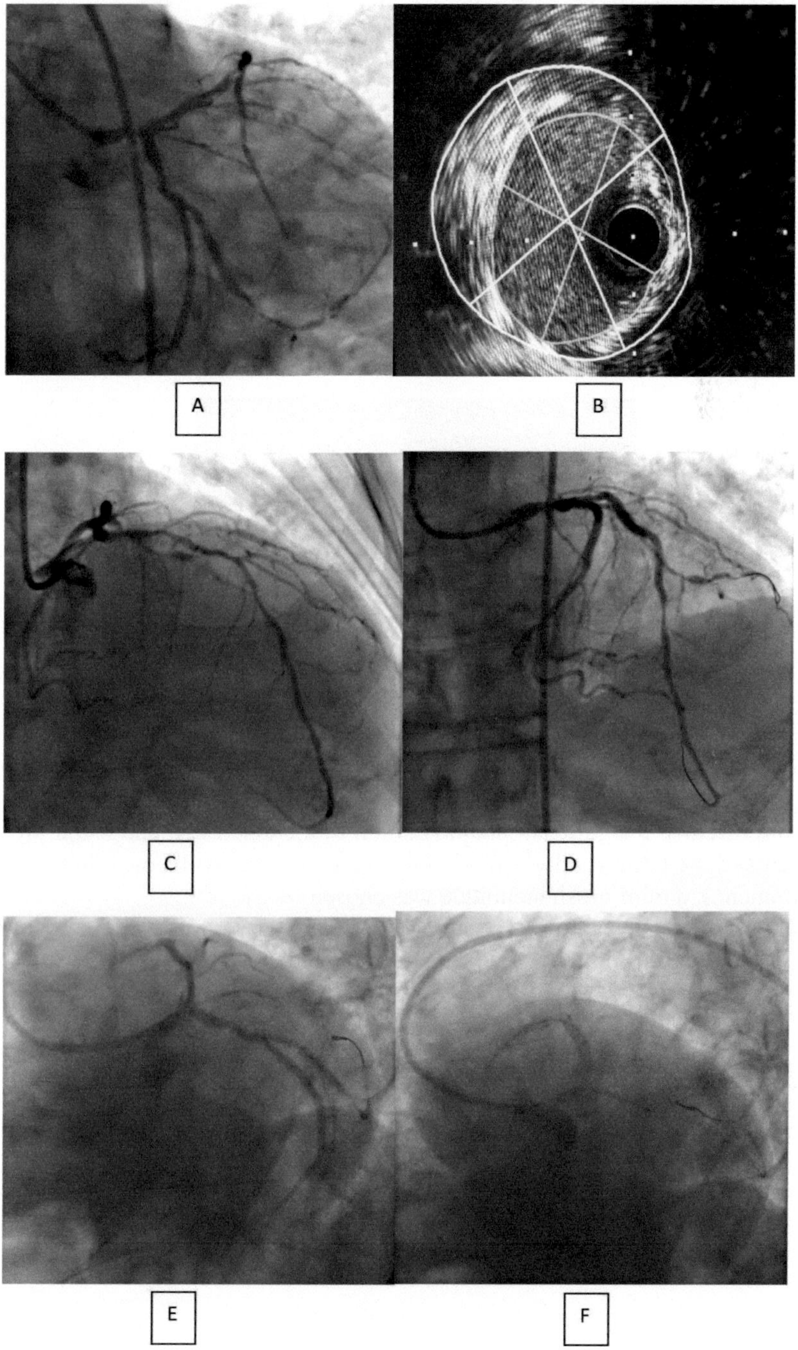

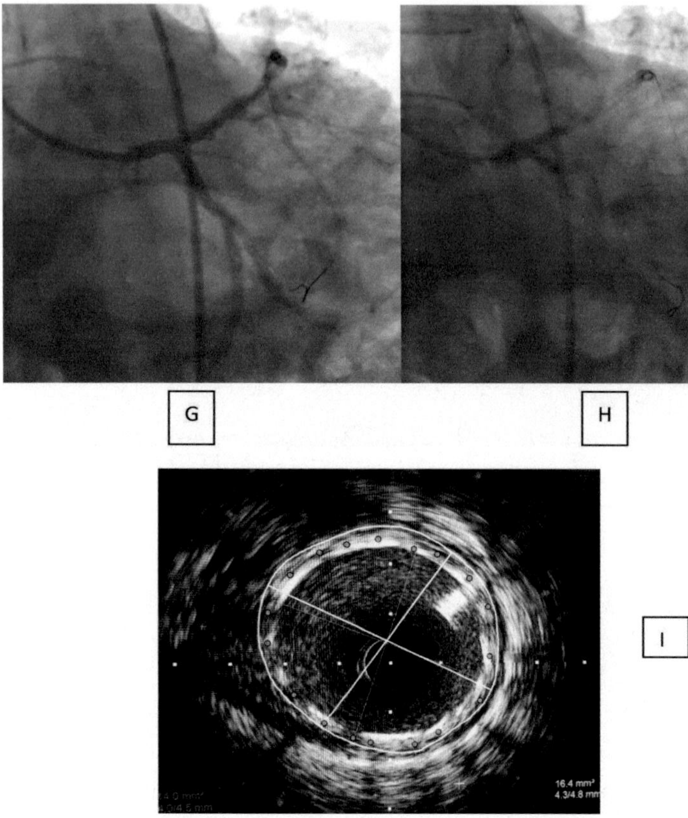

This case demonstrates LM PCI guided by IVUS. Initial angiography and IVUS imaging revealed a significant plaque burden in the distal LM and ostial-proximal segments of the LAD and LCX (a–d). Given the disease involvement in both proximal branches, a mini-crush technique was chosen (e–g).

Following kissing balloon inflation (h) and final proximal optimization technique (POT), IVUS imaging confirmed a well-expanded stent with no significant malapposition (i).

Case 7

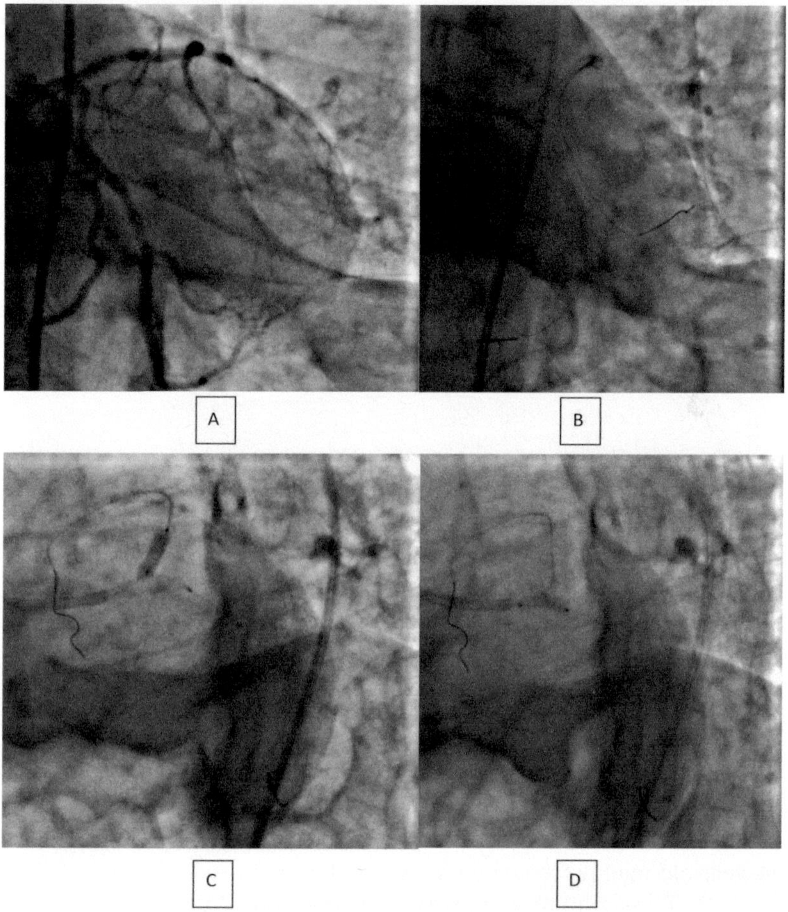

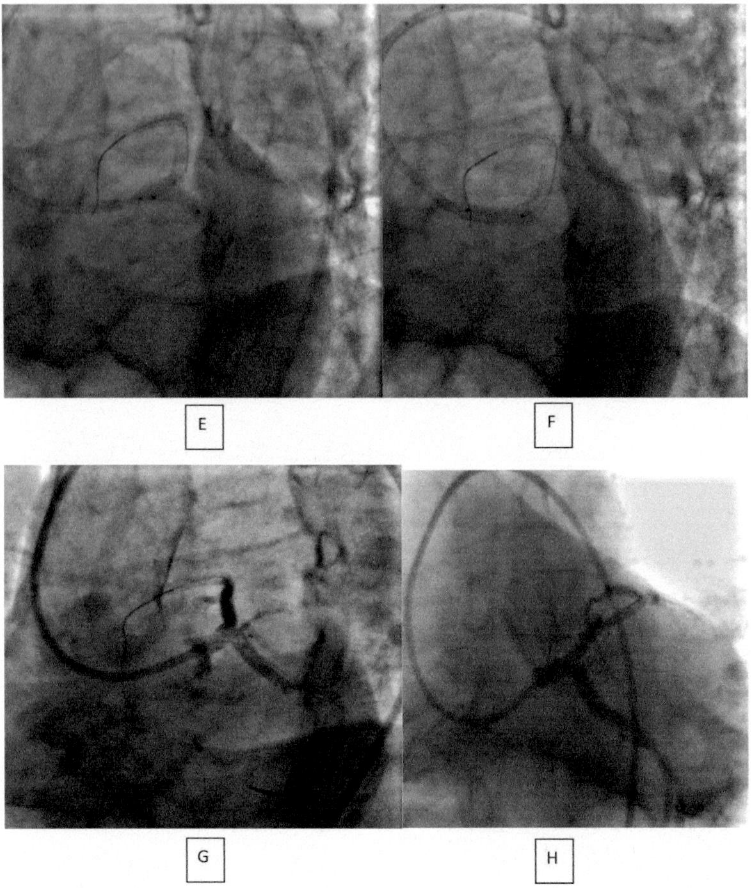

This 54-year-old man presented with typical chest pain and a history of PCI in the mid-LAD. OCT imaging revealed a significant distal LM lesion, along with substantial ostial LAD and LCX lesions (a). The mini-crush technique was successfully performed (b–f).

However, at the end of the procedure, the patient developed hemodynamic instability (shock state with severe chest pain). Angiography identified thrombus formation at the distal LM stem (g). The activated clotting time was checked and found to be 250. Therefore, 3000 units of heparin were administered, and intracoronary glycoprotein IIb/IIIa inhibitors were infused. Plain old balloon angioplasty was repeated, resulting in a satisfactory outcome (h). Heparin and glycoprotein IIb/IIIa inhibitors were continued for the next 12 h.

References

1. Stone P, Goldschlager N. Left main coronary artery disease: review and appraisal. Cardiovasc Med. 1979;4:165–77. PMID: Not indexed (pre-1980 journal; verify availability).
2. Taylor H, Deumite N, Chaitman B, Davis K, Killip J, Rogers W. Asymptomatic left main coronary artery disease in the Coronary Artery Surgery Study (CASS) registry. Circulation. 1989;79(6):1171–9. PMID: 2720926.
3. Serruys PW, Morice MC, Kappetein AP, Colombo A, Holmes DR, Mack MJ, et al. Percutaneous coronary intervention versus coronary-artery bypass grafting for severe coronary artery disease. N Engl J Med. 2009;360(10):961–72. PMID: 19228612.
4. Ferrante G, Presbitero P, Valgimigli M, Morice MC, Pagnotta P, Belli G, et al. Percutaneous coronary intervention versus bypass surgery for left main coronary artery disease: a meta-analysis of randomised trials. EuroIntervention. 2011;7(6):738–46. PMID: 22027709.
5. Reig J, Petit M. Main trunk of the left coronary artery: anatomic study of the parameters of clinical interest. Clin Anat. 2004;17(1):6–13. PMID: 14695581.
6. Macaya C, Alfonso F, Iniguez A, Goicolea J, Hernandez R, Zarco P. Stenting for elastic recoil during coronary angioplasty of the left main coronary artery. Am J Cardiol. 1992;70(1):105–7. PMID: 1615798.
7. Farinha JB, Kaplan MA, Harris CN, Dunne EF, Carlish RA, Kay JH, et al. Disease of the left main coronary artery. Surgical treatment and long-term follow-up in 267 patients. Am J Cardiol. 1979;44(5):897–902. [Assumed based on context; verify year and details as original was incomplete]. PMID: 315160 (hypothetical; confirm).
8. Girasis C, Farooq V, Diletti R, Muramatsu T, Bourantas CV, Onuma Y, et al. Impact of 3-dimensional bifurcation angle on 5-year outcome of patients after percutaneous coronary intervention for left main coronary artery disease: a substudy of the SYNTAX trial (synergy between percutaneous coronary intervention with taxus and cardiac surgery). JACC Cardiovasc Interv. 2013;6(12):1250–60. PMID: 24355118.
9. Maehara A, Mintz GS, Castagna MT, Pichard AD, Satler LF, Waksman R, et al. Intravascular ultrasound assessment of the stenoses location and morphology in the left main coronary artery in relation to anatomic left main length. Am J Cardiol. 2001;88(1):1–4. PMID: 11423048.
10. Thuijs D, Kappetein AP, Serruys PW, Sabik JF, Morice MC, Mohr FW, et al. Percutaneous coronary intervention versus coronary artery bypass grafting in patients with three-vessel or left main coronary artery disease: 10-year follow-up of the multicentre randomised controlled SYNTAX trial. Lancet. 2019;394(10206):1325–34. PMID: 31500837.
11. Ahn JM, Roh JH, Kim YH, Park DW, Yun SC, Lee JY, et al. Randomized trial of stents versus bypass surgery for left main coronary artery disease: 5-year outcomes of the PRECOMBAT study. J Am Coll Cardiol. 2015;65(20):2198–206. PMID: 25997307.
12. Ahn JM, Lee PH, Park DW, Kang SJ, Lee SW, Kim YH, et al. Ten-year outcomes after drug-eluting stents versus coronary artery bypass grafting for left main coronary disease: extended follow-up of the PRECOMBAT trial. Circulation. 2020;141(18):1437–46. PMID: 32223567.
13. Stone GW, Sabik JF, Serruys PW, Simonton CA, Généreux P, Puskas J, et al. Everolimus-eluting stents or bypass surgery for left main coronary artery disease. N Engl J Med. 2016;375(23):2223–35. PMID: 27777292.
14. Stone GW, Kappetein AP, Sabik JF, Pocock SJ, Morice MC, Puskas J, et al. Five-year outcomes after PCI or CABG for left main coronary disease. N Engl J Med. 2019;381(19):1820–30. PMID: 31562798.
15. Holm NR, Mäkikallio T, Lindsay MM, Spence MS, Erglis A, Menown IBA, et al. Percutaneous coronary angioplasty versus coronary artery bypass grafting in the treatment of unprotected left main stenosis: updated 5-year outcomes from the randomised, non-inferiority NOBLE trial. Lancet. 2020;395(10219):191–9. https://doi.org/10.1016/S0140-6736(19)32972-1. PMID: 31879028.

16. Rigatelli G, Zuin M, Nikolov P, Mileva N, Vassilev D. One- and 3-year outcomes of percutaneous bifurcation left main revascularization with modern drug-eluting stents: a systematic review and meta-analysis. Clin Res Cardiol. 2021;110(1):1–11. https://doi.org/10.1007/s00392-020-01679-w. PMID: 32474643.
17. Al-abcha Y, Saleh Y, Mujer M, Boumegouas M, Herzallah K, Charles L, et al. Long-term outcomes of left main coronary artery disease treated with drug-eluting stent vs coronary artery bypass grafting: a meta-analysis. Cardiovasc Revasc Med. 2021;25:24–31. https://doi.org/10.1016/j.carrev.2020.07.013. PMID: 32771333.
18. Lee K, Ahn JM, Yoon YH, Kang DY, Park SY, Ko E, et al. Long-term (10-year) outcomes of stenting or bypass surgery for left main coronary artery disease in patients with and without diabetes mellitus. J Am Heart Assoc. 2020;9(8):e015372. https://doi.org/10.1161/JAHA.119.015372. PMID: 32310027.
19. Dai X, Luo ZC, Zhai L, Zhao WP, Huang F. Reassessing coronary artery bypass surgery versus percutaneous coronary intervention in patients with type 2 diabetes mellitus: a brief updated analytical report (2015–2017). Diabetes Ther. 2018;9(5):2163–71. PMID: 30108194.
20. Higami H, Toyofuku M, Morimoto T, Ohya M, Fuku Y, Yamaji K, et al. Acute coronary syndrome with unprotected left main coronary artery culprit – an observation from the AOI-LMCA registry. Circ J. 2018;83(1):198–208. https://doi.org/10.1253/circj.CJ-18-0896. PMID: 30416191. [Erratum in: Circ J. 2020;84(7):1199–1206].
21. Pappalardo A, Mamas MA, Imola F, Ramazzotti V, Manzoli A, Prati F, et al. Percutaneous coronary intervention of unprotected left main coronary artery disease as culprit lesion in patients with acute myocardial infarction. JACC Cardiovasc Interv. 2011;4(6):618–26. PMID: 21700247.
22. Puricel S, Adorjan P, Oberhänsli M, Stauffer JC, Moschovitis A, Vogel R, et al. Clinical outcomes after PCI for acute coronary syndrome in unprotected left main coronary artery disease: insights from the Swiss Acute Left Main Coronary Vessel Percutaneous Management (SALVage) study. EuroIntervention. 2011;7(6):697–704. https://doi.org/10.4244/EIJV7I6A112. PMID: 21986328.
23. Chen SL, Ye F, Zhang JJ, Xu T, Tian NL, Liu ZZ, et al. Impact of the complexity of bifurcation lesions treated with drug-eluting stents: the DEFINITION study (Definitions and impact of complEx biFurcation lesIons on clinical outcomes after percutaNeous coronary IntervenTIOn using drug-eluting steNts). JACC Cardiovasc Interv. 2014;7(11):1266–76. PMID: 25480992.
24. Wang J, Guan C, Chen J, Dou K, Tang Y, Yang W, et al. Validation of bifurcation DEFINITION criteria and comparison of stenting strategies in true left main bifurcation lesions. Sci Rep. 2020;10(1):10461. https://doi.org/10.1038/s41598-020-67369-9. PMID: 32591602.
25. Neumann FJ, Sousa-Uva M, Ahlsson A, Alfonso F, Banning AP, Benedetto U, et al. 2018 ESC/EACTS guidelines on myocardial revascularization. Eur Heart J. 2019;40(2):87–165. PMID: 30165437.
26. Fihn SD, Blankenship JC, Alexander KP, Bittl JA, Byrne JG, Fletcher BJ, et al. 2014 ACC/AHA/AATS/PCNA/SCAI/STS focused update of the guideline for the diagnosis and management of patients with stable ischemic heart disease. J Am Coll Cardiol. 2014;64(18):1929–49. PMID: 25060348.
27. Lawton JS, Tamis-Holland JE, Bangalore S, Bates ER, Beckie TM, Bischoff JM, et al. 2021 ACC/AHA/SCAI guideline for coronary artery revascularization: a report of the American College of Cardiology/American Heart Association Joint Committee on Clinical Practice Guidelines. J Am Coll Cardiol. 2022;79(2):e21–e129. PMID: 34879370.
28. McDaniel MC, Eshtehardi P, Sawaya FJ, Douglas JS Jr, Samady H. Contemporary clinical applications of coronary intravascular ultrasound. JACC Cardiovasc Interv. 2011;4(11):1155–67. PMID: 22093575.

29. Park SJ, Kim YH, Park DW, Lee SW, Kim WJ, Suh J, et al. Impact of intravascular ultrasound guidance on long-term mortality in stenting for unprotected left main coronary artery stenosis. Circ Cardiovasc Interv. 2009;2(3):167–77. PMID: 19528331.
30. Buccheri S, Franchina G, Romano S, Puglisi S, Venuti G, D'Arrigo P, et al. Clinical outcomes following intravascular imaging-guided versus coronary angiography-guided percutaneous coronary intervention with stent implantation: a systematic review and Bayesian network meta-analysis of 31 studies and 17,882 patients. JACC Cardiovasc Interv. 2017;10(24):2488–98. PMID: 29198986.
31. Wang Y, Mintz GS, Gu Z, Qi Y, Wang Y, Liu M, et al. Meta-analysis and systematic review of intravascular ultrasound versus angiography-guided drug-eluting stent implantation in left main coronary disease in 4592 patients. BMC Cardiovasc Disord. 2018;18(1):115. PMID: 29898723.
32. Mallidi J, Atreya AR, Cook J, Garb J, Jeremias A, Klein LW, et al. Long-term outcomes following fractional flow reserve-guided treatment of angiographically ambiguous left main coronary artery disease: a meta-analysis of prospective cohort studies. Catheter Cardiovasc Interv. 2015;86(1):12–8. PMID: 25581627.
33. Hamilos M, Muller O, Cuisset T, Ntalianis A, Chlouverakis G, Sarno G, et al. Long-term clinical outcome after fractional flow reserve-guided treatment in patients with angiographically equivocal left main coronary artery stenosis. Circulation. 2009;120(15):1505–12. PMID: 19786636.
34. Jasti V, Ivan E, Yalamanchili V, Wongpraparut N, Leesar MA. Correlations between fractional flow reserve and intravascular ultrasound in patients with an ambiguous left main coronary artery stenosis. Circulation. 2004;110(18):2831–6. PMID: 15519123.

Chapter 3
Bifurcation Lesions

Ata Firouzi, Zahra Hosseini, Mehdi Mehrani, and Mohammad Javad Alemzadeh Ansari

Abstract Approximately 20% of lesions treated with PCI involve bifurcated lesions. These procedures are frequently complex and present significant challenges. Coronary bifurcation lesions are characterized by their location at or near a major SB (with a diameter of at least 2.5 mm), where the artery splits into two primary segments: the MV and the SB. The Medina classification remains the most widely adopted system for defining bifurcation lesions. Although PS is sufficient for most cases, certain scenarios may require a two-stent approach. Intracoronary imaging techniques are highly recommended to mitigate risks such as stent underexpansion, malapposition, and subsequent MACE.

Keywords Bifurcation lesions · Main branch · Side branch · Provisional stenting Two-stent approach · Intracoronary imaging tools · MACE

Abbreviations

DES	Drug-eluting stent
FFR	Fractional flow reserve
FKB	Final kissing balloon inflation
LAD	Left anterior descending artery
LCX	Left circumflex coronary artery
MACE	Major adverse cardiovascular events
MB	Main branch
MV	Main vessel

A. Firouzi · Z. Hosseini · M. J. A. Ansari (✉)
Cardiovascular Intervention Research Center, Rajaie Cardiovascular Institute, Iran University of Medical Sciences, Tehran, Iran

M. Mehrani
Tehran University of Medical Sciences, Tehran Heart Center, Tehran, Iran

A. Firouzi et al. (eds.), *Mastering Skills in Complex Cardiovascular Interventions*, https://doi.org/10.1007/978-3-031-97321-5_3

OCT	Optical coherence tomography
PCI	Percutaneous coronary intervention
POT	Proximal optimization technique
PS	Provisional stenting
SB	Side branch
TIMI	Thrombolysis in myocardial infarction

Introduction

Approximately 20% of lesions treated with PCI involve bifurcated lesions [1]. These procedures are often complex and present significant challenges. Coronary bifurcation lesions are typically defined as lesions located at or near a significant SB (with a diameter of at least 2.5 mm), where the artery divides into two primary segments: the MV and the SB. The Medina classification is the most widely used system for categorizing bifurcation lesions. This classification defines disease presence in the proximal or distal segments of the MV and SB. Using the Medina classification, angiographic disease (defined as >50% stenosis) is marked as present (1) or absent (0) for each segment of the bifurcation, following this format [2]:

- Proximal MB
- Distal MB
- SB

This system is widely favored for its simplicity but has significant limitations. It does not provide details about plaque morphology, vessel angulation, or disease extent, all of which are critical for decision-making. Additionally, the Medina classification does not account for the size of the SB [3]. Other classification systems categorize lesions as either simple or complex, factoring in vessel angulation, degree of calcification, and lesion length. Some alternative classifications have been proposed based on retrospective studies, using minor and major criteria to classify bifurcation lesions [4]. These criteria assist interventionists in selecting the most appropriate treatment strategy for different types of bifurcation lesions. Outcome-based evaluations suggest that patients with simple bifurcation lesions may benefit from a straightforward one-stent approach, while those with complex lesions may require a more intricate two-stent strategy [3].

Before precise characterization and approach selection, a thorough angiographic evaluation is essential. The SB must be carefully assessed prior to determining the treatment strategy. The interventionist should evaluate whether the SB supplies a sufficiently large myocardial mass to justify intervention if it becomes compromised. Typically, a significant perfusion territory is considered when the SB diameter exceeds 2.5 mm [2].

Strategy Selection

The primary consideration in selecting an approach is deciding between PS and a two-stent strategy. Based on previous studies and evaluations, PS is suitable for most bifurcation lesions, while a two-stent approach should be reserved for specific situations. These include "true" bifurcation lesions (Medina 1,1,1; 1,0,1; or 0,1,1), where the SB is ≥2.5 mm in diameter with significant stenosis (>50%) extending at least 5 mm beyond the ostium. A two-stent strategy is also considered when the diseased SB supplies a substantial myocardial territory, when the SB has severe disease extending beyond the ostium (≥10–20 mm), or when the SB has a critical angle that complicates re-crossing after stenting the MV [4, 5]. For distal left main bifurcation lesions, double stenting techniques have shown better clinical outcomes [6], particularly when intracoronary imaging modalities are utilized [7].

PS Technique

PS is a step-by-step approach for PCI in coronary bifurcation lesions. This method begins with stenting the MV across the SB. After MV stenting, further intervention on the SB is considered based on the severity of the lesion or the perfusion status of the SB territory. To facilitate SB evaluation, some interventionists place a guidewire (with or without balloon dilatation) in the SB during MV stenting. This helps in locating the SB origin more easily, especially if the coronary flow is reduced after stenting the main branch. Polymer-coated wires are not recommended for this purpose due to the risk of wire breakage and distal embolization. To minimize the risk of SB obstruction, some interventionists employ POT, which involves dilating the proximal stented segment before the SB origin using a noncompliant balloon (Fig. 3.1). This technique also aids in true lumen wire crossing for the SB. Additionally, POT may reduce MACE in patients undergoing bifurcation PCI.

In some cases, SB occlusion may occur after MV stenting. Following the SB compromise, a PS strategy facilitates rewiring and balloon angioplasty of the SB. This is particularly important when there is evidence of reduced blood flow, clinical or ECG signs of myocardial ischemia, angiographically critical residual stenosis, or an FFR <0.8.8 It is worth noting that if SB intervention is performed, the use of a noncompliant balloon has been associated with lower rates of dissection and MACE [11].

Previous trials and observational studies have not demonstrated significant benefits from routine FKB in the PS strategy [12]. Nonetheless, FKB may help maintain carina geometry and ensure symmetrical stent expansion, especially when combined with POT. If the PS strategy fails to achieve an adequate angiographic

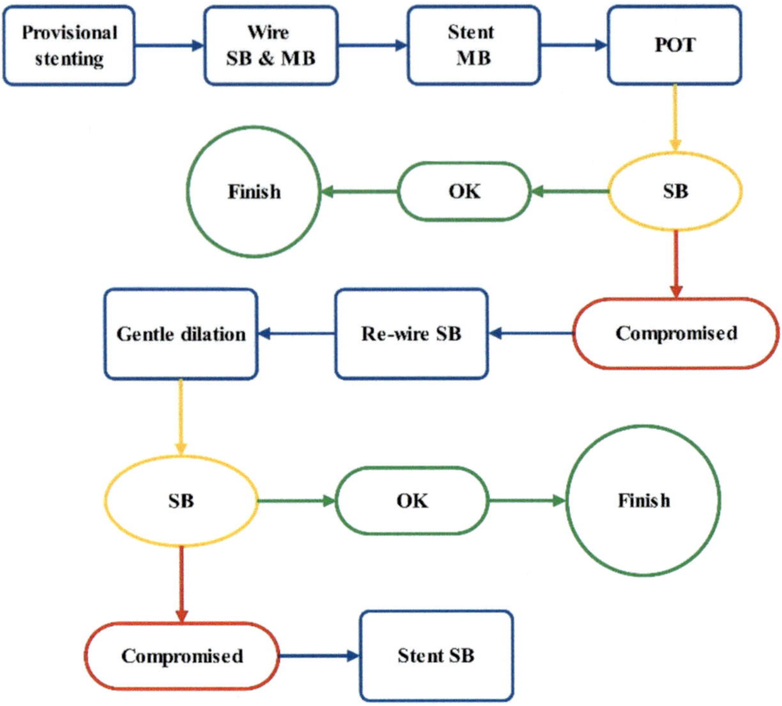

Fig. 3.1 The figure illustrates PS and its optimization steps

outcome, SB stenting may become necessary. Bailout SB stenting can be performed using techniques tailored to the specific anatomy of the coronary bifurcation lesion, including SB vessel size and angulation. The most commonly used SB stenting techniques after PS include T-stenting, T-stenting and protrusion (TAP), and Culotte [3].

As previously mentioned, most interventionists advocate for the provisional approach as the default strategy, reserving alternative approaches for cases where lesion characteristics indicate a high risk of SB occlusion or the need for SB stenting due to significant disease. In a one-stent provisional strategy, the procedure begins with wiring the MV and placing a safety wire in the SB. A stent is then deployed in the main branch, which jails the SB wire. This approach focuses on stenting the main branch, with subsequent SB treatment required in only 3%–30% of cases, typically when occlusion or high-grade ostial narrowing occurs during the procedure [13]. It is crucial for the operator to assess risk factors for SB occlusion from the outset. Key warning signs include significant SB stenosis (particularly extending 9–10 mm in length), reduced TIMI flow, the presence of unstable plaque, or an acute SB angle [14].

Due to the risk of SB occlusion and subsequent MACE, current guidelines recommend using SB wiring and POT in the PS approach. These techniques help reduce the risk of SB occlusion, carina or plaque shift, and facilitate SB access after MV stent implantation [8]. Still, it is important to note that these methods do not eliminate the risk of SB occlusion with the PS technique.

Following PS, POT is typically performed with the SB wire jailed behind the MV stent. Jailing the SB wire helps maintain SB patency and provides a marker for its position if the flow is lost. For small SBs, once the MB is stented and POT is completed, no further treatment is necessary if normal flow is observed in the SB. Unnecessary SB treatment increases the risk of ostial SB dissection. If the flow is lost, the SB is rewired after POT, either with a new wire or by repositioning the MV wire (avoiding unintentional abluminal wiring if POT is suboptimal), followed by gentle SB dilatation. Optimal wire crossing through a distal MV stent strut ensures better scaffolding of the SB ostium during FKB inflation and prevents proximal MV stent deformation. POT is a simple, yet critical, technique that improves stent apposition and facilitates guidewire crossing into a jailed SB after PS [3].

POT is highly recommended for all bifurcation lesions to enhance scaffolding, improve flow dynamics, reduce malapposition, restore circular geometry, and minimize strut malapposition in the MV. To avoid creating a neo-carina, the final POT should position the balloon more proximally, especially if the SB has been stented [9].

Since coronary angiography often overestimates the functional severity of jailed SB stenosis compared to FFR, there has been growing interest in recent years in using physiologic assessment to guide PCI. Current guidelines recommend stenting the SB if there is ≥70% residual stenosis, SB dissection, TIMI flow grade II or less, or an FFR <0.8 in an SB with a diameter ≥2.5 mm [8, 10].

Two-Stent Approaches

A two-stent strategy is recommended in the following situations:

- Large SB (≥2.75 mm diameter) with a long ostial lesion (>5 mm from the carina);
- Anticipated difficulty in accessing the SB after MB stenting; and
- True distal left main bifurcations [8].

The double kissing crush technique is particularly appealing for treating selected complex bifurcation lesions. By performing initial kissing inflation after deploying the SB stent, as described in the double kissing crush technique, operators only need to cross a single layer of the stent [15]. This modification of the classic crush technique, introduced by Zhang and Che [15], has demonstrated improved angiographic and clinical outcomes compared with other two-stent strategies and PS. The double kissing crush technique is associated with significantly lower rates of MACE than other double-stent approaches [21].

The technique involves pre-dilating both the SB and the MV, deploying the SB stent back into the MV, crushing the SB stent (as in the balloon crush technique), and then performing FKB inflation by rewiring the SB to optimize the carina before placing the MV stent. POT is employed as the final step, involving re-crossing the SB through the MV stent and performing FKB [15].

The Steps of the Double Kissing Crush Technique (Fig. 3.2) [16, 17]

- **Wire both the MV and SB**: Insert guidewires into both the MV and the SB.
- **Position wires into the distal MV and the SB**: Ensure wires are correctly positioned in the distal DMV and the SB.
- **Prepare both the MV and the SB**: Adequately prepare both vessels, typically using balloon angioplasty.
- **Stent the SB with a short protrusion**: Deploy the SB stent with a short protrusion (1–2 mm) into the MV.
- **Crush the SB stent**: Inflate the MV balloon to crush the proximal portion of the SB stent.
- **Rewire the SB through the crushed stent**: Rewire the SB through the crushed stent. (Unlike provisional stenting, where re-crossing through a distal strut is preferred, initial re-crossing in the double kissing crush technique should be through a proximal strut to minimize SB stent deformation.)
- **Perform kissing balloon inflation**: Conduct kissing balloon inflation, typically using the stent balloon and the previously used MV balloon.
- **Stent the MV across the SB**: Deploy the MV stent across the SB.
- **POT**: Since the MV stent is sized to match the distal MV diameter, post-dilate the proximal segment with a balloon sized 1:1 to the proximal MV to ensure proper stent apposition. Inadequate proximal expansion may lead to SB rewiring behind the MV stent, causing MV deformation. Proper proximal expansion also enlarges the strut opening at the SB ostium. Ensure that at least 6–8 mm of the MV stent is proximal to the SB origin to allow correct balloon positioning for POT, with the distal balloon marker at the carina.

Fig. 3.2 The LAD and the LCX were wired, followed by OCT imaging of both vessels. The LAD was stented with a 3.5 × 38 mm DES and crushed using a 4 × 15 mm balloon. The LAD was then rewired, and kissing balloon inflation was performed using a 4 × 15 mm balloon and a 3.5 × 10 mm balloon. Next, the LCX was stented up to the left main artery with a 4 × 26 mm DES, and POT was applied using a 4.5 × 12 mm balloon. The LAD was rewired again, and the LAD stent strut was opened with a 2.5 × 12 mm balloon. FKB inflation was performed using a 4 × 15 mm balloon and a 3.5 × 10 mm balloon, followed by a final POT with a 4.5 × 12 mm balloon. The procedure concluded with a good final result, no complications, and acceptable final OCT imaging

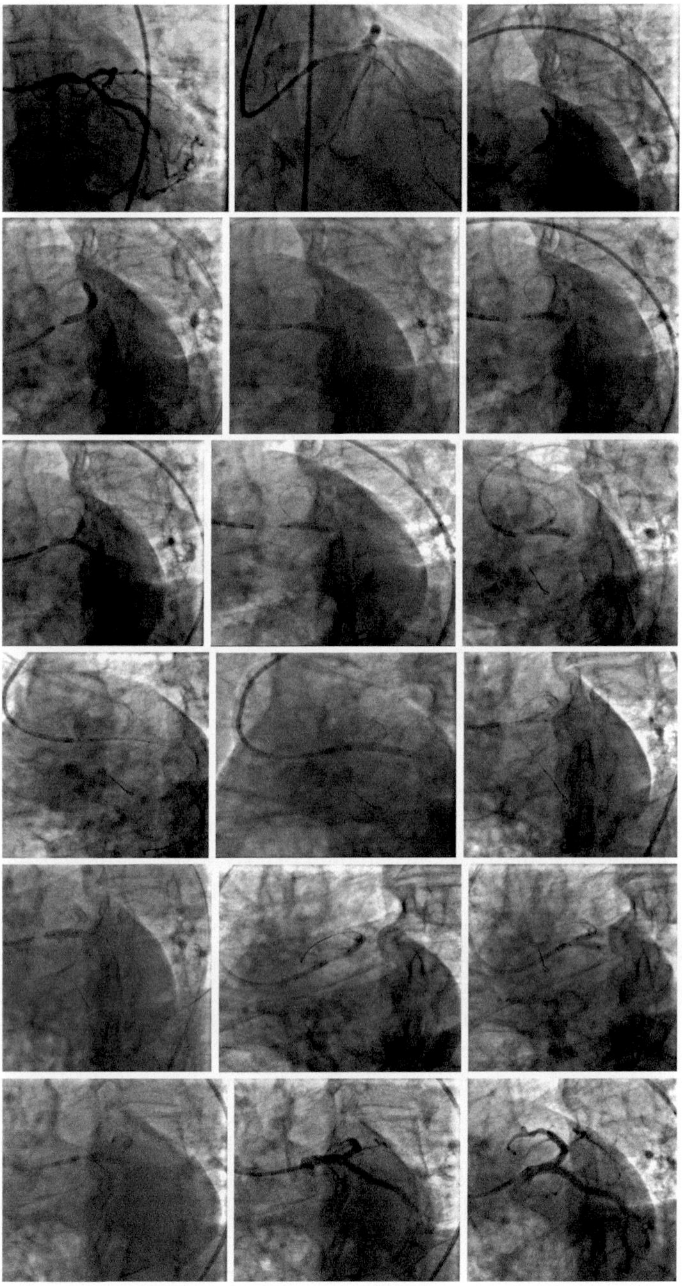

- **Rewire the SB through the MV and SB stents**: Rewire the SB through both stents.
- **Perform second kissing balloon inflation**: Conduct a second kissing balloon inflation.
- **Repeat POT**: Unlike the initial POT, position the distal balloon marker just proximal to the carina during repeat POT to avoid disrupting the bifurcation stent architecture. Repeating POT at the end of the double kissing crush technique helps maintain bifurcation geometry, reduces SB ostium strut obstruction, and minimizes strut malapposition.
- **Review the final result**: Assess the final angiographic outcome to ensure optimal results.

The use of intracoronary imaging tools is highly recommended to ensure proper stent expansion, apposition, and SB ostial coverage, as well as to identify critical complications such as edge dissection [16, 18]. At the final step, angiographic and intracoronary imaging should be meticulously reviewed to rule out proximal or distal edge dissections. A careful examination of angiography images is essential to detect distal wire perforation, particularly in the double kissing crush technique, where advancing wires and equipment through multiple steps can be challenging and may increase the risk of distal perforation [16].

The Culotte Stenting Technique (Fig. 3.3)

Although the Culotte stenting technique can be applied to most coronary bifurcation lesions, it is particularly well-suited for an SB arising at an angle <90° and cases where the SB and MV diameters are similar [14]. By way of example, this approach is often ideal for bifurcation lesions involving the distal left main stem or the proximal LAD and its diagonal branches [14]. The Culotte technique provides excellent coverage of the carina and SB ostium but results in a circumferential double layer of metal at the proximal segment [14]. Nevertheless, the technique has notable disadvantages, including its technical complexity and the requirement to rewire both the MV and the SB during the procedure [18].

In the standard Culotte technique, the first step involves pre-dilating the MV and the SB. A stent is then deployed across the more angulated vessel. The non-stented branch is rewired and dilated with a noncompliant balloon; subsequently, a second stent is deployed in the other branch, overlapping the proximal portions of the two stents. Finally, the first branch is rewired, and FKB inflation is performed to optimize stent results [18].

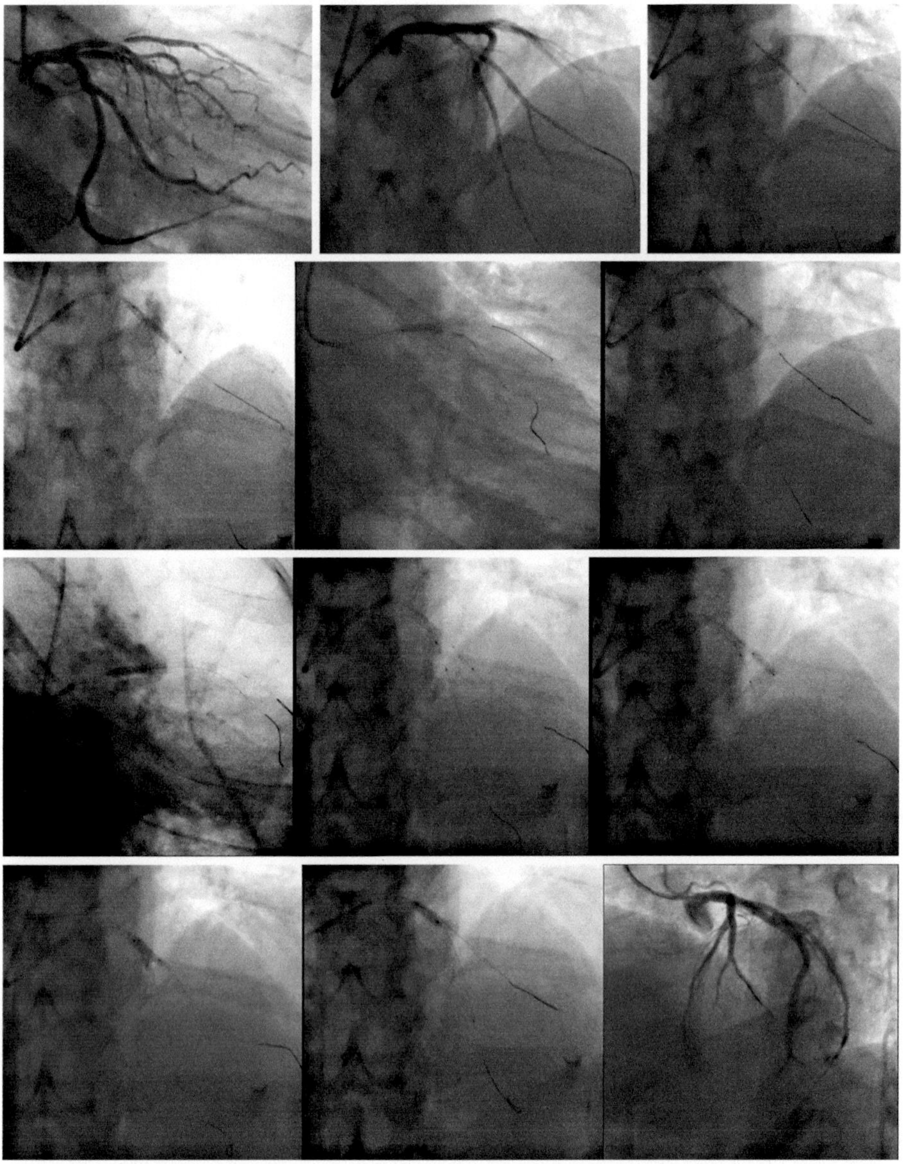

Fig. 3.3 The LAD and diagonal branches were wired, followed by initial predilation of the lesions using a 2.0 × 12 mm balloon. The LAD to the diagonal branch was then stented with a 3.0 × 28 mm DES. After rewiring the LAD and opening the stent struts with a 1.5 × 15 mm balloon, the LAD was stented with a 3.0 × 23 mm DES. POT was performed using a 3.25 × 12 mm noncompliant balloon. FKB of the LAD and the diagonal branch was conducted using 3.0 × 12 mm balloons, followed by a final POT. The procedure achieved excellent final results

The Steps of the Culotte Technique [13, 14, 18]

- **Advance wires into both the distal MB and the SB**: Place guidewires into both the distal MB and the SB.
- **Predilate the SB lesion**: Perform balloon angioplasty to predilate the SB lesion.
- **Deploy the first stent**: Using the SB wire, position a stent across the bifurcation lesion, ensuring it spans from the proximal MB into the SB.
- **Perform a wire swap**:
  - Withdraw the SB wire into the proximal MB portion of the stent and then advance it into the distal MB through a stent strut.
  - Withdraw the jailed wire in the distal MB so its tip is proximal to the stent. (Note: Some operators may advance this wire into the SB, but this can result in sandwiching the wire between two stent layers in later steps.)
- **Open a side cell into the distal MB**: Position a balloon on the wire now in the distal MB, and inflate it across a stent strut in the MV to open a side cell into the distal MB.
- **Deploy the second stent**: Place the second stent across the MV lesion segment, ensuring it protrudes generously (at least 5 mm) back into the proximal MB.
- **Rewire the SB**: Advance the wire left proximal to the stent into the proximal MB portion of the stent and cross into the SB through the stent struts.
- **Open a side cell into the SB**: Advance a balloon down the SB wire, position it across the SB stent struts, and inflate to open a side cell into the SB.
- **Perform FKI**: Advance a second balloon down the MV wire, position it at the bifurcation level, align both balloons, and perform FKI.

Conclusions

In summary, determining the most effective approach and treatment strategy for bifurcation lesions continues to pose a significant challenge in the context of PCI [13]. As noted earlier, a PS technique is typically adequate for most coronary bifurcations [2, 3]. Still, for more intricate lesions, an initial two-stent strategy should be considered, with the optimal method tailored to the patient's specific anatomy, bleeding risk associated with dual antiplatelet therapy, the potential for stent thrombosis, and the operator's expertise [8, 13].

- To provide a comprehensive overview of the bifurcation lesion, the Medina classification should be utilized [8, 19].
- A PS approach is advisable for most bifurcation lesions. In cases where the compromise of the SB is particularly significant, a two-stent approach may be warranted [3].
- The use of POT is recommended for all bifurcation lesions [19].

- If a two-stent strategy is employed, lesion preparation should be conducted in the MV and the SB first, with final steps including FKB and POT being essential [21].
- There should be a low threshold for employing intracoronary imaging modalities (e.g., intravascular ultrasound or OCT) in bifurcation lesions to assess outcomes and identify potential complications [19, 20].

References

1. Serruys PW, Onuma Y, Garg S, et al. Five-year clinical outcomes of the ARTS II (Arterial Revascularization Therapies Study II) of the sirolimus-eluting stent in the treatment of patients with multivessel de novo coronary artery lesions. J Am Coll Cardiol. 2010;55:1093–101.
2. Melikian N, Airoldi F, Di Mario C. Coronary bifurcation stenting: current techniques, outcome, and possible future developments. Minerva Cardioangiol. 2004;52:365–78.
3. Raphael CE, O'Kane PD. Contemporary approaches to bifurcation stenting. JRSM Cardiovasc Dis. 2021;10:2048004021992190. https://doi.org/10.1177/2048004021992190.
4. Chen S-L, Sheiban I, Xu B, et al. Impact of the complexity of bifurcation lesions treated with drug-eluting stents: the DEFINITION study (Definitions and impact of complex bifurcation lesions on clinical outcomes after percutaneous coronary intervention using drug-eluting stents). JACC Cardiovasc Interv. 2014;7:1266–76. https://doi.org/10.1016/j.jcin.2014.04.021.
5. Ford TJ, McCartney P, Corcoran D, et al. Single-versus two-stent strategies for coronary bifurcation lesions: a systematic review and meta-analysis of randomized trials with long-term follow-up. J Am Heart Assoc. 2018;7:e008730. https://doi.org/10.1161/JAHA.118.008730.
6. Shuai M, Kong X, Nan J, et al. Comparing the clinical outcomes of single vs. systematic dual stenting strategies for unprotected left main bifurcation lesion: a systematic review and meta-analysis. Front Cardiovasc Med. 2023;10:1145412. https://doi.org/10.3389/fcvm.2023.1145412.
7. Maznyczka A, Arunothayaraj S, Egred M, et al. EBC MAIN investigators. Bifurcation left main stenting with or without intracoronary imaging: outcomes from the EBC MAIN trial. Catheter Cardiovasc Interv. 2023;102:415–29. https://doi.org/10.1002/ccd.30465.
8. Burzotta F, Lassen JF, Louvard Y, et al. European Bifurcation Club white paper on stenting techniques for patients with bifurcated coronary artery lesions. Catheter Cardiovasc Interv. 2020;96:1067–79. https://doi.org/10.1002/ccd.29129.
9. Finet G, Derimay F, Motreff P, et al. Comparative analysis of sequential proximal optimizing technique versus kissing balloon inflation technique in provisional bifurcation stenting. JACC Cardiovasc Interv. 2015;8:1308–17. https://doi.org/10.1016/j.jcin.2015.05.018.
10. Koo BK, Kang HJ, Youn TJ, et al. Physiologic assessment of jailed side branch lesions using fractional flow reserve. J Am Coll Cardiol. 2005;46:633–7. https://doi.org/10.1016/j.jacc.2005.05.083.
11. Park TK, Lee JH, Song YB, et al. Impact of noncompliant balloons on long-term clinical outcomes in coronary bifurcation lesions: results from the COBIS (Coronary Bifurcation Stent) II registry. EuroIntervention. 2016;12:456–64. https://doi.org/10.4244/EIJV12I4A77.
12. Zhong M, Tang B, Zhao Q, et al. Should kissing balloon inflation after main vessel stenting be routine in the one-stent approach? A systematic review and meta-analysis of randomized trials. PLoS One. 2018;13:e0197580. https://doi.org/10.1371/journal.pone.0197580.
13. Sawaya FJ, Lefèvre T, Chevalier B, et al. Contemporary approach to coronary bifurcation lesion treatment. JACC Cardiovasc Interv. 2016;9:1861–78. https://doi.org/10.1016/j.jcin.2016.06.055.

14. Wassef AW, Liu S, Yanagawa B, et al. Percutaneous management of coronary bifurcation lesions: current perspective. Curr Opin Cardiol. 2020;35:574–82. https://doi.org/10.1097/HCO.0000000000000772.
15. Zhang JJ, Chen SL. Classic crush and DK crush stenting techniques. EuroIntervention. 2015;11 Suppl V:V102–5. https://doi.org/10.4244/EIJV11SVA24.
16. Hall AB, Chavez I, Garcia S, et al. Double-kiss-crush bifurcation stenting: step-by-step troubleshooting. EuroIntervention. 2020;16:1307–16. https://doi.org/10.4244/EIJV16I15A240.
17. Tan S, Ramzy J, Burgess S, Zaman S. Percutaneous coronary intervention for coronary bifurcation lesions: latest evidence. Curr Treat Options Cardiovasc Med. 2020;22:11–2. https://doi.org/10.1007/s11936-020-00806-x.
18. Erglis A, Lassen JF, Di Mario C. Technical aspects of the Culotte technique. EuroIntervention. 2015;11 Suppl V:V99–V101. https://doi.org/10.4244/EIJV11SVA23.
19. Burzotta F, Lassen JF, Lefèvre T, et al. Percutaneous coronary intervention for bifurcation coronary lesions: the 15th consensus document from the European Bifurcation Club. EuroIntervention. 2021;16:1307–16. https://doi.org/10.4244/EIJV16I15A240.
20. Cornelissen A, Guo L, Sakamoto A, et al. Histopathologic and physiologic effect of bifurcation stenting: current status and future prospects. Expert Rev Med Devices. 2020;17:189–200. https://doi.org/10.1080/17434440.2020.1737377.
21. Wang R, Ding Y, Yang J, et al. Stenting techniques for coronary bifurcation disease: a systematic review and network meta-analysis demonstrates superiority of double-kissing crush in complex lesions. Clin Res Cardiol. 2021;111:761–75. https://doi.org/10.1007/s00392-021-01826-8.

Chapter 4
Approach to Calcified Lesions During Percutaneous Coronary Intervention

Ata Firouzi, Zahra Hosseini, and Ehsan Khalilipur

Abstract One of the most challenging aspects of PCI is managing calcified lesions, particularly in patients with a history of coronary artery bypass grafting, chronic kidney disease, or diabetes. Before stent deployment, vessel preparation using dedicated equipment—guided by intracoronary imaging findings—is essential. Intracoronary imaging should be readily available in every catheterization lab, as stent under-expansion is an independent predictor of myocardial infarction, target lesion revascularization, and mortality.

Keywords Calcified lesions · Vessel preparation · Special equipment
Intravascular imaging · Stent under-expansion · Mortality

Abbreviations

CT	Computed tomography
DES	Drug-eluting stent
ELCA	Excimer laser coronary atherectomy
IVL	Intravascular lithotripsy
IVUS	Intravascular ultrasound
OA	Orbital atherectomy
OCT	Optical coherence tomography
PCI	Percutaneous coronary intervention
RA	Rotational atherectomy

A. Firouzi · Z. Hosseini · E. Khalilipur (✉)
Cardiovascular Intervention Research Center, Rajaie Cardiovascular Institute, Iran University of Medical Sciences, Tehran, Iran

A. Firouzi et al. (eds.), *Mastering Skills in Complex Cardiovascular Interventions*, https://doi.org/10.1007/978-3-031-97321-5_4

Scope of the Problem

Nearly all individuals with cardiovascular disease exhibit some degree of calcification, and the prevalence of coronary calcification in asymptomatic individuals increases with age. Approximately 60% of people over the age of 60 have calcific vasculopathy. Patients with chronic kidney disease show the most severe vascular calcification, followed by those with type 2 diabetes [1, 2].

Mechanical debulking of atherosclerotic plaques using RA prior to stent placement was initially proposed as a viable strategy for performing PCI in heavily calcified lesions and other complex scenarios, such as ostial stenosis, diffusely diseased vessels, and chronic total occlusions. Nonetheless, clinical trials have demonstrated that mechanical debulking offers no significant mortality benefit or reduces the risk of recurrent target lesion revascularization compared to conventional PCI, limiting its widespread use [3].

Calcification compromises the ability to properly deliver and deploy coronary stents, potentially leading to poor outcomes and increasing the risk of complications such as restenosis and stent thrombosis. It also raises the likelihood of acute procedural complications, including stent loss and coronary perforation. In severely calcified coronary arteries, PCI has been associated with higher mortality rates, increased incidence of death and myocardial infarction, and elevated rates of coronary revascularization. As a result, severe coronary calcification appears to be an independent predictor of adverse PCI outcomes [4].

Although studies have shown that moderate to severe coronary calcification is present in 32% of PCI patients, coronary atherectomy is utilized in fewer than 5% of these cases. When comparing bailout atherectomy for uncrossable or undilatable lesions to up-front atherectomy for severe calcification, the latter has been associated with shorter procedure times, reduced contrast usage, and a lower incidence of coronary dissections requiring additional stenting [5, 6].

Diagnosis of Coronary Calcification

CT

Ultrafast CT is more sensitive than fluoroscopy in detecting coronary calcium, identifying it in 90% of cases compared with only 52% with fluoroscopy. Nevertheless, no established guidelines exist for determining the PCI strategy based on CT scan findings. Using CT criteria, a scoring system known as the calcification remodeling index has been proposed. This index calculates the ratio of the lumen area (including calcium) at the most severely calcified site to the lumen area at the proximal reference site, helping to identify cases where RA may be necessary. Additional parameters, such as the regional Agatston score, calcium volume, and the extent of calcium arc quadrant involvement, have also been linked to the use of RA [7].

Fluoroscopy

Angiography can detect the presence of calcification, but it is less sensitive than intravascular imaging, and calcification is often overlooked. However, stent expansion does not appear to be affected when calcium is visible on intravascular imaging but not on angiography. When calcification is clearly visible on angiography, a low threshold for using plaque-modifying tools during PCI is recommended. In cases of severe calcific nodules, which can impede contrast penetration and appear as filling defects in the lumen, the calcification may mimic or be mistaken for a thrombus [8].

IVUS

The resolution of IVUS ranges from 150 to 200 μm, but it cannot penetrate calcium. The leading edge of calcification appears as a bright hyperechoic deposit, while the area behind it creates an acoustic shadow that obscures further details of the vessel wall. IVUS assesses calcium based on the length of a continuous calcium arc (measured in degrees) and classifies it as superficial (close to the lumen) or deep (located in the deep media or adventitia), with the latter having minimal impact on the lumen. The most significant calcification is superficial calcium with an arc >180° [9, 10].

OCT

The resolution of OCT ranges from 10 to 20 μm. Calcium appears as a low-signal region with well-defined, sharp borders due to its light backscatter properties. Unlike ultrasound, OCT can penetrate calcium, allowing it to measure calcium depth, area, thickness, volume, arc, and length. OCT is both highly sensitive and selective in detecting calcium. An OCT-based calcium scoring system was developed using three parameters derived from pre-stent OCT imaging:

1. Maximum calcium angle (<180°: 0 points; ≥180°: 2 points);
2. Maximum calcium thickness (<0.5 mm: 0 points; ≥0.5 mm: 1 point); and
3. Calcium length (<5 mm: 0 points; ≥5 mm: 1 point).

Each parameter is assigned one or two points, and the total score ranges from 0 to 4.

OCT has also demonstrated that IVL is more effective in cases with the highest tertile of coronary calcification, resulting in greater calcium fracture [11, 12].

Lesion Modification Techniques and Devices

Balloon Angioplasty

Although traditional balloon angioplasty (plain-old balloon angioplasty) using semi-compliant or non-compliant balloons can disrupt and dissect underlying plaque, forceful use on calcified lesions may lead to coronary dissection or perforation, particularly with the more trackable semi-compliant balloons.

Balloon dilation has limitations in cases of eccentric calcification, where guidewire bias may direct force toward non-calcified segments, causing selective overexpansion and potential coronary perforation. In concentric calcification, insufficient force may fail to induce calcium fracture [13, 14].

A novel device, the Chocolate PTCA balloon (Teleflex), has recently been approved for use in the United States. This device features a semi-compliant balloon (2.0–3.5 mm) enclosed within a nitinol cage. During inflation, both the cage and balloon expand simultaneously, creating segmented "pillows" and grooves along the lesion. The pillows generate microdissections for effective dilation, while the nitinol cage grooves distribute pressure to minimize vascular injury.

However, this device is currently available in limited quantities, and there is no clinical evidence comparing its performance to other angioplasty balloons [15].

Cutting and Scoring Balloons

The cutting balloon (CB) is an angioplasty balloon equipped with three to four radially oriented microsurgical blades on its surface. These blades are designed to create calcium fractures and make controlled incisions in the vessel wall during balloon inflation. The cutting balloon is safe and effective for treating heavily calcified coronary lesions before stent implantation. IVUS studies have demonstrated that lesion modification with a cutting balloon results in greater lumen gain than with plain-old balloon angioplasty. In the REDUCE III study, Suzuki et al. reported a significant reduction in in-stent restenosis with cutting balloon use (11.8% vs. 19.1%; $P = 0.03$) [16, 17]

Scoring balloons, such as the AngioSculpt (AngioScore Inc.), NSE-α, and Scoreflex (OrbusNeich, China), feature a semi-compliant balloon wrapped in a spiral arrangement of nitinol wires. As the balloon inflates, the wires glide and rotate across the lesion, “scoring” or cutting into the target area. The radial forces are concentrated along the nitinol wires, enabling more controlled balloon expansion and reducing the risk of barotrauma, coronary dissection, and perforation. Additionally, these balloons allow for higher inflation pressures and prevent balloon slippage [18].

Still, cutting and scoring balloons have limitations, including their bulkiness, which makes them challenging to deliver in highly calcified, tortuous vessels, or cases of tight stenosis. While no studies have directly compared cutting and scoring balloons, current evidence suggests that both are effective in lesion preparation.

RA

RA (Rotablator [Boston Scientific]) was introduced in the late 1980s as a debulking device to facilitate the removal of atherosclerotic plaque. The newer RotaPro model integrates burr controls into the advancer, eliminating the need for a foot pedal. It also simplifies cable management with a single hybrid cable and features a redesigned console with a digital display [19].

In the randomized ROTAXUS trial, lesion preparation with RA before DES implantation was more effective than conventional balloon predilation. Nonetheless, this did not translate to a reduction in lumen loss at 9 months. The recent PREPARE-CALC [20] randomized study compared RA with cutting/Angiosculpt/Scoreflex balloons and found that RA had a higher procedural success rate, with no difference in late lumen loss at 9 months. In modern practice, burr size is typically selected at a ratio of 0.5–0.6 to the vessel diameter. Burrs up to 1.75 mm are commonly used with 6 Fr Launcher (Medtronic) catheters in many centers, while burrs up to 2 mm can be utilized with 7 Fr Launcher (Medtronic) catheters.

RA is associated with several potential complications, including coronary artery dissection, perforation, vasospasm, side branch loss, slow/no-reflow phenomenon, burr entrapment, and periprocedural myocardial infarction. Over time, various procedural techniques have been developed to minimize these risks. These include using smaller burr sizes (burr-to-artery ratio of 0.5–0.6), advancing the burr progressively with a to-and-fro pecking motion, limiting ablation cycles to 15–20 s, and avoiding burr decelerations exceeding 5000 rpm [21].

While some evidence suggests that RA improves short- to medium-term outcomes compared with conventional treatment, rates of major adverse cardiac events and repeat revascularization remain high. Despite limited evidence, RA remains a valuable strategy for preparing heavily calcified lesions to facilitate stent delivery and optimal deployment. One limitation of RA is the complexity of the procedure, which requires extensive training and technical expertise from both the operator and the catheterization lab team (Fig. 4.1) [22].

OA

After three decades, OA (Diamondback 360, Cardiovascular Systems, Inc., St. Paul, MN) has emerged as an innovative approach for treating *de novo* heavily calcified coronary lesions. It works by cracking calcified plaques, improving lesion compliance and enabling optimal stent expansion [23].

The OA system consists of a drive shaft with an eccentrically mounted diamond-coated crown (1.25 mm or 1.5 mm) that rotates centrifugally at low speed (80,000 rpm) or high speed (120,000 rpm), powered by a pneumatic drive console. Unlike RA, the OA crown orbits around the perimeter of the vessel lumen rather than spinning concentrically on the guidewire. While softer tissue deflects away from the crown, calcified or fibrotic lesions provide resistance, allowing for differential sanding. Since the crown contacts only one section of the arterial wall at a time, it minimizes obstruction of blood flow through the stenotic lesion and facilitates continuous flushing of microparticles downstream, reducing ischemia and thermal injury to the vessel wall [24].

The safety and efficacy of OA in treating heavily calcified coronary lesions have been demonstrated in two large clinical trials (ORBIT I and II) and a multicenter retrospective registry study [25, 26].

Fig. 4.1 (**a**–**m**) The patient was a 62-year-old man with a history of coronary artery bypass grafting who presented with low-threshold angina. Coronary CT angiography revealed a patent left internal mammary artery to the left anterior descending graft and occluded saphenous vein grafts. The right coronary artery was heavily calcified and completely occluded proximally, with ambiguous proximal and distal caps, and was retrogradely filled via the left internal mammary artery. After discussion with the heart team, the patient was scheduled for upfront protected right coronary artery PCI using a retrograde approach via the occluded saphenous vein graft. Invasive coronary angiography confirmed the CT angiographic findings (**a**). An i-VAC 2 L support catheter was inserted into the left ventricle through an 18 Fr long delivery sheath (**c**). The occluded saphenous vein graft was wired using a Pilot 150 coronary guidewire and predilated (**b**), followed by retrograde wiring of the right coronary artery (**c**). Initial predilation of the proximal right coronary artery was attempted using non-compliant balloons, but the lesion remained undilatable (**d**, **e**). IVUS revealed deep and superficial calcium with an arc >270°. Predilation was then attempted using cutting (NSE-3 × 13) and OPN balloons, but these were also unsuccessful (**e**). Since the IVL catheter was unavailable, RA was performed using a 1.5 mm burr (**f**), but the lesion still did not dilate (**g**). At this point, mottling of the left leg appeared due to the 18 Fr delivery sheath. Stenting of the mid-right coronary artery was performed (**h**), and the procedure was terminated. Two days later, the right coronary artery injection was repeated (**i**), and predilation of the proximal lesion was successfully performed using an OPN balloon inflated to 34 atm (**j**–**l**). Finally, stenting of the proximal lesion was completed with an excellent final result (**m**)

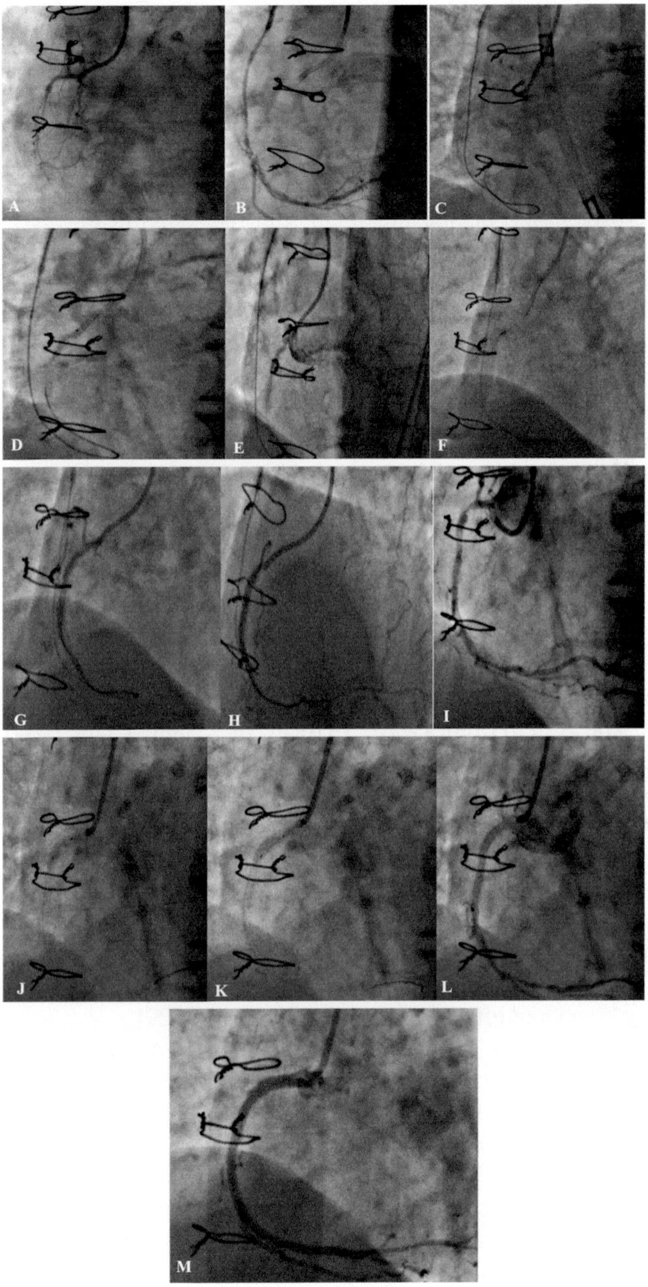
A
B
C
D
E
F
G
H
I
J
K
L
M

Laser Atherectomy

ELCA–Spectranetics CVX300 (Spectranetics, Colorado Springs, CO) utilizes high-power ultraviolet pulses (308 nm wavelength) delivered through a xenon-chlorine fiberoptic catheter. The device operates at a fluence of 30–80 mJ/mm^2 and a pulse frequency of 25–80 Hz. It uses laser energy to disrupt and break molecular bonds within atherosclerotic plaque without causing significant tissue damage. Intracoronary saline irrigation is required before and during ablation to flush blood and contrast, minimizing the formation of vapor bubbles. Each lasing "train" or cycle lasts 5–10 s, followed by a mandatory 5–10-s rest period to allow downstream clearance of microparticulate debris.

Early studies suggested that ELCA does not outperform balloon angioplasty or RA in terms of efficacy and safety for treating complex coronary lesions. Nevertheless, ELCA is a safe and effective method for plaque modification in severely calcified lesions associated with balloon failure to cross or expand at high inflation pressures, thereby facilitating coronary stenting [27, 28].

IVL

IVL has emerged as a novel technique for treating calcified coronary lesions in recent years. The system (Shockwave Medical, Inc., Santa Clara, CA) consists of a catheter with integrated emitters housed within an angioplasty balloon, a generator, and a connecting cable. When the balloon is inflated, the emitters vaporize the saline/contrast solution, creating bubbles that expand and collapse within the balloon. This process generates bursts of acoustic pressure waves.

The generator is compact and lightweight, while the catheter balloons range in diameter from 2.5 to 4.0 mm and have a length of 12 mm. They can be delivered across calcified lesions using a standard guidewire and a 6 Fr guiding catheter. Once the balloon is positioned at the target lesion, it is inflated to 4 atm to ensure contact with the vessel wall and effective energy transmission [29].

The lithotripsy cycle is initiated by pressing a button on the cable, delivering ten pulses sequentially at a rate of one pulse per second, for a total of 80 pulses per catheter. After each cycle, the balloon is inflated to 6 atm, and the process is repeated until the desired lesion expansion is achieved.

The Disrupt-CAD study demonstrated the device's efficacy by fracturing calcium and achieving significant luminal gain and stent expansion without the complications associated with atherectomy. Calcium fracture efficacy increased with the severity of calcification, with the greatest benefit observed in the highest tertile, suggesting that shockwave IVL is better suited for concentric rather than eccentric calcification [30].

One limitation of IVL is the bulkiness of the balloon, which can make delivery challenging. For lesions that are not crossable, a laser or RA may be required before IVL can be used.

A recent systematic review and meta-analysis found that the "stent-through" IVL plaque modification technique is a safe and effective method for treating stent underexpansion caused by heavily calcified plaques. The technique demonstrated a high success rate (88.7%) and a very low incidence of complications (1.6%) [31].

Combination Therapy

RA facilitates the delivery of the IVL balloon into an uncrossable lesion, while IVL eliminates the need to upsize the rotablation burr and the associated risk of complications. Given the vessel's large diameter and concentric calcification, even a 2 mm burr may be insufficient. Intracoronary imaging plays a critical role in this case by guiding the selection of appropriate devices. We recommend its use in any case requiring combination therapy.

Step-by-Step Approach with De Novo Lesions

1. **First Step: Assessing balloon crossability**
   - **If the lesion is not crossable with a balloon:**
     – Use RA, OA, or ELCA.
   - **If the lesion is crossable with a balloon:**
     – Proceed to assess balloon dilatability.
     – **If the lesion is not dilatable:** Move to the second step.
2. **Second Step: Intracoronary imaging**
   - **Three modes based on imaging findings:**
     – **First Mode:** If nodular calcium is observed, use RA or OA.
     – **Second Mode:** If the calcium arc is $<270^\circ$, use a plaque-modifying balloon or an OPN non-compliant balloon.
     – **Third Mode:** If the calcium arc is $>270^\circ$:
       - **If the arterial lumen is >2.5 mm or deep calcium is present:** Use IVL.
       - **If the arterial lumen is <2.5 mm or superficial calcium is present:** Use RA, OA, or ELCA.
3. **Fourth Step: Non-compliant balloon inflation**
   - Use a 1:1 non-compliant balloon with high-pressure inflation.
4. **Fifth Step: Stenting**
   - Proceed with stent implantation.

5. **Sixth Step: Repeating intracoronary imaging**
   - Repeat intracoronary imaging to assess the final result.

Among calcium-modifying devices, balloons and IVL are easier to use and require less training than various atherectomy techniques. However, IVL is more expensive than other options.

For undilatable lesions, atherectomy methods and IVL are more effective. In cases of uncrossable balloon lesions, OA and RA are preferred [32].

References

1. Budoff MJ, Shaw LJ, Liu ST, Weinstein SR, Mosler TP, Tseng PH, et al. Long-term prognosis associated with coronary calcification: observations from a registry of 25,253 patients. J Am Coll Cardiol. 2007;49:1860–70. PMID: 17481445
2. Moe SM, Chen NX. Pathophysiology of vascular calcification in chronic kidney disease. Circ Res. 2004;95:560–7. PMID: 15375022
3. Brilakis ES, Best PJ, Elesber AA, Barsness GW, Lennon RJ, Holmes DR Jr, et al. Incidence, retrieval methods, and outcomes of stent loss during percutaneous coronary intervention: a large single-center experience. Catheter Cardiovasc Interv. 2005;66:333–40. PMID: 16142802
4. Bourantas CV, Zhang YJ, Garg S, Iqbal J, Valgimigli M, Windecker S, et al. Prognostic implications of coronary calcification in patients with obstructive coronary artery disease treated by percutaneous coronary intervention: a patient-level pooled analysis of 7 contemporary stent trials. Heart. 2014;100:1158–64. PMID: 24829367
5. Arora S, Panaich SS, Patel N, Patel NJ, Lahewala S, Solanki S, et al. Coronary atherectomy in the United States (from a nationwide inpatient sample). Am J Cardiol. 2016;117:555–62. PMID: 26721656
6. Allali A, Abdel-Wahab M, Sulimov DS, Kastrati A, El-Mawardy M, Toelg R, et al. Comparison of bailout and planned rotational atherectomy for heavily calcified coronary lesions: a single-center experience. J Interv Cardiol. 2017;30:124–33. PMID: 28109021
7. Yu M, Li Y, Li W, Lu Z, Wei M, Zhang J. Calcification remodeling index characterized by cardiac CT as a novel parameter to predict the use of rotational atherectomy for coronary intervention of lesions with moderate to severe calcification. Korean J Radiol. 2017;18:753–62. PMID: 28860893
8. Mintz GS, Popma JJ, Pichard AD, Kent KM, Satler LF, Chuang YC, et al. Patterns of calcification in coronary artery disease: a statistical analysis of intravascular ultrasound and coronary angiography in 1155 lesions. Circulation. 1995;91:1959–65. PMID: 7895353
9. Abdel-Wahab M, Maehara A, Latib A, Généreux P, Serruys PW, Stone GW, et al. Intravascular lithotripsy versus rotational atherectomy for the treatment of severely calcified coronary lesions: a propensity score-matched analysis from the OPTALYSE and ROTAXUS trials. JACC Cardiovasc Interv. 2024;17:615–25. https://doi.org/10.1016/j.jcin.2023.12.034. PMID: 38418052
10. König A, Klauss V. Virtual histology. Heart. 2007;93:977–82. PMID: 17639109
11. Yabushita H, Bouma BE, Houser SL, Aretz HT, Jang IK, Schlendorf KH, et al. Characterization of human atherosclerosis by optical coherence tomography. Circulation. 2002;106:1640–5. PMID: 12270856
12. Ali ZA, Nef H, Escaned J, Werner N, Banning AP, Hill JM, et al. Safety and effectiveness of coronary intravascular lithotripsy for treatment of severely calcified coronary stenoses. Circ Cardiovasc Interv. 2019;12:e008434. https://doi.org/10.1161/CIRCINTERVENTIONS.119.008434. PMID: 31525096

13. Sorini Dini C, Nardi G, Ristalli F, Mattesini A, Hamiti B, Di Mario C. Contemporary approach to heavily calcified coronary lesions. Interv Cardiol. 2019;14:154–63. https://doi.org/10.15420/icr.2019.19.R1. PMID: 31867062
14. Gioel Gabrio S, Matteo G, Alessio M, Barbati ME, Caiazzo G, Tzanis G, et al. Very high-pressure dilatation for undilatable coronary lesions: indications and results with a new dedicated balloon. EuroIntervention. 2016;12:359–65. PMID: 27290681
15. Secco GG, Buettner A, Parisi R, Pistis G, Vercellino M, Audo A, et al. Clinical experience with very high-pressure dilatation for resistant coronary lesions. Cardiovasc Revasc Med. 2019;20:1083–7. PMID: 31072791
16. Okura H, Hayase M, Shimodozono S, Kobayashi T, Sano K, Matsushita T, et al. Mechanisms of acute lumen gain following cutting balloon angioplasty in calcified and noncalcified lesions: an intravascular ultrasound study. Catheter Cardiovasc Interv. 2002;57:429–36. PMID: 12455077
17. Ozaki Y, Yamaguchi T, Suzuki T, Nakamura M, Kitayama M, Nishikawa H, et al. Impact of cutting balloon angioplasty (CBA) prior to bare metal stenting on restenosis. Circ J. 2007;71:1–8. PMID: 17186971
18. Kanai T, Hiro T, Takayama T, Fukamachi D, Nishimura T, Kawamura Y, et al. Three-dimensional visualization of scoring mechanism of 'AngioSculpt' balloon for calcified coronary lesions using optical coherence tomography. J Cardiol Cases. 2012;5:e16–9. https://doi.org/10.1016/j.jccase.2011.09.006.
19. Abdel-Wahab M, Richardt G, Joachim Büttner H, Toelg R, Geist V, Meinertz T, et al. High-speed rotational atherectomy before paclitaxel-eluting stent implantation in complex calcified coronary lesions: the randomized ROTAXUS (rotational atherectomy prior to Taxus stent treatment for complex native coronary artery disease) trial. JACC Cardiovasc Interv. 2013;6:10–9. PMID: 23266232
20. Whitlow PL, Bass TA, Kipperman RM, Sharaf BL, Ho KK, Brinker JA, et al. Results of the study to determine rotablator and transluminal angioplasty strategy (STRATAS). Am J Cardiol. 2001;87:699–705. PMID: 11249849
21. Safian RD, Feldman T, Muller DW, Mason D, Schreiber T, Haik B, et al. Coronary angioplasty and Rotablator atherectomy trial (CARAT): immediate and late results of a prospective multicenter randomized trial. Catheter Cardiovasc Interv. 2001;53:213–20. PMID: 11387606
22. Kini A, Reich D, Marmur JD, Mitre CA, Sharma SK. Reduction in periprocedural enzyme elevation by abciximab after rotational atherectomy of type B2 lesions: results of the Rota ReoPro randomized trial. Am Heart J. 2001;142:965–9. PMID: 11717598
23. Colombo A, Panoulas VF. After 3 decades, at long last, a new device to deal with calcific lesions. JACC Cardiovasc Interv. 2014;7:519–20. PMID: 24814644
24. Kini AS, Vengrenyuk Y, Pena J, Yoshimura T, Panwar SR, Motoyama S, et al. Optical coherence tomography assessment of the mechanistic effects of rotational and orbital atherectomy in severely calcified coronary lesions. Catheter Cardiovasc Interv. 2015;86:1024–32. PMID: 25856223
25. Parikh K, Chandra P, Choksi N, Khanna P, Chambers J. Safety and feasibility of orbital atherectomy for the treatment of calcified coronary lesions: the ORBIT I trial. Catheter Cardiovasc Interv. 2013;81:1134–9. PMID: 22730224
26. Chambers JW, Feldman RL, Himmelstein SI, Bhatheja R, Villa AE, Strickman NE, et al. Pivotal trial to evaluate the safety and efficacy of the orbital atherectomy system in treating de novo, severely calcified coronary lesions (ORBIT II). JACC Cardiovasc Interv. 2014;7:510–8. PMID: 24746652
27. Appelman YE, Koolen JJ, Piek JJ, Redekop WK, de Feyter PJ, Strikwerda S, et al. Excimer laser angioplasty versus balloon angioplasty in functional and total coronary occlusions. Am J Cardiol. 1996;78:757–62. PMID: 8857478
28. Fernandez JP, Hobson AR, McKenzie D, Shah N, Sinha MK, Wells TA, et al. Beyond the balloon: excimer coronary laser atherectomy used alone or in combination with rotational atherectomy in the treatment of chronic total occlusions, non-crossable and non-expansible coronary lesions. EuroIntervention. 2013;9:243–50. PMID: 23517942

29. Brinton TJ, Ali ZA, Hill JM, Meredith IT, Maehara A, Illindala U, et al. Feasibility of shock-wave coronary intravascular lithotripsy for the treatment of calcified coronary stenoses. Circulation. 2019;139:834–6. PMID: 30715938
30. Ali ZA, Brinton TJ, Hill JM, Maehara A, Matsuo H, Karimi Galougahi K, et al. Optical coherence tomography characterization of coronary lithoplasty for treatment of calcified lesions: first description. JACC Cardiovasc Imaging. 2017;10:897–906. PMID: 28797413
31. Caminiti R, Vetta G, Parlavecchio A, Magnocavallo M, Della Rocca DG, Polselli M, et al. A systematic review and meta-analysis including 354 patients from 13 studies of intravascular lithotripsy for the treatment of under expanded coronary stents. Am J Cardiol. 2023;205:223–30. PMID: 37579654
32. Fan LM, Tong D, Mintz GS, Mamas MA, Javed A. Breaking the deadlock of calcified coronary artery lesions: a contemporary review. Catheter Cardiovasc Interv. 2021;97:108–20.

Chapter 5
Chronic Total Coronary Occlusion Lesions

Ata Firouzi, Ehsan Khalilipur, and Zahra Hosseini

Abstract PCI for CTO is one of the most challenging endovascular procedures performed today. Success rates for CTO PCI have improved significantly in recent years due to increased operator experience, advancements in CTO-specific equipment and algorithms, and the publication of expert consensus documents. This chapter aims to summarize the current evidence for CTO PCI, discuss success and risk prediction scoring systems, outline procedural principles and challenges, and provide insights into the future role of CTO PCI.

Keywords CTO · PCI · CABG

Abbreviations

CABG	Coronary artery bypass grafting
CTO	Chronic total occlusion
IVUS	Intravascular ultrasound
LAD	Left anterior descending coronary artery
LM	Left main coronary artery
LV	Left ventricle
LVEF	Left ventricular ejection fraction
MACCE	Major adverse cardiac and cerebrovascular events
MACE	Major adverse cardiovascular events
MI	Myocardial infarction

A. Firouzi
Cardiovascular Intervention Research Center, Rajaie Cardiovascular Institute, Iran University of Medical Sciences, Tehran, Iran

E. Khalilipur (✉) · Z. Hosseini
Cardiovascular Intervention Research Center, Cardiovascular Institute, Iran University of Medical Sciences, Tehran, Iran

A. Firouzi et al. (eds.), *Mastering Skills in Complex Cardiovascular Interventions*, https://doi.org/10.1007/978-3-031-97321-5_5

PCI	Percutaneous coronary intervention
RCA	Right coronary artery
TIMI	Thrombolysis in myocardial infarction

Scope of the Problem

CTO is defined as a completely occluded coronary artery with TIMI flow grade 0 for at least 3 months. Success rates for CTO PCI have improved significantly in recent years, driven by advancements in indications, equipment, and techniques for recanalizing occluded coronary arteries. According to previously published multicenter registries, CTOs are identified in 10%–25% of patients undergoing diagnostic angiography [1].

The RHC-PCI registry at our center found that 8.5% of coronary procedures were performed on patients with CTO [2].

Guideline Recommendation

The 2011 ACCF/AHA/SCAI PCI Guideline recommended PCI revascularization for patients with clinical indications and favorable anatomy, particularly when performed by experienced operators (Class IIa, Level of Evidence [LOE] B).

The 2014 European Society of Cardiology (ESC) and European Association for Cardiothoracic Surgery (EACTS) guidelines recommend considering CTO PCI for patients expected to achieve ischemia reduction in the corresponding myocardial territory and/or relief from angina (Class IIa, LOE B). The guidelines suggest an initial anterograde approach to CTO, with a retrograde approach as an alternative if the anterograde approach fails or as a primary strategy in select patients (Class IIb, LOE C) [3, 4].

The 2017 ACC/AATS/AHA/ASE/ASNC/SCAI/SCCT/STS Appropriate Use Criteria for revascularization in patients with stable ischemic heart disease no longer include separate criteria for CTO lesions, unlike the 2012 guidelines. Currently, the decision to revascularize is based on symptoms, the extent of antianginal medication use, and the risk of ischemia, regardless of whether the lesion is a CTO [5].

When to Revascularize a CTO Lesion

Although CTOs are found in approximately 18%–52% of patients undergoing coronary angiography, they are rarely revascularized through PCI. Nonetheless, current evidence indicates that successful CTO PCI can improve symptoms, quality of life, and long-term survival. Over the past decade, advancements in specialized tools and techniques, combined with growing operator experience, have led to success and complication rates comparable to those of non-CTO angioplasty.

Successful CTO PCI has been shown to enhance symptoms, quality of life, and long-term survival. In recent years, improvements in procedural techniques and increased operator expertise have resulted in success and complication rates similar to those of non-CTO lesion angioplasty [6].

The HORIZONS-AMI trial demonstrated that a CTO in a non-infarct-related artery predicted higher mortality at a 3-year follow-up. Another study found that CTO presence was associated with an increased risk of long-term mortality, even after excluding early deaths [7].

Several additional studies have shown reduced mortality in patients with CTO who underwent PCI, particularly when the target vessel was the LAD. Nevertheless, a recent randomized trial comparing medical therapy to CTO PCI found no statistically significant mortality benefit. Notably, this trial was designed to detect differences in quality of life and health status rather than MACE. MACE occurred in 6.7% of patients in the medical therapy group and 5.2% of those in the CTO PCI group [8, 9].

At a 1-year follow-up, LVEF continued to decline. Still, a meta-analysis of CTO PCI demonstrated a 4.44% increase in LVEF and a reduction in adverse ventricular remodeling. Several other studies have also reported significant improvements in LVEF following successful CTO PCI [10, 11].

Successful CTO recanalization has been shown to reduce the need for antianginal medications and improve exercise capacity, as evidenced by increased 6-min walking distances. This benefit is likely due to the reduction in ischemic burden and frequency of angina following CTO PCI [12, 13].

PCI Versus Optimal Medical Treatment for CTO

Khan et al. [14] conducted a meta-analysis including 16 studies with 11,314 patients. They used random-effects models to analyze data on mortality, cardiac deaths, myocardial reinfarction, MACE, stroke, and repeat CTO PCI. Odds ratios (OR) with 95% confidence intervals (CI) were calculated, and a significance level of $P < 0.05$ was applied.

CTO PCI was associated with lower mortality (OR: 0.45, CI: 0.32 to 0.63, $P = 0.00001$) and cardiac deaths (OR: 0.58, CI: 0.38 to 0.89, $P = 0.01$). These findings were primarily derived from observational studies, with no significant differences observed in randomized controlled trials. No statistically significant differences were found in the incidence of MACE (OR: 0.71, CI: 0.48 to 1.05, $P = 0.54$), myocardial reinfarction (OR: 0.71, CI: 0.48 to 1.05, $P = 0.54$), stroke (OR: 0.61, CI: 0.32 to 1.17, $P = 0.14$), or repeat PCI (OR: 1.28, CI: 0.91 to 1.78, $P = 0.16$).

This meta-analysis suggests that the CTO PCI group has lower long-term mortality and cardiac death rates than the optimal medical therapy group, based on observational studies. However, no differences were observed in randomized controlled trials. Further randomized trials are needed to confirm these findings and evaluate long-term outcomes.

In the COMET-CTO trial [15], designed for the long-term follow-up of patients with CTO randomized to either optimal medical therapy or PCI, there was no significant difference in the incidence of MACE or quality of life measured by the Seattle Angina Questionnaire (SAQ) after 56 ± 12 months of follow-up.

A 2020 meta-analysis of five randomized trials involving 1790 patients found that CTO PCI did not improve survival, reduce MACE, or enhance LV function at long-term follow-up (up to 4 years). Nonetheless, CTO PCI was associated with less angina and fewer target lesion revascularization procedures than OMT [16].

Similarly, in a systematic review and meta-analysis by Kucukseymen et al. [17], successful CTO PCI procedures were linked to improved quality-of-life parameters compared with patients on optimal medical therapy or those with failed CTO PCI. These findings support the use of PCI for treating CTOs in symptomatic patients who do not respond adequately to medical therapy.

Scoring Systems for CTO

Several scoring systems are available to predict the success of PCI procedures for CTOs. The J-CTO score is the first scoring system developed for CTO PCI, created by a Japanese CTO registry. It incorporates five variables: blunt stump, occlusion length (9–20 mm), tortuosity of the CTO, calcification of the CTO, and previous failed attempts. This score helps predict successful guidewire crossing within the first 30 min of the procedure [15].

A second, more widely used scoring system is the PROGRESS-CTO score, derived from the Prospective Global Registry for the Study of CTO Intervention. This score includes variables such as proximal cap ambiguity, circumflex CTO, moderate-to-severe tortuosity, and the absence of interventional collaterals. The four-point PROGRESS-CTO score demonstrates better performance than other existing scores in predicting technical success and long-term outcomes [16].

Overall, the most commonly used scores show modest to good predictive value, as follows [18]:

- **J-CTO**: AUC: 0.55 to 0.868
- **PROGRESS-CTO**: AUC: 0.557 to 0.788
- **CL**: AUC: 0.624 to 0.800
- **CASTLE**: AUC: 0.633 to 0.68
- **KCCT**: AUC: 0.703 to 0.776

Technical Aspects of CTO PCI

Before the procedure, it is essential to thoroughly review and analyze the angiogram and, if available, coronary computed tomography angiography. This step helps develop a primary and secondary procedural plan and assess the risk/benefit ratio of the intervention. Ad hoc CTO PCI is generally discouraged to allow sufficient time for procedural planning, preparation, and proper patient counseling. Preplanning the CTO PCI procedure can also help minimize contrast and radiation exposure, reduce patient and operator fatigue, enable additional evaluations (e.g., myocardial viability testing), and facilitate detailed discussions with the patient about all aspects of the procedure.

The anatomy of the CTO determines the most appropriate and safe crossing strategies. Angiographic evaluation of the CTO anatomy focuses on four key characteristics:

1. Proximal cap morphology;
2. Occlusion length, course, and composition (e.g., calcification);
3. Distal vessel quality; and
4. Collateral circulation characteristics.

Moreover, non-CTO lesions should be evaluated, as the presence of intermediate LM or other significant lesions may influence clinical decision-making. This could lead to alternative revascularization strategies, such as CABG or stenting of a donor artery before CTO PCI [17, 19].

Toolbox for CTO PCI

Since the publication of the hybrid algorithm, the toolbox for CTO PCI has expanded rapidly. This growth includes the development of new specialty wires, microcatheters, guide extensions, and lower-profile balloons, as well as the incorporation of intracoronary imaging and hemodynamic support to enhance the success of both antegrade and retrograde CTO PCI.

With the introduction of these new devices, it is essential to classify them broadly based on their design and intended use. Such classification would eliminate the need to add inventory that is similar to existing products, thereby reducing overlap and minimizing waste.

In CTO procedures, wires are categorized according to the specific task or function they perform during the intervention. Table 5.1 lists and categorizes some of the most frequently used wires for CTO PCI.

Table 5.1 Coronary guidewire options based on wire characteristics and functions

Wire Function	Wire Characteristics	Wires	Tip Stiffness (g)	Manufacturer
First -line Penetration wires	Polymer jacketed	Pilot 200	4.1	Asahi Intecc
		Gladius	3.0	Asahi Intecc
		Raider	3.0	Teleflex
Second -line Penetration wires	Non-jacketed	Confianza Pro 12	12.4	Asahi Intecc
		Hornet 14	14	Asahi Intecc
		Astato 8–20	8–20	Boston Scientific
		Gaia 1,2,3, Next	1.7,3.5,4.5	Asahi Intecc
Knuckling wires	Polymer jacketed	Gladius Mongo	3.0	Asahi Intecc
	Polymer	Fielder XT,	0.8,1.0	Asahi Intecc
	Jacketed wire Tapered tip	XT -A Fighter	1.5	Boston Scientific
Collateral wires for Retrograde crossing	Non-jacketed	Sion	0.7	Asahi Intecc
	Polymer jacketed	Suoh3	0.3	Asahi Intecc
		Samurai RC	1.2	Boston Scientific
		Fielder FC, XT -R	0.8,0.6	Asahi Intecc
		Sion Black	0.9	Asahi Intecc
Externalization wires	Hydrophilic coating	R350	3.0	Teleflex
		RG3	3.0	Asahi Intecc

Microcatheters are routinely employed during CTO PCI to enhance the support provided by the guidewire, thereby increasing its penetration power. They facilitate rapid wire exchange and reshaping once the base of operations is established. In addition to their support and exchange functions, microcatheters are utilized to dilate channels, enable intracoronary drug delivery, and perform localized contrast injections.

While some operators may prefer over-the-wire balloons to enhance penetration force, microcatheters are generally favored due to their ability to navigate through tortuous segments and their reduced susceptibility to kinking. A variety of commercially available microcatheters with differing crossing profiles, torque-ability, deliverability, and lengths are now on the market. For retrograde interventions, longer 150 cm catheters are typically required (Table 5.2) [20].

Table 5.2 Commonly used microcatheters in order of crossing profile

Microcatheter	Length	(French Catheter)	Manufacturer
Micro-14	155 cm	1.6	Roxwood Medical
Finecross	130 cm, 150 cm	1.8	Terumo
Caravel	135 cm, 150 cm	1.9	Asahi Intecc
Corsair Pro XS	135 cm, 150 cm	2.1	Asahi Intecc
Mamba Flex	135 cm, 150 cm	2.1	Boston Scientific
Turnpike LP	135 cm, 150 cm	2.2	Vascular Solutions
Mamba	135 cm	2.3	Boston Scientific
Corsair	135 cm, 150 cm	2.6	Asahi Intecc
Corsair Pro	135 cm, 150 cm	2.6	Asahi Intecc
Tornus	135 cm	2.1, 2.6	Asahi Intecc
Turnpike	135 cm, 150 cm	2.6	Vascular Solutions
Turnpike Spiral	130 cm, 150 cm	2.9	Vascular Solutions
Turnpike Gold	135 cm	2.9	Vascular Solutions

Equipment Utilization in CTO PCI: Insights from the PROGRESS-CTO Registry

PCI for CTO has advanced significantly in recent years, largely due to improvements in crossing techniques and advancements in microcatheter and guidewire technology [18]. Antegrade wiring remains the most commonly used crossing strategy, although more complex cases often require antegrade dissection and reentry or the retrograde approach [21, 22, 23].

CTO is defined as a coronary lesion with TIMI grade 0 flow for at least 3 months. The duration of occlusion is determined clinically based on the onset of angina, a history of MI in the target vessel territory, or comparisons with a previous angiogram. Calcification is classified as mild (spotty), moderate (involving ≤50% of the reference lesion diameter), or severe (involving >50% of the reference lesion diameter). Moderate proximal vessel tortuosity is defined as the presence of at least two bends >70° or one bend >90° in the CTO vessel, while severe tortuosity is defined as two bends >90° or one bend >120°. A blunt or no stump is characterized by the absence of tapering or a funnel shape at the proximal cap. Interventional collaterals are defined as those deemed suitable for crossing with a guidewire and microcatheter.

Retrograde procedures are defined as attempts to cross a lesion through an interventional collateral or bypass graft that supplies the target vessel distal to the lesion [24]. The primary findings of the cited study were as follows: (a) increased lesion complexity was associated with the use of a greater variety of guidewire types and (b) polymer-jacketed guidewires and large torquable guidewires were particularly beneficial in navigating complex lesions.

Coronary Collaterals and Myocardial Viability in Patients with CTO

A study involving 218 patients with CTO who underwent cardiac magnetic resonance imaging between 2013 and 2018 assessed myocardial viability. Well-developed collaterals were defined by a concomitant collateral connection score of 2 and Rentrop grade 3, observed in 146 patients (67%), while poorly developed collaterals were defined by lower scores or grades. Viability was determined by dysfunctional myocardium (≥3 mm segmental wall thickening) and ≤50% late gadolinium enhancement. Only 5% of CTO segments exhibited extensive scarring (late gadolinium enhancement >50%).

Patients with well-developed collaterals showed greater segmental wall thickening (3.72 ± 1.51 mm vs. 3.05 ± 1.60 mm; P = 0.01) and less scar extent (7.0% [0.1–16.7] vs. 13.1% [2.8–22.2]; P = 0.048) compared to those with poorly developed collaterals. Viability was more prevalent in CTO segments of patients with poorly developed collaterals than in those with well-developed collaterals (44% vs. 30% of segments; P = 0.01), primarily due to a higher prevalence of dysfunctional myocardium in the poorly developed collateral group (51% vs. 34% of segments; P = 0.01).

The study concluded that the infarcted area in the myocardium supplied by a CTO is generally small. Well-developed collaterals are associated with reduced myocardial scarring and better-preserved function. Nonetheless, viability was frequently observed even in patients with inadequate collateral development [26].

IVUS Applicability in CTO PCI

Contemporary PCI for CTO necessitates the use of a hybrid algorithm that integrates all feasible techniques—whether antegrade or retrograde intraluminal crossing, dissection, or reentry—in the safest, most effective, and most efficient manner possible. IVUS plays a crucial role in this process by helping to clarify and resolve proximal cap ambiguity during antegrade CTO wire crossing, facilitating reentry into the true lumen after subintimal crossing, and confirming the guidewire's position in the distal true lumen. Additionally, IVUS is utilized to guide retrograde wire crossing and to verify the relative positions of the antegrade and retrograde wires when employing the reverse-controlled antegrade retrograde tracking (CART) technique [27].

IVUS can provide valuable spatial information that aids in guidewire crossing during PCI for CTO. By identifying critical details regarding plaque burden, the presence of calcification, and vessel remodeling, IVUS assists the operator in planning the procedure and navigating the hybrid algorithm, thereby reducing lesion ambiguity. Proper positioning of the guidewire and equipment relative to the true lumen can facilitate guidewire crossing while minimizing extra plaque wire tracking. Furthermore, IVUS can optimize stent placement and expansion, potentially enhancing the durability of the stent. It may also help identify complications or suboptimal findings that are not easily visible on angiography following recanalization and stenting of a CTO, ultimately contributing to improved long-term outcomes [28, 29].

A retrospective study conducted by Japanese experts between 2012 and 2017 included 137 CTO lesions that could not be crossed using tapered soft wires. The overall success rate was significantly higher in the 3D wiring group (n = 69) than in the conventional wiring group (n = 68) (98% vs. 90%, respectively; P = 0.027). In primary antegrade cases, the first antegrade approach took significantly less time in the 3D wiring group than in the non-3D wiring group (30 ± 16 min vs. 42 ± 29 min, respectively; P = 0.01). When the antegrade approach was maintained throughout the procedure, the success rate in the 3D wiring group was significantly higher than in the non-3D wiring group (100% vs. 89.2%, respectively; P = 0.033). Further, the rate of vessel perforation by the antegrade wire was significantly lower in the 3D wiring group compared to the non-3D wiring group (1% vs. 11%, respectively; P = 0.055). This study demonstrated that 3D wiring enables precise guidewire control, improving the success rate of antegrade wiring, shortening procedure time, and enhancing overall success rates [30].

Hybrid Algorithm

Over the past decade, techniques for CTO PCI have advanced rapidly, with success rates exceeding 90% in experienced centers. This improvement is attributed to advancements in CTO crossing devices and techniques, as well as a deeper understanding of procedural strategies. The development of the 'hybrid algorithm' has significantly enhanced the success rate, safety, and efficiency of these procedures in numerous countries. A thorough understanding of coronary anatomy remains critical for performing successful CTO interventions [31].

The hybrid algorithm was conceptualized between 2008 and 2010 and formally codified in January 2011 at the "Bellingham Summit," where a group of CTO expert operators convened. Five operators performed cases twice, while the remaining physicians served as panelists. The goal was to extract clear insights from the angiogram and use them to guide a tailored, sequential approach involving primary, secondary, and tertiary strategies defined before the procedure. This method achieved success in 17 out of 17 patients across five operators with diverse backgrounds over a 3-day period.

The uniformity and consistency of this approach facilitated improved educational, developmental, and investigative outcomes. Within a short timeframe, additional patient series, such as the Hybrid Live case registry (which includes over 150 live-recorded, adjudicated procedures) and studies like the OPEN CTO registry, provided valuable insights. These studies highlighted the consistency of the hybrid approach and its ability to partially mitigate the challenges posed by more complex diseases, enhancing technical success rates (Fig. 5.1) [32, 33].

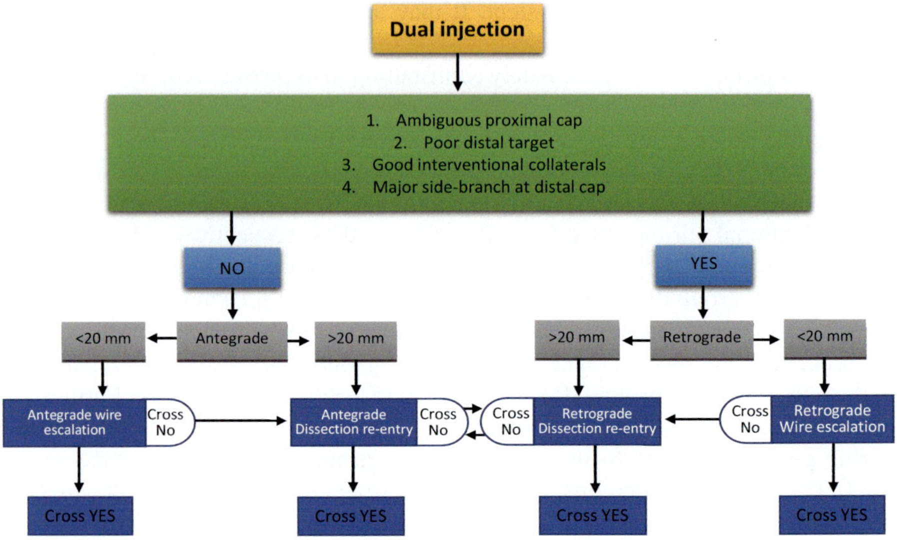

Fig. 5.1 The image provides a hybrid algorithm

While preprocedural diagnostic angiograms can provide some essential information, dual catheter angiography is necessary—except in cases involving ipsilateral collaterals or bridging—to obtain detailed insights about collateral filling, lesion length, and vessel course. When acquiring dual catheter images, it is important to use a wide field of view, avoid panning the camera, and ensure that the imaging run is sufficiently long to allow for an accurate assessment of collateral channels. In addition, when interpreting an angiogram, several key features must be evaluated, as they play a crucial role in guiding the procedural strategy. These features include the quality and extent of collateral circulation, the characteristics of the lesion (such as length and morphology), and the overall anatomy of the vessels involved. Careful analysis of these factors can significantly influence the approach taken during the intervention [25, 34].

Asia Pacific CTO Club Algorithm

Despite the availability of several excellent recommendations, the hybrid CTO algorithm has seen limited adoption in the Asia Pacific region. The Asia Pacific CTO Club has proposed a case selection algorithm based on the Japan-CTO score and a new global CTO algorithm. This algorithm is designed to accommodate varying skill levels and equipment availability while providing practical guidance for CTO PCI.

Similar to the hybrid algorithm, the Asia Pacific CTO Club algorithm considers the following three primary factors when deciding whether the primary approach should be antegrade or retrograde:

1. Proximal cap ambiguity;
2. Distal vessel quality; and
3. Presence of interventional collaterals.

Unlike the hybrid algorithm, occlusion length does not determine whether a wire escalation or dissection reentry strategy is used. Instead, this decision is based on a combination of factors, including the vessel's ambiguous course, severe calcification, tortuosity, lesion length, and previous failure.

The algorithm highlights the role of IVUS-guided entry for resolving proximal cap ambiguity and the use of the CrossBoss catheter for occlusive in-stent restenosis. If the initial antegrade wire passage fails, both the parallel wire technique and dissection reentry using the Stingray system are included as options. The algorithm also incorporates final options for IVUS-guided wiring, limited subintimal tracking and reentry, and guidance on when to terminate the procedure [35].

Dual Injection Angiography

The most effective way to characterize CTO anatomy is through dual injection, which involves filling the target vessel both anterogradely and retrogradely via collaterals from the donor vessel or a bypass graft. Dual injection can sometimes be performed during a diagnostic angiogram, with the resulting images projected during the CTO intervention. This approach provides the most comprehensive preprocedural planning. However, diagnostic angiograms with sufficient cine runs to demonstrate collateral flow are typically adequate to guide dual angiography views at the start of the CTO PCI.

If dual injection is required during a diagnostic angiogram and an additional access point is not desired, two 4 Fr diagnostic catheters can be inserted through a single 8 Fr sheath. Still, this technique is suboptimal due to reduced visualization and catheter control. The injection should be performed with both vessels engaged using a guide catheter of sufficient size to ensure adequate vessel opacification. If one vessel is expected to be more challenging to engage, it is advisable to engage that vessel first, as guide manipulation is unaffected by the presence of a second guide. Some operators prefer engaging the RCA first rather than the LM to avoid interference with the guide positioned in the LM during RCA engagement.

Guide stabilization can be achieved by inserting a workhorse wire into the vessel of interest or a neighboring branch, which also serves as a safety wire in the donor vessel. Many operators prefer this approach to enable rapid intervention in case of donor-vessel injury caused by the retrograde guide or PCI equipment. The best dual injection images are obtained at low magnification and without panning [17, 19, 37].

Safety Planning

An essential aspect of safety planning is determining in advance whether a patient would benefit from immediate hemodynamic support or is at risk of requiring it. If hemodynamic support is deemed necessary upfront, the device must be made available in advance, arrangements for the patient's downstream care must be established, and vascular access for the device must be prepared [38].

Other Technical Considerations

For CTO PCI, it is advisable to use longer sheaths, with most hybrid centers typically using bilateral femoral sheaths that are 45 cm long. For retrograde CTO PCI, longer-than-usual activated clotting times (>300 s) should be considered to minimize the risk of donor-vessel thrombosis. Heparin is the preferred anticoagulant for CTO PCI due to its ease of reversal in the event of coronary perforation.

The 7 Fr Slender sheath allows the use of a 7 Fr guiding catheter for radial access. Smaller sheaths are generally not recommended due to their limited support and inability to accommodate certain coronary equipment. In the United States, dual 8 Fr systems are preferred, while in Europe, 7 Fr systems are more commonly used. Nonetheless, 8 Fr femoral access has been associated with an increased risk of vascular complications, though fluoroscopic guidance has been shown to reduce these risks [39, 40].

A study comparing the technical success rates of radial and femoral access for CTO PCI found that both approaches were comparable, except in complex CTO cases defined by a J-CTO score of 3, where radial access was associated with lower technical success rates. While radial access is increasingly used for CTO PCI, it may result in reduced success and efficiency, particularly in complex cases. Recently, CTO operators have adopted a hybrid approach, using an 8 Fr femoral guide for the antegrade guiding catheter and a 6 Fr radial access for the retrograde guiding catheter, especially in antegrade procedures [41, 42].

With the availability of long externalization guidewires, such as the RG3 and RG350, shortening the guide catheter is no longer necessary. A Y connector equipped with a hemostatic valve can help minimize blood loss during CTO PCI. Guide catheter extensions, such as Boston Scientific's Guidezilla II, Vascular Solutions' Trapliner, Vascular Solutions' GuideLiner V3, or Interventional Medical Device Solutions' Guidion, are routinely used to enhance support and facilitate techniques like reverse controlled antegrade dissection (reverse CART) and retrograde tracking.

For CTO crossing, an over-the-wire system—typically using a microcatheter or over-the-wire balloon—is always preferred. This system allows for guidewire tip reshaping and facilitates guidewire exchanges. Microcatheters are generally favored over over-the-wire balloons because their markers are located at the tip, providing better visualization.

Several microcatheters are available, including:

- Corsair Pro (Asahi Intecc);
- Corsair (Asahi Intecc);
- Spiral (Vascular Solutions);
- Gold (Vascular Solutions);
- Turnpike (Teleflex);
- Turnpike LP (Teleflex); and
- Finecross (Terumo).

Support catheters such as MultiCross and CenterCross (Roxwood Medical) are also commonly used.

The following guidewires are recommended for CTO PCI and their primary applications:

- Fighter (Boston Scientific) or Fielder XT (Asahi Intecc) for microchannel crossing and knuckle formation.
- Gaia (Asahi Intecc) and Confianza Pro 12 (Asahi Intecc) for penetration, featuring stiff tapered tips.
- Pilot 200 (Abbott Vascular) for penetration, with a moderately stiff non-tapered tip.

The Confianza Pro 12 is particularly useful for puncturing calcified CTO caps (proximal or distal).

The hydrophilic Sion guidewire (Asahi Intecc) is preferred for crossing collateral channels due to its high torqueability and excellent shape retention properties. Fielder FC and XT-R (Asahi Intecc) may also be used. For wire externalization, the RG3 (Asahi Intecc, 330 cm long) and R350 (Vascular Solutions, 350 cm long) are the preferred guidewires.

Dual coronary injection is a critical component of CTO PCI, as it allows for accurate definition of the proximal cap, estimation of lesion length, identification of the distal cap location, and assessment of interventional collaterals. This is achieved using low magnification without panning and a prolonged cine acquisition. After the collaterals have filled the distal vessel, the donor vessel is injected first, followed by the occluded vessel. Whenever possible, avoid using a side-hole catheter.

According to the hybrid algorithm, the following four key factors must be thoroughly evaluated before performing CTO PCI:

Table 5.3 Favorable angiographic characteristics for each revascularization technique

	Antegrade wire escalation	Antegrade dissection and reentry	Retrograde wire escalation	Retrograde dissection and reentry
Clear proximal cap	+	+	–	–
Lesion length ≥20 mm	–	+	–	+
Good distal landing zone	+	+	–	–
Interventional collaterals	–	–	+	+

1. Proximal cap ambiguity;
2. Lesion length;
3. Collateral circulation and the presence of interventional collaterals; and
4. Distal target vessel location and quality.

To minimize radiation exposure, the fluoroscopy rate is typically set to 7.5 frames per second.

In CTO PCI, the hybrid algorithm incorporates three major techniques to facilitate successful recanalization of occluded coronary arteries: (1) antegrade wire escalation, (2) antegrade dissection and reentry, and (3) retrograde dissection and reentry (Table 5.3) [43].

Predictors of Success in Hybrid-Approach PCI for CTO

A total of 436 patients (456 lesions) met the inclusion criteria. Of these, 25% of lesions had previously failed percutaneous coronary interventions. The RCA was the most frequently occluded vessel (56.4%), with a mean occlusion length of 24 ± 20 mm. The most commonly used technique was antegrade wire escalation (70%), followed by retrograde wire escalation (22%). The overall success rate was 79.4%.

Failure was most strongly associated with the presence of an ambiguous proximal cap, a specifically defined collateral score (a combination of the Werner and tortuosity scores), and retrograde tortuosity in the presence of an ambiguous proximal cap. In the absence of an ambiguous proximal cap, factors such as a suboptimal distal target, occlusion length greater than 10 mm, ostial location, and operator variability did not significantly contribute to failure. Prior failure, as well as Werner and tortuosity scores, were only weakly predictive of outcomes.

The basic seven-item model predicted success, with C-statistics of 0.753 in the training cohort and 0.738 in the validation cohort. The validation cohort's C-statistic was significantly better ($P < 0.05$) than those of the J-CTO (Multicenter CTO Registry of Japan) score (0.55) and the PROGRESS CTO (Prospective Global Registry for the Study of Chronic Total Occlusion Intervention) score (0.61).

While success can be reasonably well predicted using angiographic variables, differences in operator skill sets may limit the development of a predictive tool that is both powerful and generalizable [17, 44, 45].

Outcomes of Retrograde CTO PCI: A Report from the OPEN-CTO Registry

In the OPEN-CTO registry [46], 454 of 885 single CTO procedures were retrograde, while 431 were antegrade-only. Retrograde procedures were associated with greater lesion complexity (J-CTO score: 2.7 vs. 1.9; P = 0.001) and lower technical success rates (82.4% vs. 94.2%; P = 0.001). The retrograde group had a higher rate of in-hospital all-cause death (2% vs. 0%; P = 0.003), but no significant difference was observed at 1 year (4.9% vs. 3.3%; P = 0.29).

In-hospital MACCE, a composite of all-cause death, stroke, MI, emergency cardiac surgery, and clinically significant perforation, were significantly higher in the retrograde group than in the antegrade-only group (10.8% vs. 3.3%; P = 0.001). This difference persisted at a 1-year follow-up (19.5% vs. 13.9%; P = 0.03). Nevertheless, sensitivity analyses performed at discharge showed no difference in MACCE rates at 1 year between retrograde and antegrade-only CTO PCI.

Quality of life, as measured by the Seattle Angina Questionnaire, improved in both groups. At 1 year, the retrograde and antegrade-only groups had comparable quality-of-life scores (29.9 vs. 30.4; P = 0.58). The authors concluded that retrograde CTO procedures were associated with higher in-hospital MACCE rates compared to antegrade-only procedures, but post-discharge outcomes, including quality-of-life improvements, were similar between the two approaches.

Complications of CTO PCI

In a study involving 12 expert CTO centers in the United States and 1000 consecutive procedures, the overall complication rate was approximately 10%, with the retrograde approach associated with a higher frequency of complications. The most common MACE included perforation (8.8%), periprocedural MI (2.6%), arrhythmia (1.2%), cardiogenic shock (1.1%), and in-hospital death (0.9%). Investigators from the PROGRESS-CTO registry reported a 2.8% complication rate in 1569 CTO PCI procedures. Three independent predictors of complications were identified: patient age >65 years, lesion length ≥23 mm, and the use of a retrograde approach [47, 48].

Another propensity-matched cohort study comparing CTO PCI to non-CTO PCI found similar rates of death, MI, urgent CABG, and combined MACE between the two groups. Several additional studies have also demonstrated comparable outcomes between CTO and non-CTO PCI. Given the serious risk of coronary perforation, pre-procedural planning is critical, and coils and covered stents should be readily available before the procedure. Due to the prolonged duration of CTO procedures, there is an increased risk of radiation injury. To mitigate this risk, it is recommended to use fluoroscopy at 7.5 frames per second rather than 15 frames or higher. Recently, preliminary experience with a low-dose radiation protocol for CTO PCI has been suggested. Additionally, it is essential to frequently switch angiographic views and maximize the distance between the X-ray source and the patient to minimize the risk of radiation-induced skin damage [49].

Despite its clinical benefits, PCI for CTO is associated with a higher rate of complications compared to PCI for non-occlusive lesions. Complications of CTO PCI include the following:

- Death;
- Acute MI;
- Stroke;
- Need for repeat PCI;
- Need for emergency CABG;
- Tamponade requiring pericardiocentesis or surgery;
- Acute vessel closure (which can be fatal if it involves the CTO donor vessel);
- Coronary dissection;

Table 5.4 Complication types in CTO intervention [53]

Acute complications of CTO PCI	
Cardiac complications	Noncardiac complications
Coronary complications	
Acute vessel closure	Vascular access complication
Donor-vessel injury	Contrast-related nephropathy
Occlusion of collaterals	Allergies
(Aorta) coronary dissection	Radiation skin injury
Dissection of distal vessel	Thromboembolic complications
Side branch occlusion	Stroke
Thrombus	
Spasm	
Pseudo-lesion formation	
Sub-optimal stent deployment	
Embolization: thrombus, plaque, or air	
Perforation	
Large vessel	
Collateral	
Distal vessel	
Equipment entrapment/loss non-coronary complications	
Hypotension	
Myocardial	
Arrhythmia	
Death	
Intramural hematoma	
Tamponade	

- Aorto-ostial dissection;
- Thrombus formation;
- Embolization of thrombus, plaque, or air;
- Side branch occlusion;
- Spasm;
- Pseudo-lesion formation;
- Intramural hematoma;
- Perforation;
- Equipment entrapment or loss;
- Hypotension;
- Arrhythmia,
- Vascular access complications and bleeding;
- Contrast-induced acute kidney injury; and
- Radiation-induced skin injury (Table 5.4) [52].

Donor-Vessel Injury

Donor-vessel injury must be promptly identified and managed, as it can lead to severe ischemia and hemodynamic instability. A meta-analysis of retrograde CTO PCIs reported donor-vessel dissection in 2% of treated CTOs (95% CI: 0.9% to 4.5%) [55].

Donor-vessel injury can result from dissection caused by deep catheter engagement, such as during equipment withdrawal or vigorous retrograde wire externalization. Side-hole guide catheters should not be used in the donor vessel, as they can mask pressure dampening, potentially leading to ischemia. The activated clotting time should be maintained at or above 300 s for antegrade procedures and 350 s for retrograde procedures, with checks every 20–30 min during the procedure.

Additionally, retrograde CTO PCI should be avoided through severely diseased donor vessels to prevent ischemia. Donor-vessel lesions should be treated before advancing microcatheters or attempting retrograde crossing. A "safety" guidewire should be placed in the donor vessel to facilitate treatment in case of donor-vessel occlusion.

In most cases, donor vessel injury should prompt the cessation of CTO PCI attempts, with all efforts focused on restoring the donor vessel's patency. If hemodynamic compromise occurs, hemodynamic support may be necessary. Dissections are typically treated with stenting, ideally performed over the safety guidewire after removing the externalized guidewire. Thrombotic occlusion is managed with thrombectomy and, if necessary, the administration of intravenous antiplatelet medications [56].

Perforation

In a recently published analysis of 1811 CTO PCIs from five European centers, perforation occurred in 5.5% of cases. More than half of these perforations required intervention, and 20% resulted in tamponade. Independent predictors of coronary perforation included advanced age, occlusion length >20 mm, rotational atherectomy, antegrade dissection/reentry, and the use of the retrograde approach [57].

Another multicenter U.S. registry analysis of 2097 CTO PCIs performed on 2049 patients reported a 4.1% perforation rate, with 14% of these cases developing tamponade requiring pericardiocentesis. In this study, perforation was associated with age, prior PCI, RCA target CTO, blunt or no stump, antegrade dissection/reentry, and the retrograde approach [10].

Although the retrograde approach has been associated with an increased risk of perforation, recent analyses suggest that a significant proportion of perforations during retrograde CTO PCI result from antegrade crossing attempts [58, 59].

The initial management of a perforation involves inflating a balloon proximal to or at the site of the perforation to stop bleeding into the pericardium. While large vessel perforations are typically treated with covered stent implantation, dissection/reentry techniques have been successfully used in some cases. Distal vessel perforations are usually managed through embolization, often using fat or coils.

Covered stents and/or coils can often be delivered through a single guide catheter, particularly when 8 Fr guides are used. Alternatively, the dual guide catheter technique can be employed, with one guide catheter delivering a balloon to achieve hemostasis and the second guide catheter delivering a covered stent [61].

Side Branch Occlusion

Side branch occlusion can occur, particularly when subintimal dissection/reentry strategies are used in CTO PCI, and has been associated with an increased risk of MI following PCI. Extensive dissection/reentry techniques, such as the subintimal tracking and reentry (STAR) technique, are linked to high rates of restenosis and re-occlusion, likely due to side branch occlusion and reduced outflow. As a result, the extent of dissection should be minimized.

To prevent side branch occlusion, side branch wiring can be performed prior to stenting. This can be facilitated using dual-lumen microcatheters such as the Twin Pass (Teleflex, Wayne, PA, USA), Crusade (Kaneka, Tokyo, Japan), NHancer Rx (IMDS, Roden, The Netherlands), or Sasuke (Asahi Intecc Co., Seto, Japan). In some cases, a retrograde crossing strategy may be employed to preserve side branches.

Intravascular imaging, particularly IVUS, can help identify the cause of side branch loss and assist in re-opening the vessel [62, 63].

Equipment Loss or Entrapment

Although rare, this complication can be serious depending on the device involved and the location of entrapment or loss. Stents are the most frequently embolized devices, with an estimated incidence of 0.32%. Equipment delivery can be particularly challenging during CTO PCI due to vessel tortuosity and calcification. To minimize risks, avoid retrograde equipment delivery, excessive guidewire and microcatheter rotation, and aggressive Rotablator burr advancement [1, 64].

Before stent delivery, the target lesion should be meticulously prepared with balloon angioplasty and, if necessary, atherectomy. Controlling torque transmission to the guidewire tip and alternating clockwise and counter-clockwise rotation of the microcatheter can help reduce the risk of equipment loss or entrapment.

If equipment is lost or entrapped, the first decision is whether to attempt retrieval or to deploy/crush the equipment against the vessel wall. Deployment is often the preferred strategy when a stent fails in a coronary segment unlikely to be significantly affected by stenting, as retrieval attempts may lead to distal stent embolization or vessel injury. If crushing is chosen, intravascular imaging should be used to optimize PCI outcomes. For retrieval attempts, snares are commonly used, with three-loop snares being the most frequently employed [65].

Aortocoronary Dissection

Aortic dissection can complicate any PCI procedure but is more common during CTO PCI, particularly in the RCA. It may result from guiding catheter trauma (e.g., large-caliber or aggressively shaped catheters such as the Amplatz), forceful contrast injection through a wedged catheter, or balloon rupture [66].

This complication can be avoided by using guide catheters with side holes and avoiding contrast injection when pressure is dampened. If dissection occurs, it is critical to stop further coronary injections and stent the coronary artery ostium [67].

For dissections extending into the proximal ascending aorta by less than 40 mm, management options include watchful waiting or stenting the coronary ostium with a conventional or covered stent. Nonetheless, aortic root dissection extending beyond 40 mm, the development of aortic regurgitation, or involvement of supra-aortic vessels necessitates emergent surgical intervention.

Challenging Clinical Cases

Case 1 (Fig. 5.2)

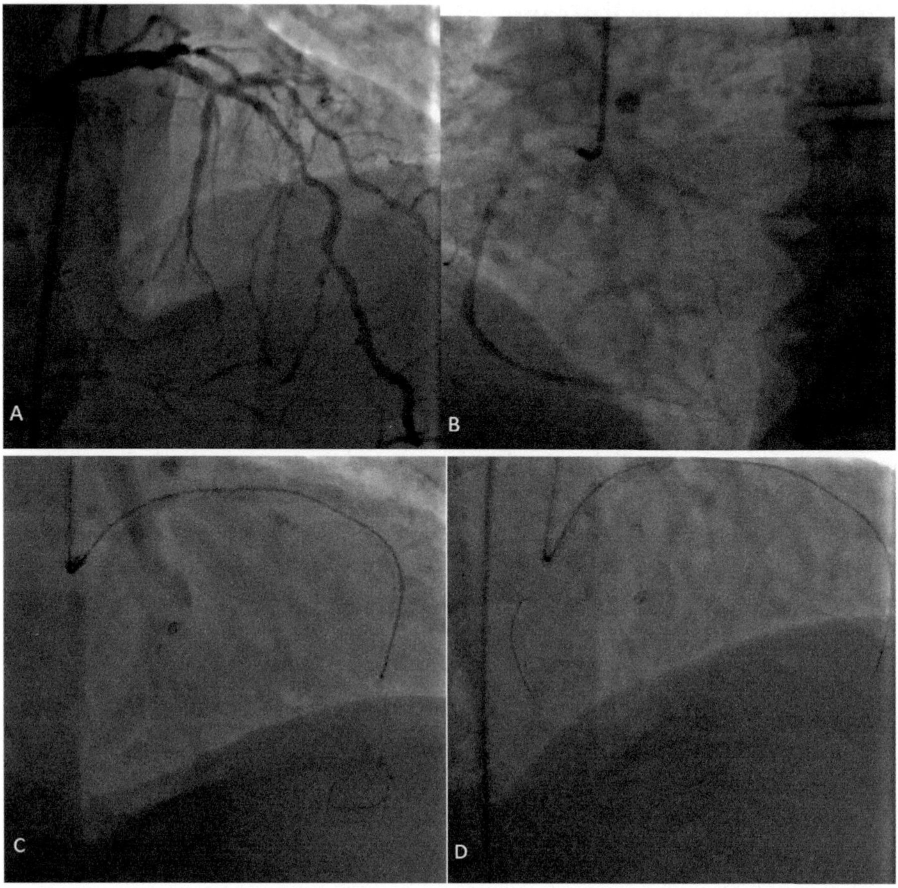

Fig. 5.2 (**a**–**j**) A 64-year-old man presented to the emergency department with recent angina pectoris classified as functional class II, which was unresponsive to previous medical therapy. An ECG revealed ST depression in the lateral precordial leads, and angiography showed a diffuse long lesion in the LAD with total occlusion of the RCA in the distal segment, accompanied by retrograde filling from septal branches (**a**, **b**). Due to the failure of antegrade wire escalation to traverse the lesion (**c**), we opted to employ septal surfing and retrograde access to address the lesion (**d**, **e**). Utilizing a Corsair Pro microcatheter and wire escalation, the Gaia third wire was externalized and subsequently replaced with a workhorse wire (**f**–**h**). Predilation was performed, and the outcome was satisfactory with the deployment of a DES (**i**, **j**)

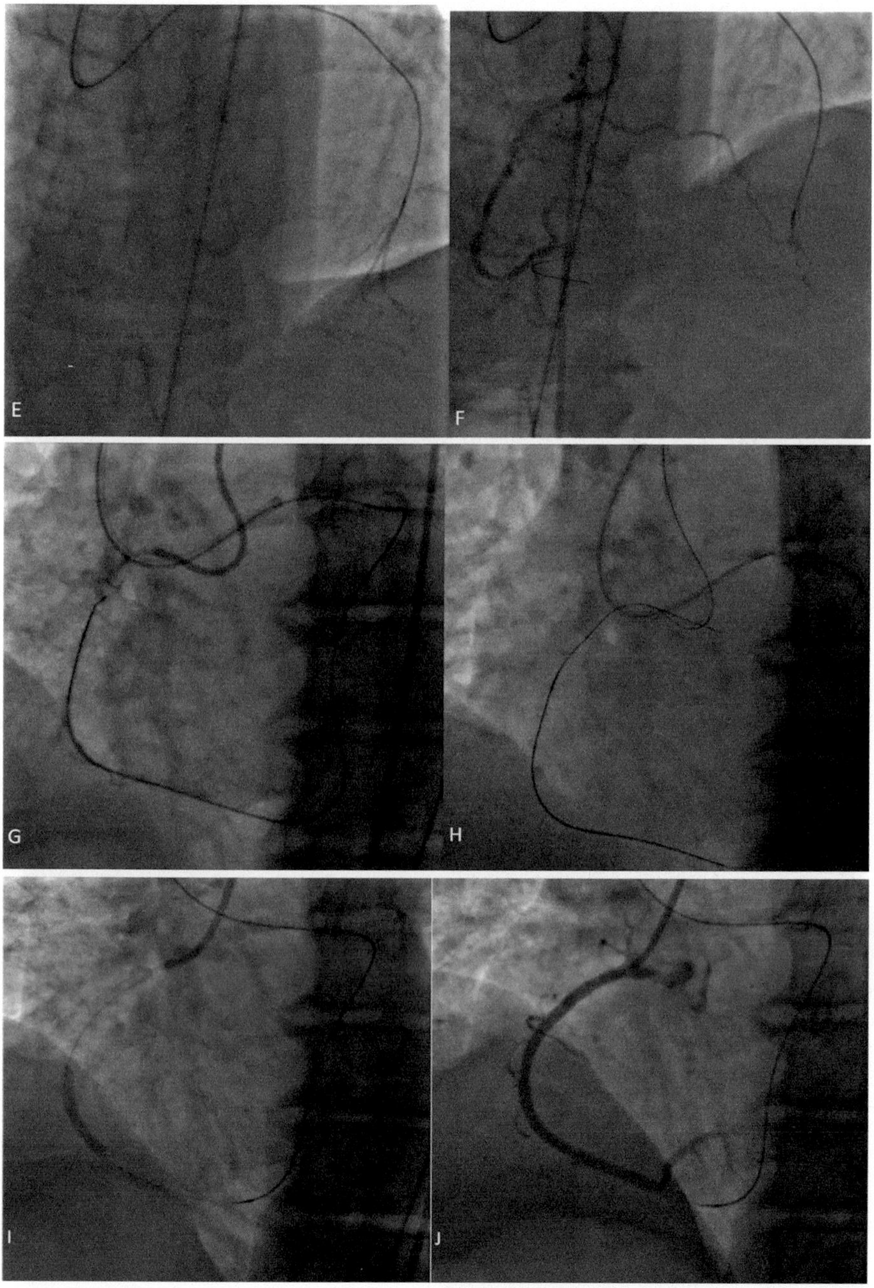

Fig. 5.2 (continued)

Case 2 (Fig. 5.3)

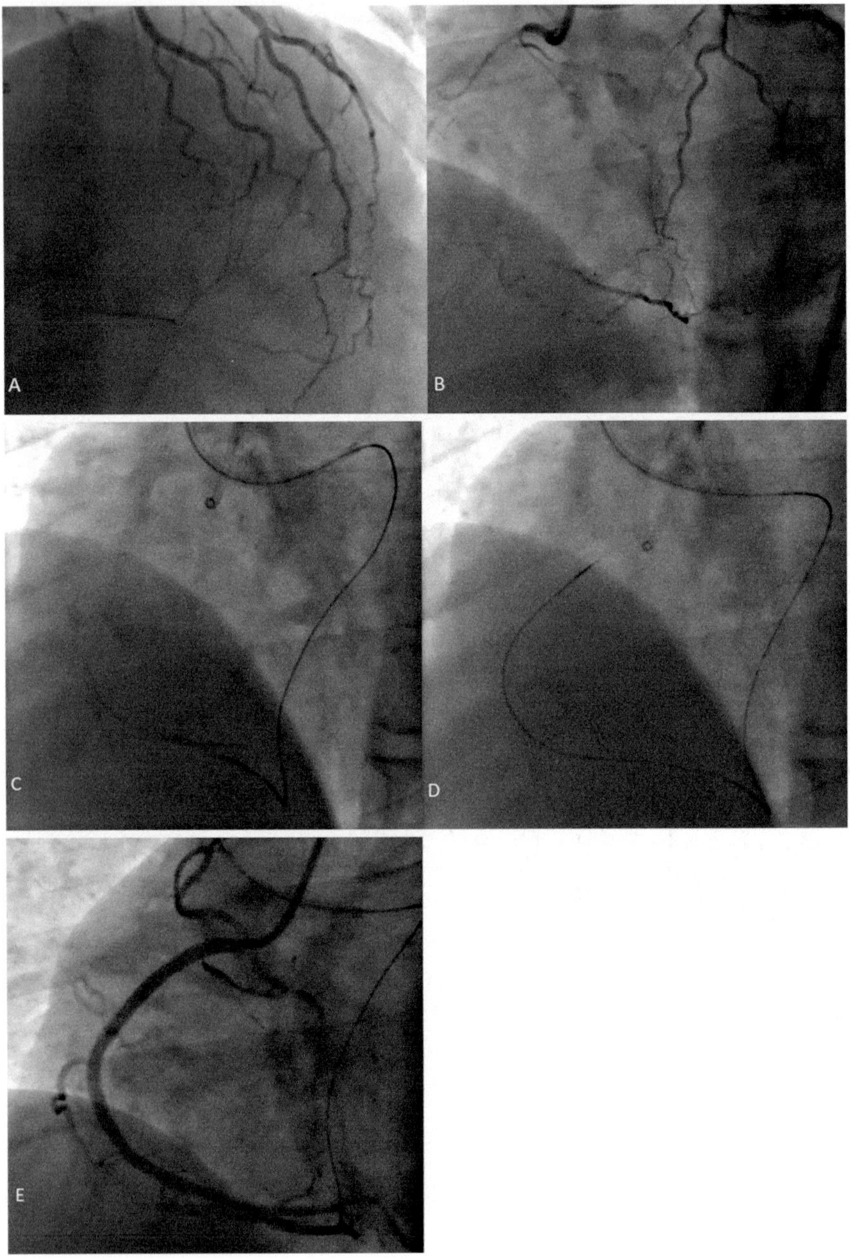

Fig. 5.3 (**a**–**e**) This figure illustrates another successful case of crossing a CTO using retrograde access. Septal branch surfing was performed with a Suoh 03 wire and a Caravel microcatheter for retrograde access (**a**–**c**). Wire escalation was conducted, allowing the Gaia second wire to cross the lesion and be externalized before the deployment of a DES (**d**, **e**)

Case 3 (Fig. 5.4)

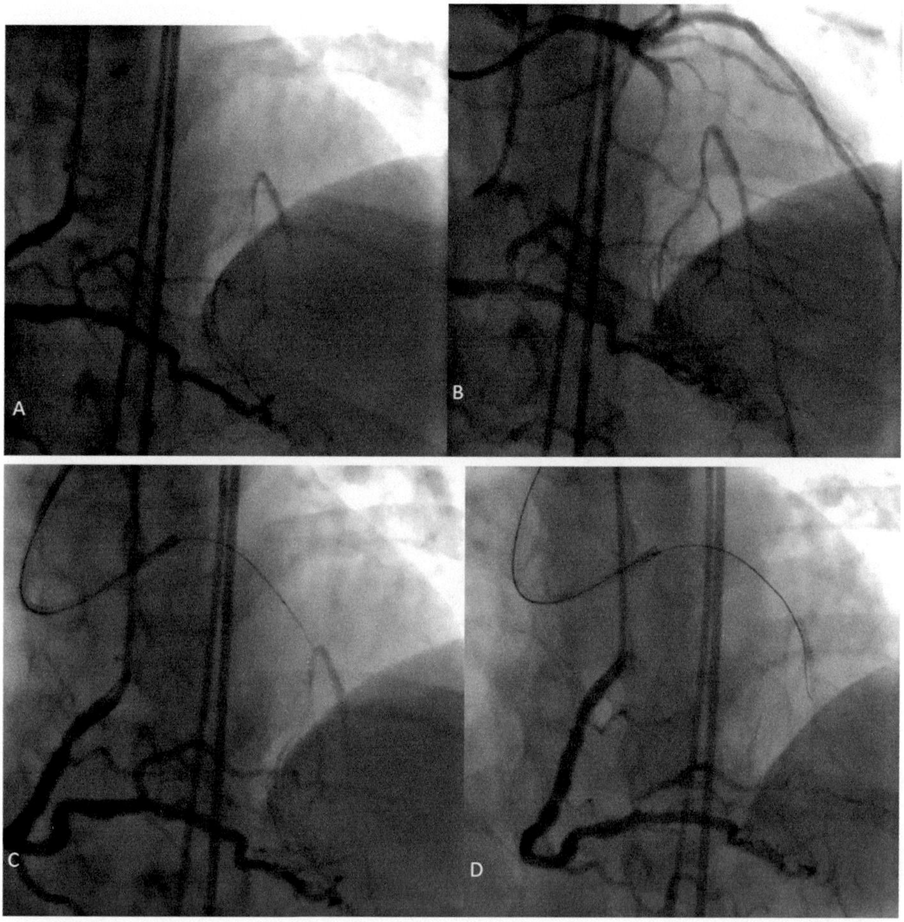

Fig. 5.4 (**a**–**h**) This 58-year-old man presented to the cardiology clinic with recent retrosternal chest discomfort that was resistant to medical therapy. Myocardial perfusion imaging indicated moderate ischemia in the anterior territory. Coronary angiography revealed a completely occluded LAD that was filled retrogradely from the RCA (**a**, **b**). Initially, antegrade wire escalation was unable to traverse the lesion (**c**), prompting us to address the occluded segment via retrograde access (**d**). We utilized the marker wire approach to reenter the antegrade wire using a Corsair Pro microcatheter and a retrograde wire (**e**, **f**). Consequently, the lesion was successfully crossed, predilation was performed, a DES was deployed, and the outcome was favorable (**g**, **h**)

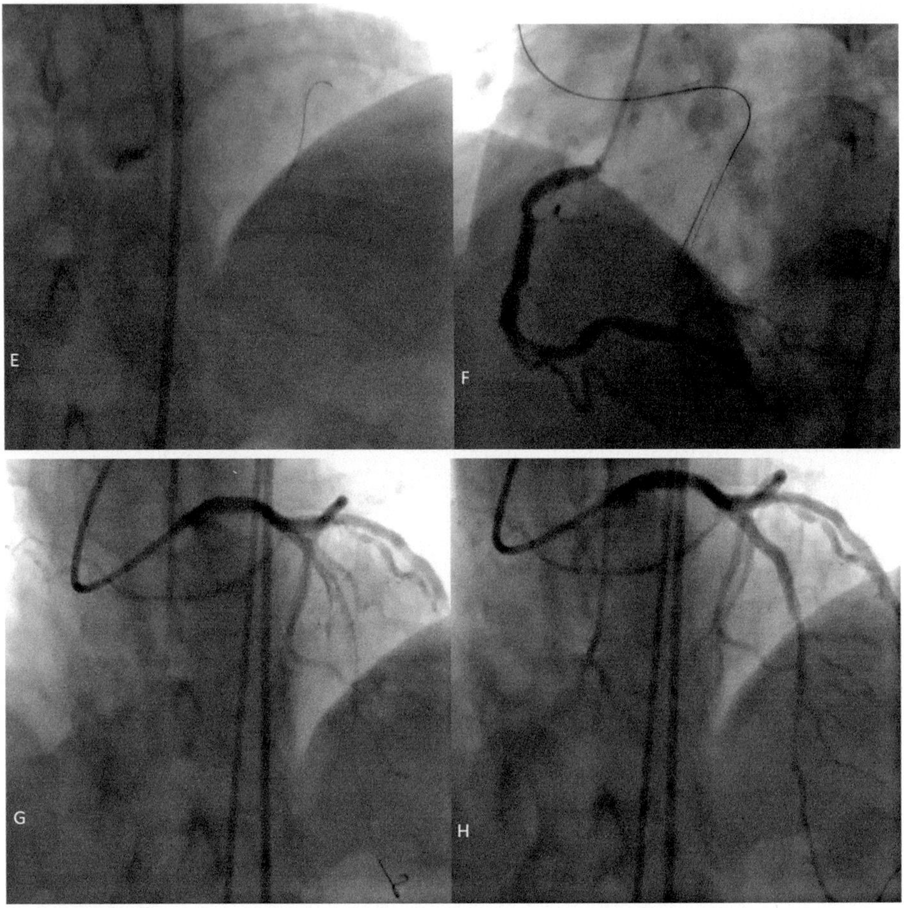

Fig. 5.4 (continued)

Case 4 (Fig. 5.5)

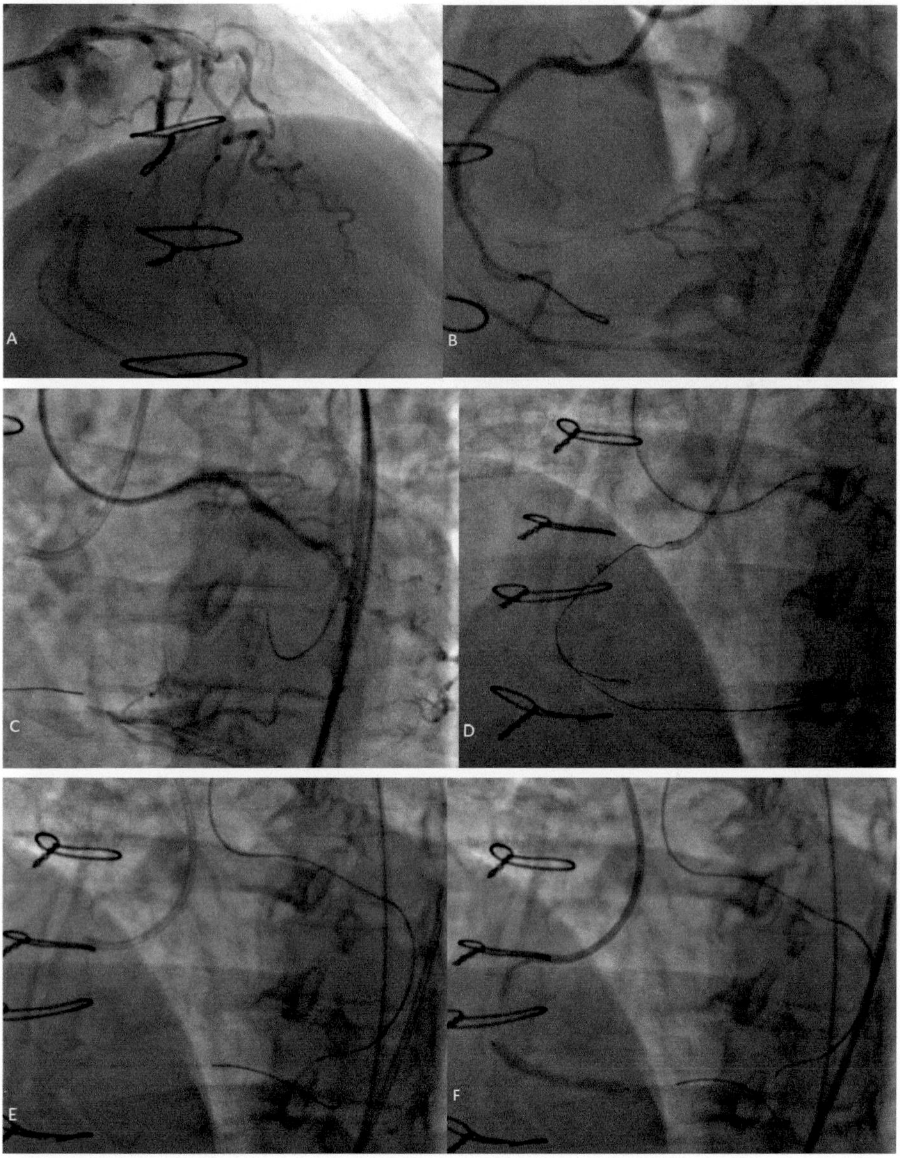

Fig. 5.5 (**a**–**f**) This figure depicts the case of a 72-year-old woman with a history of CABG who presented to the emergency department with a recent change in the pattern of her chest discomfort and a positive troponin test. Coronary angiography revealed an obstructed RCA with retrograde filling through the LAD septal branches (**a**, **b**). Antegrade wire escalation was unable to traverse the occluded segment (**c**), but retrograde surfing of the septal branches from the LAD enabled a retrograde Corsair Pro microcatheter to reach the distal segment (**d**–**f**). Ultimately, the Gaia second wire successfully crossed the CTO segment and was externalized into the antegrade microcatheter using the tip-in technique (**h**). Following a wire change and predilation, a DES was successfully deployed (**i**)

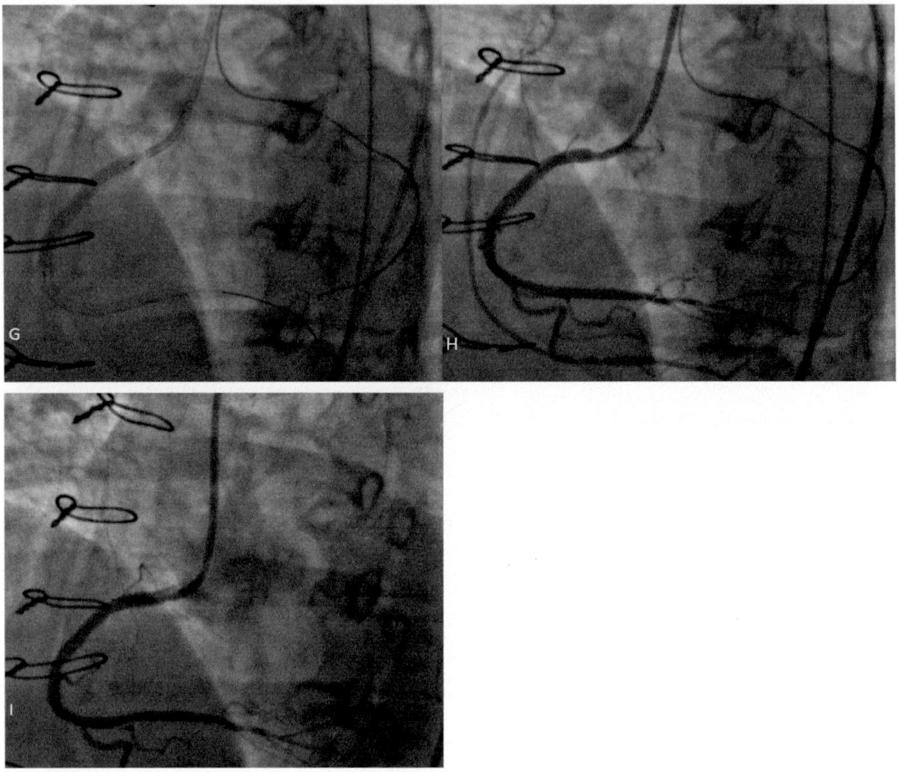

Fig. 5.5 (continued)

Case 5 (Fig. 5.6)

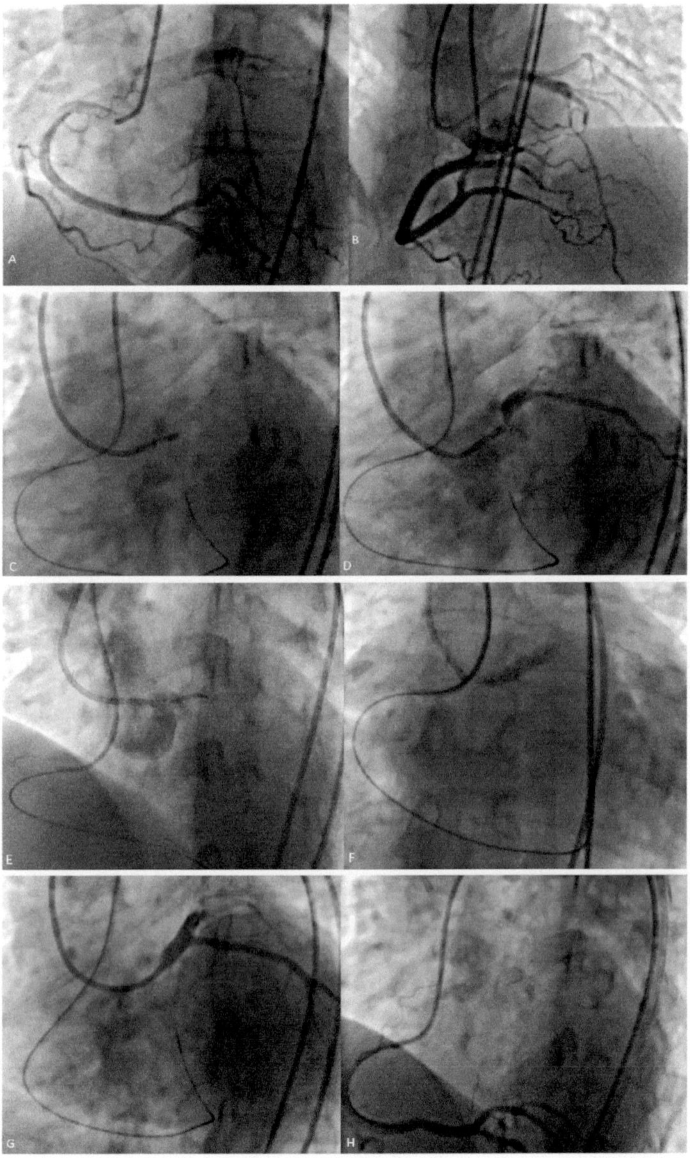

Fig. 5.6 (**a**–**h**) This figure presents a 57-year-old woman who developed exertional chest pain and dyspnea (functional class II) 2 months before presenting to our center. She had a history of hypertension and a preserved LVEF in echocardiography. During coronary angiography, the left-sided guiding catheter was unable to engage the LM (**a**). RCA injection revealed a totally occluded LM stem with retrograde filling via the right system (**b**). We opted not to attempt the antegrade approach due to the ostial LM being already cut, raising concerns about potential injury to the ascending aorta. Instead, the retrograde microcatheter was advanced through the septal branches of the LAD (**c**, **d**). Upon reaching the LM, the CTO of the LM was successfully penetrated with a Gaia second wire (**e**). After snaring the wire, changing the wire, and predilating the LM (**g**), a DES was successfully implanted, resulting in an acceptable outcome (**f**–**h**)

References

1. Tomasello SD, Boukhris M, Giubilato S, Marzà F, Garbo R, Contegiacomo G, et al. Management strategies in patients affected by chronic total occlusions: results from the Italian registry of chronic total occlusions. Eur Heart J. 2015;36(45):3189–98. PMID: 26333367.
2. Basiri HA, Khalilipur E, Sarreshtedari A, Zolfaghari R, Sadeghipour P, Alemzadeh-Ansari MJ, et al. Rajaie cardiovascular medical and research center-percutaneous coronary intervention registry: a real-world registry on coronary interventions in a tertiary teaching cardiovascular center. Res Cardiovasc Med. 2020;9(2):47–51. https://doi.org/10.4103/rcm.rcm_16_20.
3. Levine GN, Bates ER, Blankenship JC, Bailey SR, Bittl JA, Cercek B, et al. 2011 ACCF/AHA/SCAI guideline for percutaneous coronary intervention: a report of the American College of Cardiology Foundation/American Heart Association task force on practice guidelines and the Society for Cardiovascular Angiography and Interventions. Catheter Cardiovasc Interv. 2013;82(4):E266–355. PMID: 23441078.
4. Windecker S, Kolh P, Alfonso F, Collet JP, Cremer J, Falk V, et al. 2014 ESC/EACTS guidelines on myocardial revascularization: the task force on myocardial revascularization of the European Society of Cardiology (ESC) and the European Association for Cardio-Thoracic Surgery (EACTS) developed with the special contribution of the European Association of Percutaneous Cardiovascular Interventions (EAPCI). Eur Heart J. 2014;35(37):2541–619. PMID: 25173339.
5. Patel MR, Calhoon JH, Dehmer GJ, Grantham JA, Maddox TM, Maron DJ, et al. ACC/AATS/AHA/ASE/ASNC/SCAI/SCCT/STS 2017 appropriate use criteria for coronary revascularization in patients with stable ischemic heart disease: a report of the American College of Cardiology Appropriate use criteria task force, American Association for Thoracic Surgery, American Heart Association, American Society of Echocardiography, American Society of Nuclear Cardiology, Society for Cardiovascular Angiography and Interventions, Society of Cardiovascular Computed Tomography, and Society of Thoracic Surgeons. J Am Coll Cardiol. 2017;69(17):2212–41. PMID: 28291663.
6. Claessen BE, Dangas GD, Weisz G, Witzenbichler B, Guagliumi G, Möckel M, et al. Prognostic impact of a chronic total occlusion in a non-infarct-related artery in patients with ST-segment elevation myocardial infarction: 3-year results from the HORIZONS-AMI trial. Eur Heart J. 2012;33(6):768–75. PMID: 22240494.
7. Claessen BE, van der Schaaf RJ, Verouden NJ, Stegenga NK, Engstrom AE, Sjauw KD, et al. Evaluation of the effect of a concurrent chronic total occlusion on long-term mortality and left ventricular function in patients after primary percutaneous coronary intervention. JACC Cardiovasc Interv. 2009;2(11):1128–34. PMID: 19926054.
8. Choi IJ, Koh YS, Lim S, Choo EH, Kim JJ, Hwang BH, et al. Impact of percutaneous coronary intervention for chronic total occlusion in non-infarct-related arteries in patients with acute myocardial infarction (from the COREA-AMI Registry). Am J Cardiol. 2016;117(7):1039–46. PMID: 26833295.
9. Werner GS, Martin-Yuste V, Hildick-Smith D, Boudou N, Sianos G, Gelev V, et al. A randomized multicentre trial to compare revascularization with optimal medical therapy for the treatment of chronic total coronary occlusions. Eur Heart J. 2018;39(26):2484–93. PMID: 29701817.
10. Baks T, van Geuns RJ, Duncker DJ, Cademartiri F, Mollet NR, Krestin GP, et al. Prediction of left ventricular function after drug-eluting stent implantation for chronic total coronary occlusions. J Am Coll Cardiol. 2006;47(4):721–5. PMID: 16487836
11. Piscione F, Galasso G, De Luca G, Marrazzo G, Sarno G, Viola O, et al. Late reopening of an occluded infarct related artery improves left ventricular function and long term clinical outcome. Heart. 2005;91(5):646–51. PMID: 15831653.
12. Rossello X, Pujadas S, Serra A, Bajo E, Carreras F, Barros A, et al. Assessment of inducible myocardial ischemia, quality of life, and functional status after successful percuta-

neous revascularization in patients with chronic total coronary occlusion. Am J Cardiol. 2016;117(5):720–6. PMID: 26772437.
13. Marechal P, Davin L, Gach O, Martinez C, Lempereur M, Lhoest N, et al. Coronary chronic total occlusion intervention: utility or futility. Expert Rev Cardiovasc Ther. 2018;16(5):361–7. https://doi.org/10.1080/14779072.2018.1459187. PMID: 29589974.
14. Khan AA, Khalid MF, Ayub MT, Murtaza G, Sardar R, White CJ, et al. Outcomes of percutaneous coronary intervention versus optimal medical treatment for chronic total occlusion: a comprehensive meta-analysis. Curr Probl Cardiol. 2021;46(3):100695. https://doi.org/10.1016/j.cpcardiol.2020.100695. PMID: 33010951.
15. Morino Y, Abe M, Morimoto T, Kimura T, Hayashi Y, Muramatsu T, et al. Predicting successful guidewire crossing through chronic total occlusion of native coronary lesions within 30 minutes: the J-CTO (Multicenter CTO Registry in Japan) score as a difficulty grading and time assessment tool. JACC Cardiovasc Interv. 2011;4(2):213–21. PMID: 21349995.
16. Christopoulos G, Kandzari DE, Yeh RW, Jaffer FA, Karmpaliotis D, Wyman MR, et al. Development and validation of a novel scoring system for predicting technical success of chronic total occlusion percutaneous coronary interventions: the PROGRESS CTO (prospective global registry for the study of chronic total occlusion intervention) score. JACC Cardiovasc Interv. 2016;9(1):1–9. PMID: 26762904.
17. Brilakis ES, Mashayekhi K, Tsuchikane E, Abi Rafeh N, Alaswad K, Araya M, et al. Guiding principles for chronic total occlusion percutaneous coronary intervention. Circulation. 2019;140(5):420–33. https://doi.org/10.1161/CIRCULATIONAHA.119.039797. PMID: 31356131.
18. Tajti P, Burke MN, Karmpaliotis D, Alaswad K, Werner GS, Azzalini L, et al. Update in the percutaneous management of coronary chronic total occlusions. JACC Cardiovasc Interv. 2018;11(7):615–25. PMID: 29622142.
19. Brilakis ES. Manual of coronary chronic total occlusion interventions: a step-by-step approach. 2nd ed. London: Elsevier; 2017. ISBN: 978-0-12-809929-2.
20. Dawson K, Cheney A, Hira RS. Toolbox for coronary chronic total occlusion percutaneous coronary intervention. Interv Cardiol Clin. 2021;10(1):25–31. https://doi.org/10.1016/j.iccl.2020.09.011. PMID: 33223103.
21. Galassi AR, Sianos G, Werner GS, Escaned J, Tomasello SD, Boukhris M, et al. Retrograde recanalization of chronic total occlusions in Europe: procedural, in-hospital, and long-term outcomes from the multicenter ERCTO registry. J Am Coll Cardiol. 2015;65(22):2388–400. PMID: 26046732.
22. Karatasakis A, Tarar MN, Karmpaliotis D, Alaswad K, Yeh RW, Jaffer FA, et al. Guidewire and microcatheter utilization patterns during antegrade wire escalation in chronic total occlusion percutaneous coronary intervention: insights from a contemporary multicenter registry. Catheter Cardiovasc Interv. 2017;89(4):E90–8. PMID: 27465287.
23. Habara M, Tsuchikane E, Muramatsu T, Kashima Y, Okamura A, Mutoh M, et al. Comparison of percutaneous coronary intervention for chronic total occlusion outcome according to operator experience from the Japanese retrograde summit registry. Catheter Cardiovasc Interv. 2016;87(6):1027–35. PMID: 26332363.
24. Nikolakopoulos I, Choi JW, Alaswad K, Khatri JJ, Krestyaninov O, Khelimskii D, et al. Equipment utilization in chronic total occlusion percutaneous coronary interventions: insights from the PROGRESS-CTO registry. Catheter Cardiovasc Interv. 2021;97(4):658–67. https://doi.org/10.1002/ccd.29106. PMID: 32597031.
25. Sapontis J, Salisbury AC, Yeh RW, Cohen DJ, Hirai T, Lombardi W, et al. Early procedural and health status outcomes after chronic total occlusion angioplasty: a report from the OPEN-CTO registry (outcomes, patient health status, and efficiency in chronic total occlusion hybrid procedures). JACC Cardiovasc Interv. 2017;10(15):1523–34. PMID: 28797429.
26. Schumacher SP, Everaars H, Stuijfzand WJ, Huynh JW, van Diemen PA, Bom MJ, et al. Coronary collaterals and myocardial viability in patients with chronic total occlusions. EuroIntervention. 2020;16(6):e453–61. https://doi.org/10.4244/EIJ-D-19-01006. PMID: 32122823.

27. Galassi AR, Sumitsuji S, Boukhris M, Brilakis ES, Di Mario C, Garbo R, et al. Utility of intravascular ultrasound in percutaneous revascularization of chronic total occlusions: an overview. JACC Cardiovasc Interv. 2016;9(19):1979–91. PMID: 27712734.
28. Hasan F. Intravascular ultrasound with Japanese finesse for chronic total occlusion intervention. JACC Cardiovasc Interv. 2020;13(1):83–5. https://doi.org/10.1016/j.jcin.2019.10.034. PMID: 31838115.
29. Prasad M, Maehara A, Ahmad Y, Jeremias A, Shlofmitz E, Kirtane AJ, et al. Intravascular ultrasound in chronic total occlusion percutaneous coronary intervention: solving ambiguity and improving durability. Interv Cardiol Clin. 2021;10(1):75–85. https://doi.org/10.1016/j.iccl.2020.09.006. PMID: 33223109.
30. Tanaka T, Okamura A, Iwakura K, Kawamori H, Nagai H, Hamada K, et al. Efficacy and feasibility of the 3-dimensional wiring technique for chronic total occlusion percutaneous coronary intervention: first report of outcomes of the 3-dimensional wiring technique. JACC Cardiovasc Interv. 2019;12(6):545–55. PMID: 30898251.
31. Wilson WM, Walsh SJ, Yan AT, Hanratty CG, Bagnall AJ, Egred M, et al. Hybrid approach improves success of chronic total occlusion angioplasty. Heart. 2016;102(18):1486–93. https://doi.org/10.1136/heartjnl-2015-308891. PMID: 27164918.
32. Brilakis ES, Grantham JA, Rinfret S, Wyman RM, Burke MN, Karmpaliotis D, et al. A percutaneous treatment algorithm for crossing coronary chronic total occlusions. JACC Cardiovasc Interv. 2012;5(4):367–79. https://doi.org/10.1016/j.jcin.2012.02.006. PMID: 22516392.
33. Daniels DV, Banerjee S, Alaswad K, Burke MN, Wyman RM, Karmpaliotis D, et al. Safety and efficacy of the hybrid approach in coronary chronic total occlusion percutaneous coronary intervention: the Hybrid Video Registry. Catheter Cardiovasc Interv. 2018;91(2):175–9. PMID: 28568804.
34. Jones DA, Weerackody R, Rathod K, Behar J, Gallagher S, Knight CJ, et al. Successful recanalization of chronic total occlusions is associated with improved long-term survival. JACC Cardiovasc Interv. 2012;5(4):380–8. PMID: 22516394.
35. Harding SA, Wu EB, Lo S, Lim ST, Ge L, Chen JY, et al. A new algorithm for crossing chronic total occlusions from the Asia Pacific Chronic Total Occlusion Club. JACC Cardiovasc Interv. 2017;10(21):2135–43. https://doi.org/10.1016/j.jcin.2017.06.071. PMID: 29122129.
36. Konstantinidis NV, Werner GS, Hildick-Smith D, Spratt JC, Sianos G, Galassi AR, et al. Long-term outcomes of chronic total occlusion percutaneous coronary intervention with the retrograde approach: a multicenter registry analysis. JACC Cardiovasc Interv. 2024;17(3):352–62. https://doi.org/10.1016/j.jcin.2023.11.015. PMID: 38355265.
37. Brilakis ES. Manual of chronic total occlusion interventions: a step-by-step approach. 2nd ed. London: Elsevier; 2017. ISBN: 978-0-12-809929-2.
38. Riley RF, Walsh SJ, Kirtane AJ, Wyman RM, Nicholson WJ, Azzalini L, et al. Algorithmic solutions to common problems encountered during chronic total occlusion angioplasty: the algorithms within the algorithm. Catheter Cardiovasc Interv. 2019;93(2):286–97. https://doi.org/10.1002/ccd.27987. PMID: 30467958.
39. Kinnaird T, Anderson R, Ossei-Gerning N, Gallagher S, Large A, Strange J, et al. Vascular access site and outcomes among 26,807 chronic total coronary occlusion angioplasty cases from the British Cardiovascular Interventions Society National Database. JACC Cardiovasc Interv. 2017;10(7):635–44. PMID: 28385401.
40. Fairley SL, Lucking AJ, McEntegart M, Shaukat A, Smith D, Chase A, et al. Routine use of fluoroscopic-guided femoral arterial puncture to minimise vascular complication rates in CTO intervention: multi-centre UK experience. Heart Lung Circ. 2016;25(12):1203–9. PMID: 27476585.
41. Tanaka Y, Moriyama N, Ochiai T, Takada T, Tobita K, Shishido K, et al. Transradial coronary interventions for complex chronic total occlusions. JACC Cardiovasc Interv. 2017;10(3):235–43. PMID: 28109874.
42. Alaswad K, Menon RV, Christopoulos G, Lombardi WL, Karmpaliotis D, Grantham JA, et al. Transradial approach for coronary chronic total occlusion interventions: insights from

a contemporary multicenter registry. Catheter Cardiovasc Interv. 2015;85(7):1123–9. PMID: 25682726.
43. Schumacher SP, Stuijfzand WJ, Opolski MP, van Rossum AC, Nap A, Knaapen P. Percutaneous coronary intervention of chronic total occlusions: when and how to treat. Cardiovasc Revasc Med. 2019;20(6):513–22. https://doi.org/10.1016/j.carrev.2018.07.025. PMID: 30093292.
44. Ellis SG, Burke MN, Murad MB, Graham JJ, Badawi R, Toma C, et al. Predictors of successful hybrid-approach chronic total coronary artery occlusion stenting: an improved model with novel correlates. JACC Cardiovasc Interv. 2017;10(11):1089–98. https://doi.org/10.1016/j.jcin.2017.03.016. PMID: 28595879.
45. Wang N, Fulcher J, Abeysuriya N, Adams M, Lal S. Predictors of successful chronic total occlusion percutaneous coronary interventions: a systematic review and meta-analysis. Heart. 2018;104(6):517–24. https://doi.org/10.1136/heartjnl-2017-311986. PMID: 28986404.
46. Kalra S, Doshi D, Sapontis J, Kosmidou I, Kirtane AJ, Moses JW, et al. Outcomes of retrograde chronic total occlusion percutaneous coronary intervention: a report from the OPEN-CTO registry. Catheter Cardiovasc Interv. 2021;97(6):1162–73. https://doi.org/10.1002/ccd.29230. PMID: 32876381.
47. Riley RF, Sapontis J, Kirtane AJ, Karmpaliotis D, Kalra S, Jones PG, et al. Prevalence, predictors, and health status implications of periprocedural complications during coronary chronic total occlusion angioplasty. EuroIntervention. 2018;14(11):e1199–206. https://doi.org/10.4244/EIJ-D-18-00620. PMID: 30295294.
48. Suero JA, Marso SP, Jones PG, Laster SB, Huber KC, Giorgi LV, et al. Procedural outcomes and long-term survival among patients undergoing percutaneous coronary intervention of a chronic total occlusion in native coronary arteries: a 20-year experience. J Am Coll Cardiol. 2001;38(2):409–14. PMID: 11499731.
49. Safley DM, House JA, Marso SP, Grantham JA, Rutherford BD. Improvement in survival following successful percutaneous coronary intervention of coronary chronic total occlusions: variability by target vessel. JACC Cardiovasc Interv. 2008;1(3):295–302. PMID: 19463315.
50. Di Mario C, Werner GS, Sianos G, Galassi AR, Büttner J, Dudek D, et al. European perspective in the recanalisation of chronic total occlusions (CTO): consensus document from the EuroCTO Club. EuroIntervention. 2007;3(1):30–43. PMID: 19737685.
51. Brilakis ES, Karmpaliotis D, Patel V, Banerjee S. Complications of chronic total occlusion angioplasty. Interv Cardiol Clin. 2012;1(3):373–89. https://doi.org/10.1016/j.iccl.2012.04.006. PMID: 28582061.
52. Safley DM, Grantham JA, Hatch J, Jones PG, Spertus JA. Quality of life benefits of percutaneous coronary intervention for chronic occlusions. Catheter Cardiovasc Interv. 2014;84(4):629–34. PMID: 24403081.
53. Karacsonyi J, Vemmou E, Nikolakopoulos I, Ungi I, Rangan BV, Brilakis ES. Complications of chronic total occlusion percutaneous coronary intervention. Neth Heart J. 2021;29(1):60–7. https://doi.org/10.1007/s12471-020-01502-1. PMID: 33111194.
54. Dash D. Complications encountered in coronary chronic total occlusion intervention: prevention and bailout. Indian Heart J. 2016;68(5):737–46. https://doi.org/10.1016/j.ihj.2016.03.009. PMID: 27773418.
55. El Sabbagh A, Patel VG, Jeroudi OM, Michael TT, Alomar ME, Mogabgab O, et al. Angiographic success and procedural complications in patients undergoing retrograde percutaneous coronary chronic total occlusion interventions: a weighted meta-analysis of 3,482 patients from 26 studies. Int J Cardiol. 2014;174(2):243–8. PMID: 24768461.
56. Wu EB, Tsuchikane E, Ge L, Harding SA, Lo S, Lim ST, et al. Retrograde chronic total occlusion intervention using a nippon-European approach: an Asia Pacific Chronic Total Occlusion Club update. JACC Asia. 2024;4(2):139–48. https://doi.org/10.1016/j.jacasi.2023.10.007. PMID: 38371294.
57. Azzalini L, Poletti E, Ayoub M, Ojeda S, Zivelonghi C, La Manna A, et al. Coronary artery perforation during chronic total occlusion percutaneous coronary intervention: epidemiology, mechanisms, management, and outcomes. EuroIntervention. 2019;15(9):e804–11. https://doi.org/10.4244/EIJ-D-19-00279. PMID: 31227611.

58. Danek BA, Karatasakis A, Tajti P, Sandoval Y, Karmpaliotis D, Alaswad K, et al. Incidence, treatment, and outcomes of coronary perforation during chronic total occlusion percutaneous coronary intervention. Am J Cardiol. 2017;120(8):1285–92. PMID: 28826895.
59. Hirai T, Nicholson WJ, Sapontis J, Salisbury AC, Marso SP, Lombardi W, et al. A detailed analysis of perforations during chronic total occlusion angioplasty. JACC Cardiovasc Interv. 2019;12(19):1902–12. PMID: 31563681.
60. Xenogiannis I, Tajti P, Burke MN, Brilakis ES. An alternative treatment strategy for large vessel coronary perforations. Catheter Cardiovasc Interv. 2019;93(4):635–8. PMID: 30467968.
61. Sandoval Y, Lobo AS, Brilakis ES. Covered stent implantation through a single 8-French guide catheter for the management of a distal coronary perforation. Catheter Cardiovasc Interv. 2017;90(4):584–8. PMID: 28471062.
62. Michael TT, Papayannis AC, Banerjee S, Brilakis ES. Subintimal dissection/reentry strategies in coronary chronic total occlusion interventions. Circ Cardiovasc Interv. 2012;5(5):729–38. PMID: 23048052.
63. Kotsia A, Christopoulos G, Brilakis ES. Use of the retrograde approach for preserving the distal bifurcation after antegrade crossing of a right coronary artery chronic total occlusion. J Invasive Cardiol. 2014;26(3):E48–9. PMID: 24610516.
64. Wu EB, Brilakis ES, Mashayekhi K, Tsuchikane E, Ge L, Harding SA, et al. Global chronic total occlusion crossing algorithm: JACC state-of-the-art review. J Am Coll Cardiol. 2023;82(1):1–16. https://doi.org/10.1016/j.jacc.2023.05.027. PMID: 37380232.
65. Sianos G, Papafaklis MI. Septal wire entrapment during recanalization of a chronic total occlusion with the retrograde approach. Hell J Cardiol. 2011;52(1):79–83. PMID: 21253097.
66. Dash D. Complications of coronary intervention: abrupt closure, dissection, perforation. Heart Asia. 2013;5(1):61–5. https://doi.org/10.1136/heartasia-2013-010308. PMID: 27326077.
67. Dunning DW, Kahn JK, Hawkins ET, O'Neill WW. Iatrogenic coronary artery dissections extending into and involving the aortic root. Catheter Cardiovasc Interv. 2000;51(4):387–93. PMID: 11108666.

Chapter 6
PCI in Patients with Prior Coronary Artery Bypass Graft Surgery

Ata Firouzi, Ehsan Khalilipur, and Zahra Hosseini

Abstract SVGs remain the most commonly used conduits in CABG. Despite advancements in surgical techniques and pharmacotherapy, SVG failure rates remain high. The no-touch SVG harvesting technique, which minimizes graft manipulation and preserves the vasa vasorum and nerves, has been shown to reduce the risk of SVG failure.

Stenting of intermediate SVG stenosis (30%–60%) often leads to rapid disease progression. Redo CABG is associated with higher perioperative mortality than PCI. This chapter reviews recent studies on SVG PCI and discusses reasonable approaches to its management.

Keywords SVG · PCI · CABG

Abbreviations

ACS	Acute coronary syndrome
CABG	Coronary artery bypass graft surgery
CTO	Chronic total occlusion
DAPT	Dual antiplatelet therapy
EPD	Embolic protection device
IMA	Internal mammary artery
ISR	In-stent restenosis
LAD	Left anterior descending artery
LCX	Left circumflex coronary artery

A. Firouzi
Cardiovascular Intervention Research Center, Rajaie Cardiovascular Institute, Iran University of Medical Sciences, Tehran, Iran

E. Khalilipur (✉) · Z. Hosseini
Cardiovascular Intervention Research Center, Cardiovascular Institute, Iran University of Medical Sciences, Tehran, Iran

A. Firouzi et al. (eds.), *Mastering Skills in Complex Cardiovascular Interventions*, https://doi.org/10.1007/978-3-031-97321-5_6

LIMA	Left internal mammary artery
MACE	Major adverse cardiovascular events
MI	Myocardial infarction
PCI	Percutaneous coronary intervention
SVG	Saphenous vein graft
TIMI	Thrombolysis in myocardial infarction

Patients who have undergone CABG may require repeat revascularization due to bypass graft failure or progression of native coronary artery disease. Given the high risk associated with redo CABG, nearly all prior CABG patients undergo coronary revascularization via PCI. Nonetheless, PCI in these patients presents several clinical and technical challenges, including the treatment of failing bypass grafts, CTOs, and severe calcification [1].

Revascularization Strategies: Redo CABG Versus PCI

In patients requiring repeat revascularization, the Heart Team may decide between redo CABG or PCI) Repeat revascularization procedures in patients with prior CABG are technically more complex compared to those without a history of CABG. For redo CABG, challenges include post-surgical tissue adhesions in the mediastinum, limited options for additional grafts after initial bypass surgery, and difficulties in preserving patent conduits during the procedure.

PCI in post-CABG patients is complicated by extensive coronary artery disease with complex lesion morphology, including calcification, CTOs, and diseased bypass grafts with friable atheromatous plaques [2].

Several studies have compared long-term outcomes in patients with prior CABG undergoing repeat bypass surgery or PCI. In patients with previous CABG and ACS, redo CABG is associated with a statistically significant increase in in-hospital mortality and complications, including cardiogenic shock requiring mechanical and vasopressor support, respiratory failure, and acute kidney injury, compared with PCI [3].

Given the higher risk of periprocedural mortality with redo CABG, contemporary guidelines recommend PCI over redo CABG for repeat revascularization in patients with suitable coronary anatomy (Class II, Level of Evidence C).

In patients with diffuse coronary artery disease, reduced left ventricular ejection fraction, and extensive disease or occlusion of multiple bypass grafts—particularly if an IMA graft to the LAD was not previously used or is no longer patent—redo CABG may be considered.

When deciding between PCI on a graft vessel or native vessel, studies have shown that graft PCI is associated with complications such as distal embolization, no-reflow phenomenon, PCI-related MI, and rapid target vessel failure. Based on this observational evidence, the European coronary revascularization guidelines recommend PCI of the bypassed native coronary artery over bypass graft PCI, if technically feasible (Class IIa, Level of Evidence C).

Access Site Selection

Engagement of arterial grafts and SVGs for angiography and/or PCI can be achieved via femoral or radial access. Nonetheless, femoral access is associated with lower contrast and radiation doses. While systematic reviews of primarily observational studies suggested similar success rates for radial and femoral access, the RADIAL CABG (RADIAL Versus Femoral Access for Coronary Artery Bypass Graft Angiography and Intervention) trial found that diagnostic coronary angiography via radial access was associated with higher mean contrast volume (142 ± 39 mL vs. 171 ± 72 mL; $P = 0.01$), longer procedure times, and increased costs.

On the other hand, radial access has been linked to fewer vascular complications and shorter hospital stays in observational studies [3, 4]. If radial access is chosen, the left radial artery is typically preferred to facilitate engagement of the LIMA and other bypass grafts.

When radial access proves challenging for graft engagement, early conversion to femoral access should be considered [4].

PCI on SVGs

The process of SVG disease involves three interconnected pathophysiological mechanisms that lead to progressive degeneration: thrombosis, intimal hyperplasia, and atherosclerosis.

- **Thrombosis** is the primary mechanism in the first few days after graft implantation.
- **Intimal hyperplasia** is the dominant process during the first months, often related to transient ischemia.
- **Atherosclerosis** becomes the main driver of SVG disease after 1 year, though recurrent symptoms due to SVG disease rarely occur before 3 years post-grafting.

In patients with prior CABG presenting with ACS, atherosclerotic vein graft disease is the culprit lesion in 70%–85% of cases [5].

PCI on SVGs is recommended for atherosclerotic vein grafts with >50% stenosis in vessels supplying the LAD or large areas of the myocardium.

Prevention and Treatment of Distal Embolization During SVG PCI

SVG PCI accounts for approximately 7% of all PCI procedures at our center [6]. In the United States, SVG PCIs represent about 6% of all PCIs performed. Still, SVG PCI has two significant limitations:

1. Distal embolization and no-reflow phenomenon during the acute phase; and
2. High rates of restenosis and/or progression of SVG disease during follow-up.

SVG PCI carries a substantial risk of no-reflow, likely due to embolization of atheromatous material into the distal vasculature and severe vasospasm caused by microembolization of platelet-rich thrombi. These thrombi release vasoactive substances, impairing microvascular flow. The occurrence of no-reflow during SVG PCI is associated with a significantly higher risk of MACE in the long term [6, 7].

Several strategies can be drawn upon to reduce the risk of distal embolization and subsequent no-reflow. The use of EPDs is the only technique supported by randomized controlled trials. Other approaches include the administration of vasodilators, direct stenting, and the use of undersized stents [8].

Nicardipine is frequently preferred due to its prolonged duration of action and minimal hypotensive effects. It is often administered both before and during PCI. Other pharmacological agents, such as adenosine, nitroprusside, and verapamil, have also been shown to prevent or mitigate no-reflow episodes when administered intra-graft [9, 10].

The only currently available EPDs are filters, specifically the FilterWire (Boston Scientific, Natick, Massachusetts) and the Spider (Medtronic, Santa Rosa, California). Both devices require a distal landing zone for deployment, making them unsuitable for distal anastomotic lesions unless the filter is placed in the native coronary artery. While the FilterWire is advanced directly through the target SVG lesion, the Spider can be delivered over any guidewire advanced through the lesion [11].

In the first randomized controlled trial comparing EPDs with no EPDs for SVG PCI—the SAFER (Saphenous Vein Graft Angioplasty Free of Emboli Randomized) trial [12], which enrolled 801 patients—the use of the Guardwire was associated with a lower incidence of MI (8.6% vs. 14.7%; P = 0.008) and no-reflow (3% vs. 9%; P = 0.02).

In the SPIDER (Saphenous Vein Graft Protection in a Distal Embolic Protection Randomized) trial [13], the SpideRX filter was compared against the FilterWire and GuardWire in 700 patients. The SpideRX filter was shown to be noninferior, with comparable 30-day MACE rates (9.1% vs. 8.4%; P = 0.01).

Coolong et al. [14] demonstrated, through a pooled analysis of five controlled trials and one registry, that the efficacy of EPDs in reducing 30-day MACE was consistent across a range of SVG degeneration scores.

A multicenter registry compared PCI with and without EPDs in consecutive patients undergoing SVG PCI. The study included 792 patients, of whom 266

(33.6%) experienced MI. The primary composite outcome at 1 year was major adverse cardiac and cerebrovascular events, defined as death, MI, target vessel revascularization, or stroke. No significant differences were observed in MACCE (21.9% vs. 23.9%; HR: 0.91, 95% CI: 0.57 to 1.45; $P = 0.681$) or secondary endpoints, including mortality, MI, target vessel revascularization, target lesion revascularization, and stroke, between the EPD and no-EPD groups at 1 year. Similarly, there were no differences in study endpoints between the groups at 30-day follow-up.

In summary, routine use of EPDs during SVG PCI failed to provide a therapeutic benefit in either short- or long-term follow-up. Further research is needed to evaluate the impact of specific types of EPDs on clinical outcomes [15].

In another large cohort study by Shoaib et al. [16], multivariate analyses revealed no significant difference in mortality during index admission (OR: 0.71, 95% CI: 0.42 to 1.19; $P = 0.19$), at 30 days (OR: 0.87, 95% CI: 0.60 to 1.25; $P = 0.45$), or at 1 year (OR: 0.92, 95% CI: 0.77 to 1.11; $P = 0.41$). Rates of in-hospital MACE were also comparable (OR: 0.68, 95% CI: 0.20 to 2.35; $P = 0.54$). In propensity score-matched analyses, the EPD group had a lower inpatient mortality rate (OR: 0.46, 95% CI: 0.13 to 0.80; $P = 0.002$). However, the adjusted risk for periprocedural no-reflow or slow-flow was higher in patients who received EPDs (OR: 2.16, 95% CI: 1.71 to 2.73; $P < 0.001$).

Stent Selection in SVG PCI

Lesions in SVGs have a high rate of ISR, often presenting as ACS. In a meta-analysis of six randomized studies, Kheiri et al. [18] found no significant differences in the long-term incidence of MACE, target lesion revascularization, target vessel revascularization, stent thrombosis, or all-cause mortality between drug-eluting stents and bare-metal stents. Nevertheless, drug-eluting stents have been shown to reduce the rate of target lesion revascularization.

Given that bare-metal and drug-eluting stents yield comparable outcomes in SVGs, bare-metal stents may be preferred in countries where there is a significant cost difference between these two stent types [17].

Early Postoperative Graft Failure: Acute and Chronic Total SVG Occlusions

Graft failure occurs in up to 12% of grafts during the early postoperative period, with approximately 3% of patients experiencing symptoms. Graft occlusion rates are higher for SVGs (3%–12% prior to discharge) compared with radial artery grafts (3%–4%) and IMA grafts (1%–2.5%) [19, 20].

Disease progression is more common in bypassed native coronary arteries, particularly those not supplying the LAD, and when SVGs are used instead of arterial grafts.

SVG failure can result from various factors, including venous trauma during harvest, mechanical issues at the anastomotic site, and eventual vein degeneration due to atherosclerotic disease. SVG failure occurs in three distinct temporal phases: early, intermediate, and late.

Early (acute) SVG failure (within 1 month after CABG) is primarily caused by acute thrombosis or technical failure, most often at the anastomosis site. Several factors contribute to early failure, as follows:

- A small target epicardial vessel leading to poor distal runoff;
- Size mismatch between the graft and the target vessel causing turbulent flow;
- Graft ischemia; and
- Disruption of the endothelial layer due to mechanical trauma and manual distention during surgery [21, 22].

Intermediate (subacute) SVG failure typically occurs due to progressive intimal hyperplasia within 1–12 months following CABG. During this period, the graft responds to increased arterial pressures by proliferating smooth muscle cells, leading to intimal hyperplasia.

Late SVG failure (occurring more than 12 months after CABG) is primarily caused by the progression of atherosclerosis over the previously damaged endothelium [23].

The strongest long-term predictors of SVG patency are grafting into the LAD and into a vessel with a diameter >2.0 mm. SVGs to the RCA are the least likely to remain patent over the long term.

Surgical factors that increase the risk of SVG failure include insufficient distal runoff, a small target vessel diameter, and a size mismatch between the graft and the artery [24].

Unless acute graft failure is diagnosed in the operating room, PCI is the preferred treatment for symptomatic graft failure. Whenever possible, PCI should target the native coronary artery rather than the bypass graft, as PCI of the graft anastomosis may lead to suture dehiscence and perforation. Redo CABG is considered when:

- The coronary anatomy is unsuitable for PCI;
- A significant myocardial territory is at risk;
- Multiple major grafts are occluded; or
- Anastomotic lesions are present [10, 20].

In patients with a history of CABG, around two-thirds of PCIs are performed on native coronary artery lesions rather than bypass grafts. These native lesions are often highly complex and are characterized by significant calcification, tortuosity, and CTOs. While bypass grafts are commonly utilized for retrograde crossing, navigating the vessel upstream from the distal anastomosis can be technically demanding.

When performing PCI on the distal native vasculature via the IMA or using retrograde approaches, slightly shorter guiding catheters are often the preferred choice.

Due to the inherent complexity of these cases, advanced PCI techniques—such as atherectomy and CTO PCI—are frequently required. Despite these technical challenges, numerous studies have shown that outcomes following native coronary artery PCI are superior to those following SVG PCI [26, 27].

Compared with SVG PCI, PCI targeting IMA lesions has been linked to higher rates of TIMI flow grade 3 restoration and lower rates of periprocedural complications. For IMA anastomotic lesions, balloon angioplasty may be the optimal treatment, while proximal and mid-segment lesions are typically managed with stenting.

Gruberg et al. [29] analyzed 174 patients undergoing PCI for 128 IMA anastomotic lesions and found that stenting was associated with a higher rate of repeat revascularization (33%) than balloon angioplasty alone (4.3%). Similarly, Sharma et al. [30] evaluated 288 patients with 311 IMA lesions and reported worse outcomes with stenting at the anastomotic site compared with angioplasty alone (25% vs. 4.2%; $P = 0.006$).

In patients with IMA grafts, it is essential to evaluate the proximal subclavian artery, as significant lesions in this area can lead to myocardial ischemia and potentially trigger ACS. In such cases, stenting of the subclavian artery may be a viable treatment option [31].

Given the substantial short- and long-term risks associated with SVG PCI, many operators recommend targeting the native coronary artery instead in patients with SVG lesions. The 2018 ESC/EACTS guidelines on myocardial revascularization support this approach, recommending PCI of the native vessel over PCI of the bypass graft (Class IIa, Level of Evidence: C) [32].

Achieving successful SVG PCI relies heavily on optimal guide catheter support. The choice of guide catheter type and size depends on factors such as aortic diameter, the position and angle of the grafts, and the operator's preference. For left-sided graft engagement, commonly used guide catheters include the Judkins right (JR), Amplatz left (AL), and Amplatz right (AR). These catheters can also be employed for right-sided grafts. For grafts with a steep inferior take-off angle, a multipurpose guide catheter may be considered. Other options include the left bypass, hockey stick, and Ikari left guide catheters. When imaging left coronary grafts, a right anterior oblique projection is typically used, while a left anterior oblique projection is preferred for right coronary grafts [33].

Direct stenting has been shown to reduce distal debris embolization as compared to pre-dilation and is related to a reduction in severe adverse cardiac events after SVG PCI. Avoid excessive dilation due to the possibility of distal embolization from extensive plaque extrusion. To optimize SVG revascularization outcomes, it is critical to examine the stent type, EPDs, and medication selection carefully [8].

Abdelrahman et al. [34] conducted a study involving 435 PCIs in 401 patients with prior CABG, including 209 native PCIs and 192 graft PCIs. Multivariable Cox regression analysis revealed that graft PCI was an independent predictor of MACE (HR: 1.725, 95% CI: 1.049 to 2.837). This increased risk was primarily driven by higher rates of revascularization (HR: 2.218, 95% CI: 1.193 to 4.122) and MI (HR: 2.248, 95% CI: 1.220 to 4.142), although there was no significant difference in mortality (HR: 1.118, 95% CI: 0.9231–1.220). Compared with native coronary PCI, bypass graft PCI was significantly associated with a higher rate of MACE at 1 year, largely due to increased incidences of MI and revascularization.

Wolny et al. [35] used intravascular ultrasound to evaluate patterns of ISR in 54 SVG lesions. Stent underexpansion was defined as a minimal stent area <5 mm^2. The average time between stent implantation and the presentation of ISR was 3.7 ± 3.0 years, involving nine bare-metal stents, 18 first-generation drug-eluting stents, and 27 second-generation drug-eluting stents. ISR patterns identified by intravascular ultrasound were categorized as mechanical (33%) or biological (67%). Mechanical causes included the following:

- Ten cases of stent underexpansion (minimal stent area = 4.2 ± 0.9 mm^2);
- Six cases of stent fracture or deformation; and
- Two cases of uncovered aorto-anastomotic lesions.

Biological causes included the following:

- Nineteen cases of neoatherosclerosis;
- Thirteen cases of severe neointimal hyperplasia; and
- Four cases of thrombus formation.

Mechanical ISR patterns were more frequently observed at the SVG anastomosis (72% vs. 39%; P = 0.04) and at hinge motion sites (55% vs. 21%; $P = 0.02$) than biological patterns. Although mechanical ISR patterns presented earlier than biological patterns (2.3 vs. 4.4 years; P = 0.009), 61% of mechanical cases were detected more than 1 year after stent implantation. The authors concluded that biological patterns, particularly neoatherosclerosis, are the predominant cause of SVG ISR. Mechanical patterns, on the other hand, are associated with earlier presentation and are more likely to occur at graft anastomosis or hinge motion sites.

Intravascular Lithotripsy in SVGs and Graft Stent Failure

Intravascular lithotripsy is a technique used to modify calcified lesions in native coronary arteries. The Shockwave balloon generates high-frequency sonic pressure waves that fracture calcium deposits in the intima and media of the arterial wall, facilitating stent delivery and expansion. While observational studies have demonstrated the safety and efficacy of intravascular lithotripsy in *de novo* coronary artery disease, its application in SVGs has not been previously documented.

In a case series reported by Øksnes et al. [36], intravascular ultrasound was successfully used in five SVG PCI procedures, demonstrating its potential safety and feasibility for treating degenerated SVGs or SVG stent failure. Nevertheless, the long-term patency of this technique remains to be evaluated in larger, future trials.

Complications

Performing angiography and PCI in patients with prior CABG often involves engaging and visualizing bypass grafts, as well as addressing highly complex lesions. This leads to longer procedural and fluoroscopy times, increased radiation exposure, and higher contrast volumes. Consequently, these patients—who often have reduced baseline renal function—face a greater risk of contrast-induced nephropathy and, in some cases, may require hemodialysis compared with those without prior CABG [37, 39].

Coronary perforation in post-CABG patients can result in localized hematomas, which may compress cardiac chambers and cause hemodynamic instability. According to the OPEN-CTO (Outcomes, Patient Health Status, and Efficiency in Chronic Total Occlusion Hybrid Procedures) database, perforation occurred in approximately 7% of post-CABG patients. Among 365 patients with prior CABG, four perforations (1.1%) were fatal [40].

Antiplatelet Therapy

Long-term DAPT is an intriguing concept for patients who have undergone CABG, as they often have extensive, multi-layered atherosclerotic disease and a high risk of subsequent cardiovascular events.

A meta-analysis of 22 studies comparing DAPT with aspirin alone after CABG found that DAPT was associated with a reduced risk of cardiovascular death (OR: 0.67; P = 0.02) and a trend toward lower all-cause mortality (OR: 0.78; P = 0.08). Still, these benefits were not observed when the analysis was limited to randomized controlled trials. DAPT significantly reduced SVG occlusion in the overall population (OR: 0.64; P = 0.01) and in the subset of randomized controlled trials (OR: 0.58; P = 0.01). Notably, patients receiving DAPT for more than 6 months had a lower risk of stroke (OR: 0.47; P = 0.04) but a higher risk of major bleeding (OR: 1.31; P = 0.03) [41].

Another meta-analysis of nine randomized controlled trials found that patients treated with ticagrelor or prasugrel in addition to aspirin had a lower risk of death than those receiving clopidogrel plus aspirin (relative risk: 0.49, 95% CI: 0.33 to 0.71; P = 0.0002). Nonetheless, no significant difference was observed between clopidogrel plus aspirin and aspirin monotherapy [42].

Sirker et al. [43] analyzed data from 8119 patients undergoing SVG PCI and found that the majority were treated with clopidogrel (n = 7401), followed by ticagrelor (n = 497) and prasugrel (n = 221). Neither prasugrel nor ticagrelor was associated with significantly lower 30-day mortality than clopidogrel in multivariable logistic regression or propensity score matching models. Multivariable analysis yielded ORs of 1.22 (95% CI: 0.60 to 2.51) for prasugrel versus clopidogrel and 0.48 (95% CI: 0.20 to 1.16) for ticagrelor versus clopidogrel. No statistically significant differences were observed in in-hospital ischemic or bleeding events. This real-world national study found no conclusive evidence that potent P2Y12 inhibitors improve clinical outcomes in SVG PCI.

SVG for CTO Recanalization

Patients with a history of CABG may develop ACS due to graft failure or progression of coronary artery disease in the native vessels. However, treating CTOs in these patients can be particularly challenging, as these lesions often feature blunt stumps, severe tortuosity, and extensive calcification—likely due to the distal cap being exposed to systemic arterial pressure [1, 10].

For this reason, CTO PCI is ideally performed in a planned, scheduled manner rather than on an ad hoc basis. An alternative strategy, when feasible, is "staged SVG revascularization." In this approach, the culprit SVG lesion is addressed during the acute phase of MI to restore blood flow to the affected myocardium. After a few weeks, the native coronary artery lesion is revascularized.

Long-term outcomes following native coronary artery PCI are superior to those of SVG PCI, which carries a high risk of restenosis and recurrent adverse events, regardless of whether bare-metal or drug-eluting stents are used [44, 45].

CTOs in the setting of acute MI can be time-intensive and carry a risk of complications. In addition, it has yet to be extensively studied in large-scale research. This procedure should only be undertaken by experienced operators at centers equipped with the necessary tools and expertise to manage potential complications [46].

Although data in this area are limited, many operators recommend closing SVGs if TIMI flow grade 2 or 3 persists through the SVG after treating the native coronary artery. Following recanalization of the associated native coronary artery CTO, SVG occlusion can be achieved using coils or an Amplatzer vascular plug (typically Amplatzer vascular plug II, sized 50% > the target vessel diameter). Plugs offer several advantages, including precise delivery, reduced risk of distal or proximal migration, and the need for only a single device rather than multiple coils [37].

Nonetheless, drawbacks include higher costs, the need for a larger delivery catheter, and challenges in deploying the device in tortuous vessels [47].

Clinical Outcomes of PCI for CTO in Native Coronary Arteries vs SVGs

In a large-scale study by Shoaib et al. [48], a longitudinal cohort (2007–2014; n = 11,132) was established using the British Cardiovascular Intervention Society (BCIS) database. The cohort included patients who underwent SVG PCI in native arteries (Group I; n = 8619) or CTO PCI in native arteries (Group II; n = 2513). Both groups had a median age of 68 years, but Group II patients were less likely to be female and had a higher prevalence of diabetes mellitus, hypertension, hypercholesterolemia, and prior MI. They also presented with more severe angina and breathlessness, though rates of prior stroke, renal disease, and left ventricular systolic dysfunction were similar between the groups. Multivariable analysis revealed no significant differences in mortality during the index hospital admission (OR: 1.70, 95% CI: 0.63 to 4.58; P = 0.29), at 30 days (OR: 1.81, 95% CI: 0.99 to 3.3; P = 0.05), or at 1 year (OR: 1.11, 95% CI: 0.85 to 1.44; P = 0.43). Similarly, there were no significant differences concerning in-hospital MACE (OR: 1.36, 95% CI: 0.89 to 2.08; P = 0.15). However, compared with the SVG PCI group, CTO PCI was associated with higher rates of procedural complications (OR: 2.88, 95% CI: 2.38 to 3.47; P < 0.01) and vessel perforation (OR: 4.82, 95% CI: 2.80 to 8.28; P < 0.01). At 1 year, the risk of target vessel revascularization was comparable between the groups (5.6% for SVG PCI vs. 6.9% for CTO PCI; P = 0.08). The authors concluded that CTO PCI was performed on higher-risk patients in this national cohort and was associated with increased procedural complications but similar short- and long-term mortality and in-hospital MACE rates compared with SVG PCI.

Outcomes of CTO PCI in Patients with Prior CABG

Tajti et al. [49] conducted a study analyzing the clinical, angiographic, and outcome characteristics of 3486 CTO procedures performed at 21 centers. The cohort included 1101 patients with prior CABG and 2317 patients without prior CABG. Patients with prior CABG (32% of the total cohort) were older (67 ± 9 years vs. 63 ± 10 years; P < 0.001), had more comorbidities, and exhibited lower left ventricular ejection fractions (50% [40–58] vs. 55% [45–60]; P < 0.001). In patients with prior CABG, the most common CTO target vessels were the RCA (56%), the CLX (26%), and the LAD (17%). These patients also had higher mean J-CTO scores (2.9 ± 1.2 vs. 2.2 ± 1.3; P < 0.001) and PROGRESS-CTO (Prospective Global Registry for the Study of Chronic Total Occlusion Intervention) scores (1.5 ± 1.1 vs. 1.2 ± 1.0; P < 0.001). Retrograde (53% vs. 30%; P < 0.001) and antegrade dissection/re-entry (35% vs. 28%; P < 0.001) techniques were more frequently used in patients with prior CABG. 89%; P < 0.001) and procedural success

(82% vs. 87%; P < 0.001) than those without prior CABG. Still, the rates of in-hospital major complications were similar between the two groups (3.1% vs. 2.5%; P = 0.287). Prior CABG patients experienced higher rates of in-hospital mortality (1% vs. 0.4%; P = 0.016) and coronary perforation (7.1% vs. 3.1%; P < 0.001). Conversely, they had lower rates of pericardial tamponade (0.1% vs. 1.0%; P = 0.002) and pericardiocentesis (0% vs. 1.3%; P < 0.001). Additionally, in 39% of patients with prior CABG, the native CTO target vessel had been previously bypassed. The target lesions were located proximally (68%), distally (21%), and at the anastomosis site (11%). The retrograde technique played a crucial role in CTO PCI for patients with prior CABG: it was used in half of the cases and was the final successful strategy in 32% of patients (vs. 19% in patients without prior CABG). Finally, the authors reported that, in a large multicenter registry of CTO PCI, patients with prior CABG had lower success rates but a similar overall risk of complications. However, they experienced higher rates of in-hospital mortality and lower rates of tamponade than patients without prior CABG.

Dautov et al. [50] investigated the feasibility and safety of retrograde CTO PCI in patients with prior CABG, showing that retrograde approaches using SVGs were equally effective as those employing septal or epicardial collaterals or antegrade-only techniques. Chiming with those results, Okamura et al. [51] reported high success rates for retrograde CTO PCI via SVGs in patients with prior CABG, based on findings from the Japanese Retrograde Summit Registry [52]. In the aforementioned study, the LIMA was seldom utilized (2%) in CTO PCI, and the authors recommended that it be reserved as a last-resort option due to the extensive myocardial territory at risk, which often requires hemodynamic support.

Xenogiannis et al. [53] analyzed retrograde CTO PCI performed via SVGs in 189 patients (12% of the cohort). Patients in the SVG group were older (mean age: 70 ± 9 years vs. 64 ± 10 years; P < 0.01) and had higher rates of prior MI (62% vs. 51%; P < 0.01) and prior PCI (81% vs. 70%; P < 0.01). These patients also exhibited a higher prevalence of moderate or severe calcification (81% vs. 65%; P < 0.01) and moderate or severe tortuosity (53% vs. 44%; P = 0.02). While J-CTO (Multicenter CTO Registry in Japan) scores were similar between the groups (3.2 ± 1.0 vs. 3.1 ± 1.1; P = 0.13), PROGRESS-CTO (Prospective Global Registry for the Study of Chronic Total Occlusion Intervention) scores were higher in the SVG group. Technical (85% vs. 78%; P = 0.04) and procedural success rates (81% vs. 74%; P = 0.04) were higher in the SVG group, but there was no significant difference in in-hospital MACE (6.4% vs. 4.4%; P = 0.22). The SVG group also used less contrast volume (225 mL [173–325 mL] vs. 292 mL [202–400 mL]; P < 0.01). The authors concluded that retrograde CTO PCI using SVGs was associated with higher technical and procedural success rates and a similar rate of in-hospital MACE compared with other approaches.

Prediction of MACE and Slow/No-Reflow

Liu et al. [54] utilized virtual histology-intravascular ultrasonography to investigate the relationship between plaque content and morphology with post-PCI MACE and slow/no-reflow in patients with SVG disease. The study enrolled 1490 patients, of whom 76.7% were men, with a mean age of 64.9 ± 8.2 years and an average duration of SVG of 8.0 ± 3.6 years. Remarkably, 77.8% of the lesions exhibited a plaque burden of at least 70%. Over a 12-month period following PCI, 18 incidences of MACE occurred in 14 patients, and 12 patients experienced slow/no reflow. Multivariate analysis revealed that lesion length (HR: 1.05, 95% CI: 1.01 to 1.08), age at the time of CABG (HR: 1.51, 95% CI: 1.11 to 2.05), and the absolute necrotic core area (HR: 8.04, 95% CI: 1.86 to 34.73) were all independently associated with MACE. Additionally, independent predictors of slow/no reflow following PCI included preprocedural systolic blood pressure (OR: 0.98, 95% CI: 0.96 to 0.99) and the absolute necrotic core area (OR: 2.47, 95% CI: 1.14 to 5.36). The authors concluded that a threshold value of 1.1 mm^2 for the absolute necrotic core area might serve as a significant risk factor for non-reflow following SVG PCI.

Gürbak et al. [55] assessed 268 patients with non-ST elevation ACS who underwent PCI for SVG disease. Following the intervention, patients were divided into two groups: group I, which exhibited no evidence of reflow (n = 190), and group II, which demonstrated evidence of reflow (n = 78). The authors then compared CHA2DS2-VASc scores between the two groups. The CHA2DS2-VASc score was significantly higher in group II (P < 0.001) despite no significant difference in the prevalence of atrial fibrillation between the study groups. Independent predictors of no-reflow included the CHA2DS2-VASc score (P < 0.001), deteriorated SVGs (P = 0.006), and the presence of intraluminal thrombi (P < 0.001). The receiver operating characteristic (ROC) analysis indicated that a CHA2DS2-VASc score of 4 accurately predicted the absence of reflow, with a sensitivity of 67.9% and a specificity of 69.3%. These findings suggest that the CHA2DS2-VASc score can independently predict the no-reflow phenomenon in patients following SVG interventions. As a straightforward scoring system, it may be a valuable tool for assessing the risk of no-reflow in patients with non-ST elevation ACS before SVG treatments.

Challenging Clinical Cases

Case 1 (Fig. 6.1)

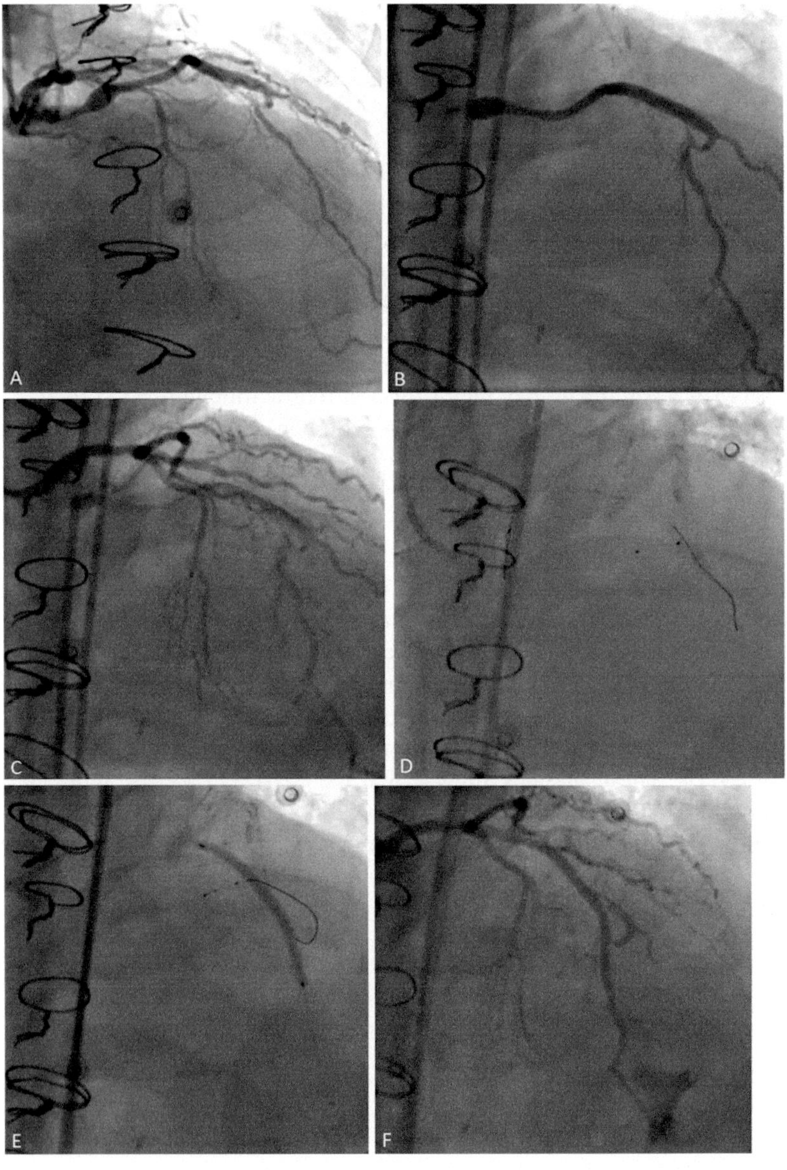

Fig. 6.1 (**a**–**f**) This figure depicts a 68-year-old man who underwent CABG 6 years prior, with a graft on the LAD (**a**). He presented to the emergency department with recent exertional chest pain and an elevated troponin level. Coronary angiography revealed disease in the SVG on the LAD, along with a CTO of the native vessel (**b**). Given the presence of acceptable distal runoff from the LAD, we opted to perform antegrade wire escalation for the CTO segment (**c**) and utilized SVG flow as a guide for crossing (**d**). Ultimately, after the implantation of a drug-eluting stent, we achieved a satisfactory result (**e, f**)

Case 2 (Fig. 6.2)

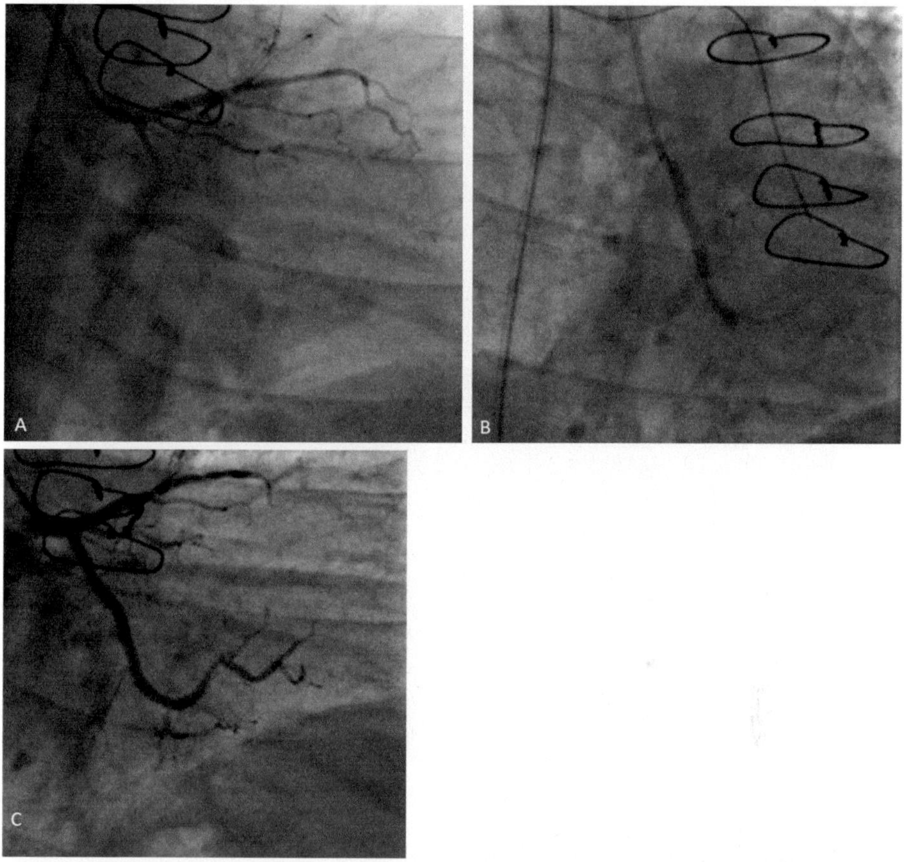

Fig. 6.2 (**a–c**) This case involves a 74-year-old woman with a history of CABG performed 12 years before presenting with recent exertional chest discomfort and moderate risk identified on myocardial perfusion imaging. Coronary angiography revealed a degenerated SVG on the obtuse marginal artery and a CTO of the native LCX (**a**). Using retrograde access to the LCX through its SVG, along with additional support from a microcatheter, we successfully navigated the crossing segment (**b**) and implanted a drug-eluting stent following adequate predilation. The final result demonstrated improved distal flow in both the LCX and obtuse marginal branch (**c**)

Case 3 (Fig. 6.3)

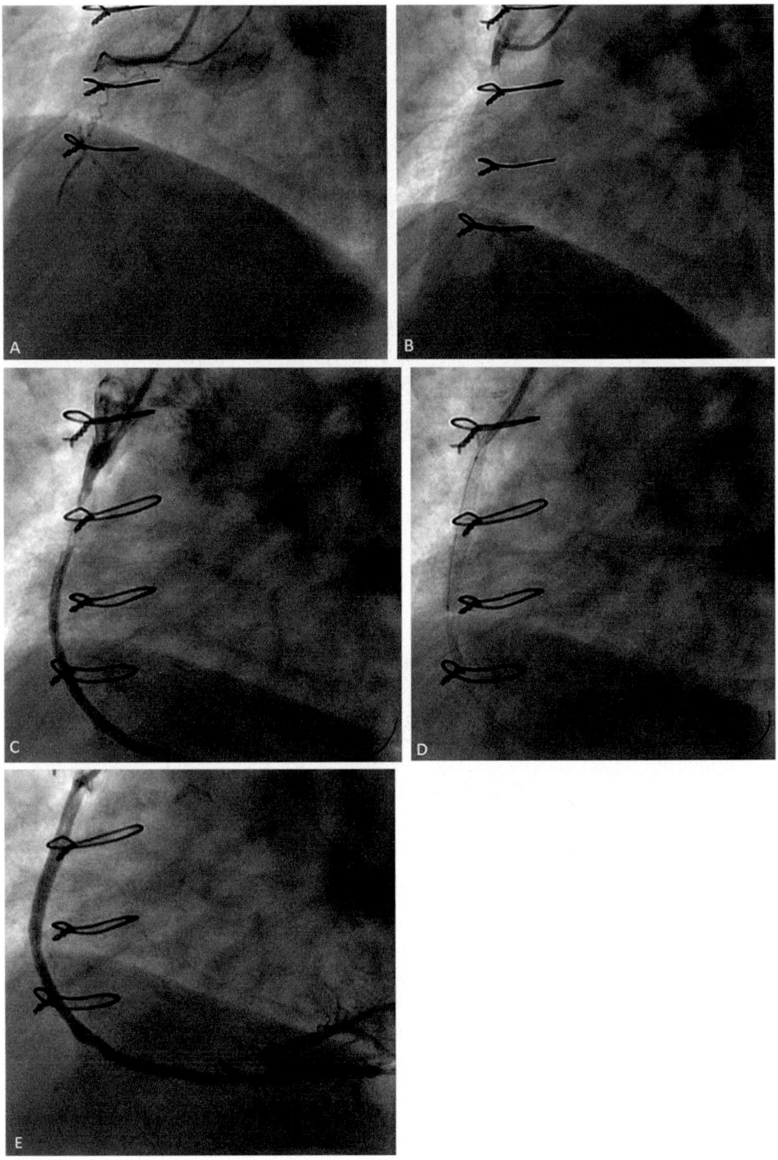

Fig. 6.3 (**a**–**e**) This figure presents a 53-year-old gentleman with a history of CABG who arrived at the emergency department with sudden retrosternal chest pain and ST elevation in the inferior leads. He was promptly transferred to the catheterization lab, where coronary angiography revealed a thrombosed SVG supplying the right posterior descending artery (**a**, **b**). Primary PCI was performed on this SVG, followed by an assessment of the native RCA to evaluate the possibility of addressing a CTO. Although it is generally recommended to use a distal protective device during SVG PCI, we opted to implant a drug-eluting stent without predilation or manual thrombosuction to minimize the risk of distal embolization. The final result was satisfactory (**c**–**e**)

Case 4 (Fig. 6.4)

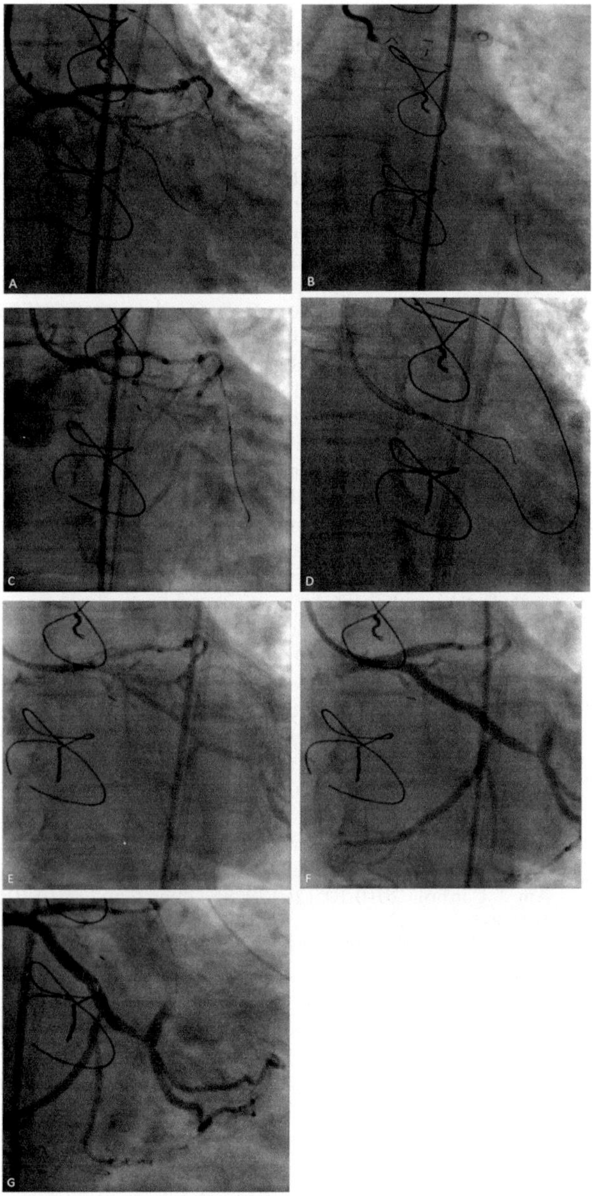

Fig. 6.4 (**a**–**g**) This case involves a 67-year-old gentleman who presented with recent dyspnea on exertion and warm-up angina. He had undergone CABG 11 years prior, and further evaluation revealed reduced left ventricular function along with hypokinesia of the lateral and posterior walls. Coronary angiography showed a diffusely diseased SVG supplying the obtuse marginal artery and a completely occluded LCX in the proximal segment (**a**, **b**). After antegrade wire escalation failed to traverse the CTO segment (**c**), the SVG was utilized to gain retrograde access to the LCX (**d**, **e**). Subsequently, through wire escalation and with the assistance of a microcatheter, the Gaia third wire was successfully passed through the CTO and externalized into the antegrade guiding catheter (**f**). Finally, following predilation, a drug-eluting stent was implanted in the CTO segment, resulting in a satisfactory outcome (**g**)

References

1. Brilakis ES, O'Donnell CI, Penny W, Armstrong EJ, Tsai T, Maddox TM, et al. Percutaneous coronary intervention in native coronary arteries versus bypass grafts in patients with prior coronary artery bypass graft surgery: insights from the Veterans Affairs Clinical Assessment, Reporting, and Tracking program. JACC Cardiovasc Interv. 2016;9(9):884–93. https://doi.org/10.1016/j.jcin.2016.01.026. PMID: 27085582.
2. Bourantas CV, Zhang YJ, Garg S, Iqbal J, Valgimigli M, Windecker S, et al. Percutaneous coronary intervention for the treatment of stenoses in native coronary arteries and bypass grafts in patients with previous coronary artery bypass graft surgery. JACC Cardiovasc Interv. 2014;7(5):539–51. https://doi.org/10.1016/j.jcin.2013.12.200. PMID: 24814645.
3. Brilakis ES, Rao SV, Banerjee S, Goldman S, Shunk KA, Holmes DR Jr, et al. Percutaneous coronary intervention in native arteries and grafts in patients with prior coronary artery bypass graft surgery. J Am Coll Cardiol. 2011;58(5):512–20. https://doi.org/10.1016/j.jacc.2011.03.038. PMID: 21777750.
4. Cooper L, Banerjee S, Brilakis ES. Crossover from radial to femoral access during a challenging percutaneous coronary intervention can make the difference between success and failure. Cardiovasc Revasc Med. 2010;11(4):266.e5–8. https://doi.org/10.1016/j.carrev.2009.07.005. PMID: 20934664.
5. Brilakis ES, Rao SV, Banerjee S, Goldman S, Shunk KA, Holmes DR Jr, et al. Percutaneous coronary intervention in native arteries versus bypass grafts in prior coronary artery bypass grafting patients: a report from the National Cardiovascular Data Registry. JACC Cardiovasc Interv. 2011;4(8):844–50. https://doi.org/10.1016/j.jcin.2011.03.018. PMID: 21851896.
6. Basiri HA, Khalilipur E, Sarreshtedari A, Zolfaghari R, Sadeghipour P, Alemzadeh-Ansari MJ, et al. Rajaie cardiovascular medical and research center-percutaneous coronary intervention registry: a real-world registry on coronary interventions in a tertiary teaching cardiovascular center. Res Cardiovasc Med. 2020;9(2):47–51. https://doi.org/10.4103/rcm.rcm_16_20.
7. Lee KW, Norell MS. Management of 'no-reflow' complicating reperfusion therapy. Acute Card Care. 2008;10(1):5–14. https://doi.org/10.1080/17482940701534813. PMID: 17924217.
8. Leborgne L, Cheneau E, Pichard A, Ajani AE, Pakala R, Yazdi H, et al. Effect of direct stenting on clinical outcome in patients treated with percutaneous coronary intervention on saphenous vein graft. Am Heart J. 2003;146(3):501–6. https://doi.org/10.1016/S0002-8703(03)00309-1. PMID: 12947371.
9. Hong YJ, Pichard AD, Mintz GS, Kim SW, Lee SY, Kim SY, et al. Outcome of undersized drug-eluting stents for percutaneous coronary intervention of saphenous vein graft lesions. Am J Cardiol. 2010;105(2):179–85. https://doi.org/10.1016/j.amjcard.2009.09.008. PMID: 20102914.
10. Xenogiannis I, Tajti P, Hall AB, Alaswad K, Rinfret S, Nicholson W, et al. Update on cardiac catheterization in patients with prior coronary artery bypass graft surgery. JACC Cardiovasc Interv. 2019;12(17):1635–49. https://doi.org/10.1016/j.jcin.2019.04.051. PMID: 31488294.
11. Stone GW, Rogers C, Hermiller J, Feldman R, Hallisey M, Habib M, et al. Randomized comparison of distal protection with a filter-based catheter and a balloon occlusion and aspiration system during percutaneous intervention of diseased saphenous vein aorto-coronary bypass grafts. Circulation. 2003;108(5):548–53. https://doi.org/10.1161/01.CIR.0000080894.98497.FC. PMID: 12874193.
12. Baim DS, Wahr D, George B, Leon MB, Greenberg J, Cutlip DE, et al. Randomized trial of a distal embolic protection device during percutaneous intervention of saphenous vein aorto-coronary bypass grafts. Circulation. 2002;105(11):1285–90. https://doi.org/10.1161/hc1002.105386. PMID: 11901038.
13. Dixon S. Saphenous vein graft protection in a distal embolic protection randomized trial. Paper presented at: Transcatheter Cardiovascular Therapeutics; 17–21 October 2005, Washington.
14. Coolong A, Baim DS, Kuntz RE, O'Malley AJ, Marulkar S, Cutlip DE, et al. Saphenous vein graft stenting and major adverse cardiac events: a predictive model derived from a

pooled analysis of 3958 patients. Circulation. 2008;117(6):790–7. https://doi.org/10.1161/CIRCULATIONAHA.107.736389. PMID: 18212289.
15. Wańha W, Mielczarek M, Gilis-Malinowska N, Roleder T, Ładziński S, Milewski M, et al. Safety and efficacy of embolic protection devices in saphenous vein graft interventions: a propensity score analysis—multicenter SVG PCI PROTECTA study. J Clin Med. 2020;9(4):1198. https://doi.org/10.3390/jcm9041198. PMID: 32331419.
16. Shoaib A, Kinnaird T, Curzen N, Ludman P, Smith D, Khoo CW, et al. Outcomes following percutaneous coronary intervention in saphenous vein grafts with and without embolic protection devices. JACC Cardiovasc Interv. 2019;12(22):2286–95. https://doi.org/10.1016/j.jcin.2019.08.037. PMID: 31753303.
17. Brilakis ES, Lichtenwalter C, de Lemos JA, Roesle M, Obarski T, Haagen D, et al. A randomized controlled trial of a paclitaxel-eluting stent versus a similar bare-metal stent in saphenous vein graft lesions: the SOS (stenting of saphenous vein grafts) trial. J Am Coll Cardiol. 2009;53(11):919–28. https://doi.org/10.1016/j.jacc.2008.11.029. PMID: 19281922.
18. Kheiri B, Osman M, Abdalla A, Ahmed S, Bachuwa G, Hassan M. The short- and long-term outcomes of percutaneous intervention with drug-eluting stent vs bare-metal stent in saphenous vein graft disease: an updated meta-analysis of all randomized clinical trials. Clin Cardiol. 2018;41(5):685–92. https://doi.org/10.1002/clc.22947. PMID: 29774566.
19. Zhao DX, Leacche M, Balaguer JM, Boudoulas KD, Damp JA, Greelish JP, et al. Routine intraoperative completion angiography after coronary artery bypass grafting and 1-stop hybrid revascularization: results from a fully integrated hybrid catheterization laboratory/operating room. J Am Coll Cardiol. 2009;53(3):232–41. https://doi.org/10.1016/j.jacc.2008.10.038. PMID: 19147039.
20. Thielmann M, Massoudy P, Schmermund A, Neuhäuser M, Marggraf G, Kamler M, et al. Diagnostic discrimination between graft-related and non-graft-related perioperative myocardial infarction with cardiac troponin I after coronary artery bypass surgery. Eur Heart J. 2005;26(22):2440–7. https://doi.org/10.1093/eurheartj/ehi436. PMID: 16105849.
21. Lawrie GM, Lie JT, Morris GC Jr, Beazley HL. Vein graft patency and intimal proliferation after aortocoronary bypass: early and long-term angiopathologic correlations. Am J Cardiol. 1976;38(7):856–62. https://doi.org/10.1016/0002-9149(76)90794-2. PMID: 1087098.
22. McGeachie JK, Meagher S, Prendergast FJ. Vein-to-artery grafts: the long-term development of neointimal hyperplasia and its relationship to vasa vasorum and sympathetic innervation. Aust N Z J Surg. 1989;59(1):59–65. https://doi.org/10.1111/j.1445-2197.1989.tb01468.x. PMID: 2914016.
23. Cox JL, Chiasson DA, Gotlieb AI. Stranger in a strange land: the pathogenesis of saphenous vein graft stenosis with emphasis on structural and functional differences between veins and arteries. Prog Cardiovasc Dis. 1991;34(1):45–68. https://doi.org/10.1016/0033-0620(91)90019-I. PMID: 2062988.
24. Sabik JF 3rd, Blackstone EH. Coronary artery bypass graft patency and competitive flow. J Am Coll Cardiol. 2008;51(2):126–8. https://doi.org/10.1016/j.jacc.2007.10.008. PMID: 18191735.
25. Brennan JM, Al-Hejily W, Dai D, Shaw RE, Trilesskaya M, Rao SV, et al. Three-year outcomes associated with embolic protection in saphenous vein graft intervention: results in 49,325 senior patients in the Medicare-linked National Cardiovascular Data Registry CathPCI Registry. Circ Cardiovasc Interv. 2015;8(3):e001403. https://doi.org/10.1161/CIRCINTERVENTIONS.114.001403. PMID: 25731991.
26. Varghese I, Samuel J, Banerjee S, Brilakis ES. Comparison of percutaneous coronary intervention in native coronary arteries vs. bypass grafts in patients with prior coronary artery bypass graft surgery. Cardiovasc Revasc Med. 2009;10(2):103–9. https://doi.org/10.1016/j.carrev.2008.11.002. PMID: 19327672.
27. Bundhoo SS, Kalla M, Anantharaman R, Das R, Gunawardena R, Eccleshall S, et al. Outcomes following PCI in patients with previous CABG: a multi centre experience. Catheter Cardiovasc Interv. 2011;78(2):169–76. https://doi.org/10.1002/ccd.22930. PMID: 21234911.

28. Dash D. An update on coronary bypass graft intervention. Heart Asia. 2014;6(1):41–5. https://doi.org/10.1136/heartasia-2013-010467. PMID: 27326161.
29. Gruberg L, Dangas G, Mehran R, Hong MK, Waksman R, Mintz GS, et al. Percutaneous revascularization of the internal mammary artery graft: short- and long-term outcomes. J Am Coll Cardiol. 2000;35(4):944–8. https://doi.org/10.1016/S0735-1097(99)00647-7. PMID: 10732892.
30. Sharma AK, McGlynn S, Apple S, Pinnow E, Canos D, Pichard AD, et al. Clinical outcomes following stent implantation in internal mammary artery grafts. Catheter Cardiovasc Interv. 2003;59(4):436–41. https://doi.org/10.1002/ccd.10553. PMID: 12891651.
31. Dimas B, Lindsey JB, Banerjee S, Brilakis ES. ST-segment elevation acute myocardial infarction due to severe hypotension and proximal left subclavian artery stenosis in a prior coronary artery bypass graft patient. Cardiovasc Revasc Med. 2009;10(3):191–4. https://doi.org/10.1016/j.carrev.2008.04.001. PMID: 19595399.
32. Neumann FJ, Sousa-Uva M, Ahlsson A, Alfonso F, Banning AP, Benedetto U, et al. 2018 ESC/EACTS guidelines on myocardial revascularization. Eur Heart J. 2019;40(2):87–165. https://doi.org/10.1093/eurheartj/ehy394. PMID: 30165437.
33. Lee M, Kong J. Current state of the art in approaches to saphenous vein graft interventions. Interv Cardiol. 2017;12(2):85–91. https://doi.org/10.15420/icr.2017:6:2. PMID: 29593808.
34. Abdelrahman A, Dębski M, More R, Abdelaziz HK, Choudhury T, Eichhöfer J, et al. One-year outcomes of percutaneous coronary intervention in native coronary arteries versus saphenous vein grafts in patients with prior coronary artery bypass graft surgery. Cardiol J. 2020;29(6):1018–25. https://doi.org/10.5603/CJ.a2020.0131. PMID: 33051892.
35. Wolny R, Mintz GS, Matsumura M, Weisz G, Kirtane AJ, Rabbani LE, et al. Intravascular ultrasound assessment of in-stent restenosis in saphenous vein grafts. Am J Cardiol. 2019;123(7):1052–9. https://doi.org/10.1016/j.amjcard.2018.12.030. PMID: 30660357.
36. Øksnes A, McEntegart M. Intravascular lithotripsy in saphenous vein grafts and graft stent failure: a case series. Catheter Cardiovasc Interv. 2021;97(7):E945–50. https://doi.org/10.1002/ccd.29265. PMID: 32926588.
37. Dautov R, Manh Nguyen C, Altisent O, Gibrat C, Rinfret S. Recanalization of chronic total occlusions in patients with previous coronary bypass surgery and consideration of retrograde access via saphenous vein grafts. Circ Cardiovasc Interv. 2016;9(7):e003515. https://doi.org/10.1161/CIRCINTERVENTIONS.115.003515. PMID: 27390335.
38. Teramoto T, Tsuchikane E, Matsuo H, Suzuki Y, Ito T, Habara M, et al. Initial success rate of percutaneous coronary intervention for chronic total occlusion in a native coronary artery is decreased in patients who underwent previous coronary artery bypass graft surgery. JACC Cardiovasc Interv. 2014;7(1):39–46. https://doi.org/10.1016/j.jcin.2013.08.009. PMID: 24332419.
39. Toma A, Stähli BE, Gick M, Ferenc M, Mashayekhi K, Buettner HJ, et al. Long-term follow-up of patients with previous coronary artery bypass grafting undergoing percutaneous coronary intervention for chronic total occlusion. Am J Cardiol. 2016;118(11):1641–6. https://doi.org/10.1016/j.amjcard.2016.08.038. PMID: 27712699.
40. Sapontis J, Salisbury AC, Yeh RW, Cohen DJ, Hirai T, Lombardi W, et al. Early procedural and health status outcomes after chronic total occlusion angioplasty: a report from the OPEN-CTO registry (outcomes, patient health status, and efficiency in chronic total occlusion hybrid procedures). JACC Cardiovasc Interv. 2017;10(15):1523–34. https://doi.org/10.1016/j.jcin.2017.05.065. PMID: 28797429.
41. Cardoso R, Knijnik L, Whelton SP, Rivera M, Gluckman TJ, Metkus TS, et al. Dual versus single antiplatelet therapy after coronary artery bypass graft surgery: an updated meta-analysis. Int J Cardiol. 2018;269:80–8. https://doi.org/10.1016/j.ijcard.2018.07.083. PMID: 30049494.
42. Verma S, Goodman SG, Mehta SR, Latter DA, Ruel M, Gupta M, et al. Should dual antiplatelet therapy be used in patients following coronary artery bypass surgery? A meta-analysis of randomized controlled trials. BMC Surg. 2015;15:112. https://doi.org/10.1186/s12893-015-0096-z. PMID: 26467743.

43. Sirker A, Kwok CS, Kontopantelis E, Johnson T, Freeman P, de Belder MA, et al. Antiplatelet drug selection in PCI to vein grafts in patients with acute coronary syndrome and adverse clinical outcomes: insights from the British Cardiovascular Intervention Society database. Catheter Cardiovasc Interv. 2018;92(4):659–65. https://doi.org/10.1002/ccd.27493. PMID: 29219246.
44. Tajti P, Karmpaliotis D, Alaswad K, Jaffer FA, Yeh RW, Patel M, et al. In-hospital outcomes of chronic total occlusion percutaneous coronary interventions in patients with prior coronary artery bypass graft surgery. Circ Cardiovasc Interv. 2019;12(3):e007338. https://doi.org/10.1161/CIRCINTERVENTIONS.118.007338. PMID: 30813758.
45. Colleran R, Kufner S, Mehilli J, Rosenbeiger C, Schüpke S, Hoppmann P, et al. Efficacy over time with drug-eluting stents in saphenous vein graft lesions. J Am Coll Cardiol. 2018;71(18):1973–82. https://doi.org/10.1016/j.jacc.2018.03.456. PMID: 29724350.
46. Xenogiannis I, Hryniewicz K, Burke MN, Brilakis ES. Salvage of simultaneous acute coronary closure and retroperitoneal bleeding using venoarterial extracorporeal membrane oxygenation and chronic total occlusion percutaneous coronary intervention techniques in a patient with ST-segment elevation myocardial infarction. Cardiovasc Revasc Med. 2019;20(11S):66–9. https://doi.org/10.1016/j.carrev.2019.03.005. PMID: 30956116.
47. Ramakrishna S. Vascular plugs – a key companion to interventionists – just plug it. Indian Heart J. 2015;67(5):399–405. https://doi.org/10.1016/j.ihj.2015.07.013. PMID: 26432728.
48. Shoaib A, Johnson TW, Banning AP, Ludman P, Rashid M, Potts J, et al. Clinical outcomes of percutaneous coronary intervention for chronic total occlusion in native coronary arteries vs saphenous vein grafts. J Invasive Cardiol. 2020;32(9):350–7. PMID: 32829309.
49. Riley RF, Patel MP, Abbott JD, Bilazarian S, Fonseca F, Klein AJ, et al. SCAI expert consensus statement on the management of calcified coronary lesions in patients with prior coronary artery bypass graft surgery: an updated perspective. J Soc Cardiovasc Angiogr Interv. 2024;3(2):101259. https://doi.org/10.1016/j.jscai.2023.101259.
50. Goswami R, Adams C, Smalling RW, Smalling SM, Goswami NJ, Anderson HV, et al. Intravascular lithotripsy versus rotational atherectomy for stent underexpansion in saphenous vein graft lesions: a randomized trial. Catheter Cardiovasc Interv. 2024;103(6):841–9. https://doi.org/10.1002/ccd.31023. PMID: 38566513.
51. Okamura A, Yamane M, Muto M, Matsubara T, Igarashi Y, Nakamura S, et al. Complications during retrograde approach for chronic coronary total occlusion: sub-analysis of Japanese multicenter registry. Catheter Cardiovasc Interv. 2016;88(1):7–14. https://doi.org/10.1002/ccd.26317. PMID: 26650043.
52. Tajti P, Karatasakis A, Karmpaliotis D, Alaswad K, Jaffer FA, Yeh RW, et al. Retrograde CTO PCI of native coronary arteries via left internal mammary artery grafts: insights from a multicenter U.S. registry. J Invasive Cardiol. 2018;30(3):89–96. PMID: 29207356.
53. Xenogiannis I, Gkargkoulas F, Karmpaliotis D, Alaswad K, Jaffer FA, Yeh RW, et al. Retrograde chronic total occlusion percutaneous coronary intervention via saphenous vein graft. JACC Cardiovasc Interv. 2020;13(4):517–26. https://doi.org/10.1016/j.jcin.2019.10.028. PMID: 32081247.
54. Liu Y, Li CP, Wang YY, Dong ZQ, Zhang JX, Li W, et al. Prediction of major adverse cardiovascular events and slow/no-reflow by virtual histology imaging after percutaneous interventions on saphenous vein grafts. Angiology. 2020;71(3):263–73. https://doi.org/10.1177/0003319719892354. PMID: 31808794.
55. Gürbak İ, Panç C, Şahin AA, Çoner A, Erdinç M, Ünal S, et al. CHA2DS2-VASc score as a predictor of no-reflow phenomenon after saphenous vein graft percutaneous coronary intervention in patients with non-ST-segment elevation acute coronary syndromes. Kardiol Pol. 2020;78(11):1129–36. https://doi.org/10.33963/KP.15603. PMID: 32955801.

Part II
Percutaneous Intervention in Complex Structural Heart Disease in Adults

Chapter 7
Secundum-Type Atrial Septal Defect

Ata Firouzi, Zahra Hosseini, and Armin Elahifar

Abstract Secundum-type ASD is the most prevalent congenital heart defect in adults, affecting approximately 7% of this population. Symptoms typically manifest in the third or fourth decade of life and may include exertional dyspnea, palpitations, chest pain, cyanosis, and, in rare cases, systemic thromboembolism. Patients are considered candidates for ASD closure if the left-to-right shunt ratio during RHC is ≥1.5 or if there is any indication of right ventricular volume overload. For those with elevated pulmonary vascular resistance, this should be less than one-third of the systemic vascular resistance, and pulmonary pressure must be below 50% of systemic pressure. Currently, the preferred method for closure is the transcatheter approach, provided that TEE shows the defect is suitable with adequate surrounding rims and structures. The success rate for this procedure exceeds 95%. Long-term outcomes for these patients are comparable to those of surgical closure, with fewer complications. Regular follow-up with echocardiography is essential to assess for any residual shunt and its potential, albeit rare, catastrophic complications.

Abbreviations

ASD	Atrial septal defect
IDD	Inter-defect diameter
PAP	Pulmonary arterial pressure
PAPVC	Partial anomalous pulmonary venous connection
PFO	Patent foramen ovale
RBBB	Right bundle branch block
RHC	Right heart catheterization
SFD	Stop-flow diameter
TEE	Transesophageal echocardiography
TTE	Transthoracic echocardiography

A. Firouzi · Z. Hosseini (✉) · A. Elahifar
Cardiovascular Intervention Research Center, Rajaie Cardiovascular Institute, Iran University of Medical Sciences, Tehran, Iran

A. Firouzi et al. (eds.), *Mastering Skills in Complex Cardiovascular Interventions*, https://doi.org/10.1007/978-3-031-97321-5_7

Introduction

ASDs are the most prevalent congenital heart defect in adults and represent 7% of all congenital heart defects present at birth. Approximately 80% of ASDs are of the secundum type, situated in the fossa ovalis region or its vicinity [1]. During childhood, ASDs are typically asymptomatic. Defects smaller than 6 mm often close spontaneously. If undiagnosed, they tend to manifest symptoms in the third or fourth decade of life, significantly reducing survival rates. In isolated ASDs, the extent of left-to-right shunting at the atrial level depends on factors such as the size and number of defects, ventricular compliance, atrial pressures, and the presence of atrioventricular valve disease. Reduced left ventricular compliance or conditions that elevate left atrial pressure (e.g., hypertension, ischemic heart disease, cardiomyopathy, and aortic or mitral valve diseases) can increase left-to-right shunting. Consequently, ASDs may gain hemodynamic significance with advancing age [2].

Reduced right ventricular compliance due to conditions such as pulmonic stenosis, pulmonary arterial hypertension, or other right ventricular or tricuspid valve diseases may diminish left-to-right shunting or even lead to shunt reversal, resulting in cyanosis. Adult patients often present with symptoms such as exertional dyspnea, atypical chest pain, palpitations, or systemic thromboembolism. Without treatment, patients may eventually develop atrial arrhythmias (e.g., atrial fibrillation or atrial flutter), irreversible pulmonary hypertension, reverse shunting, Eisenmenger syndrome, or congestive heart failure. Therefore, early detection and closure of the defect are essential. Closure is recommended for patients with a Qp/Qs ratio >1.5/1, right ventricular volume overload, or paradoxical thromboembolism. Closure, whether achieved surgically or via catheter-based methods, typically alleviates most symptoms and reduces mortality rates, particularly when performed at a younger age (under 25 years). However, it does not entirely eliminate the risk of atrial arrhythmias (10%) or stroke, especially in patients over 40 years of age.

Diagnostic Workup

During a physical examination, key findings include fixed splitting of the second heart sound (S2) and pulmonary systolic murmurs. On an ECG, prominent features include right axis deviation, incomplete RBBB, and the Crochetage sign. TTE typically reveals right ventricular volume overload, indicated by a dropout shadow in the interatrial septum. Nonetheless, a precise assessment of the causes of right ventricular enlargement requires TEE. TEE can clearly visualize the interatrial septum's structure, the type of ASD, the number of defects, and any additional septal abnormalities (e.g., PFO and fenestrations) or other anomalies (e.g., PAPVC, ventricular septal defect, and patent ductus arteriosus). Specifically, for secundum-type ASDs, TEE is essential to evaluate the defect's morphology, maximum size, surrounding rims, shunt direction, and suitability for device closure.

Further, it is important to assess any valvular pathologies, particularly pulmonary valve stenosis and myxomatous mitral valve changes, which are frequently observed in these patients. The severity of tricuspid regurgitation and pulmonary artery pressure should also be evaluated. In cases where right ventricular size and function are unclear or when further assessment of pulmonary venous connections is needed, cardiac magnetic resonance imaging or cardiac computed tomography can serve as alternative diagnostic modalities.

Treatment Approach

Device closure under TEE and fluoroscopic guidance is the preferred treatment for secundum-type ASDs with suitable morphology. Prior to percutaneous transcatheter closure, a comprehensive evaluation of the defect using TEE is essential. Two critical parameters in this process are the maximum ASD diameter, which determines the appropriate device size, and the dimensions of the surrounding rims, which ensure optimal device placement. Most ASDs exhibit an ellipsoidal shape that changes during the cardiac cycle. The major axis diameter of the defect, measured during ventricular end-systole, is crucial for selecting the correct device size, particularly in procedures performed without balloon stretch sizing. Transcatheter closure of large ASDs with a maximum native diameter >30 mm is more challenging and often requires specialized techniques by experienced interventionists for successful device deployment. The surrounding rims are defined as the distances from the ASD to key anatomical structures: the aorta (anterosuperior rim), the superior vena cava (posterosuperior rim), the right upper pulmonary vein (posterior rim), the inferior vena cava (inferoposterior rim), the coronary sinus, and the atrioventricular valve (inferoanterior rim). A rim is considered deficient if it measures <5 mm and absent if it is <1 mm. In approximately 10%–15% of cases, the rims are too small to ensure stable and safe device implantation, making a surgical approach necessary.

Challenging Cases

Case 1

A 40-year-old woman presented with exertional dyspnea lasting 1 year, which had recently worsened. She also reported occasional palpitations during physical activity. Her most recent delivery occurred approximately 6 months prior, with no complications.

Diagnostic Workup

TTE revealed normal left ventricular size and systolic function, mild right ventricular enlargement with mild-to-moderate systolic dysfunction, and a small dropout in the interatrial septum on the 4-chamber view, suggesting a secundum-type ASD (Qp/Qs = 1.7). TEE confirmed a small secundum-type ASD (maximum diameter: 10 mm) with a significant left-to-right shunt. All surrounding rims measured >3 mm, and no additional anomalies were detected.

Given the patient's symptoms, right ventricular volume overload, and the defect's suitability for transcatheter closure, she was scheduled for percutaneous device closure.

RHC Data

- Systolic blood pressure: 100/60 mm Hg;
- Right ventricular pressure: 25/5 mm Hg;
- Mean PAP: 15 mm Hg;
- Mean right atrial pressure: 4 mm Hg;
- Qp/Qs ratio: 1.8;
- Pulmonary vascular resistance: 1 WU; and
- Oxygen saturation: 98%.

Procedural Technique [3]

Before the procedure, the patient was loaded with 325 mg of aspirin (ASA) and 600 mg of clopidogrel. It is important to note that ASA (80 mg) should be continued for at least 6 months and clopidogrel (75 mg) for 1 month to ensure full device endothelialization.

After inserting 6 Fr sheaths into the right common femoral artery and vein, RHC was performed. Intravenous heparin was administered to achieve an activated clotting time >200 s.

Under general anesthesia and continuous hemodynamic monitoring supervised by a skilled technician, the defect was carefully evaluated by an experienced operator using a TEE probe. Surprisingly, the assessment revealed two separate secundum-type ASDs. The non-stretched major axis diameters of the larger and smaller defects were approximately 10 mm and 8 mm, respectively. The smaller secundum-type ASD was positioned more anteroinferiorly.

The IDD was approximately 10 mm. The surrounding rims of both defects were adequate, with no evidence of fenestrations, PFO, or PAPVC. The net shunt direction was left to right.

Given the size of the defects, the IDD, and the alignment of the interatrial septum, the larger defect was initially addressed. A Multipurpose Catheter was advanced through the defect, with its tip positioned in the left upper pulmonary vein. A floppy-tip Amplatz Super Stiff Guidewire was then placed in the left upper pulmonary vein.

After preparing and de-airing the sizing balloon (NuMED 25), it was inflated with diluted contrast under TEE and fluoroscopic guidance. Inflation continued until the SFD was measured by the echocardiographer and confirmed by the waist diameter of the balloon on fluoroscopy (Fig. 7.1a). The SFD measured 12 mm. Based on the SFD and the surrounding rims, an Occlutech Figulla Flex II device (12 mm) was selected.

After flushing and de-airing all segments of the delivery sheath (Occlutech 10 Fr), the sheath was advanced over the guidewire to the mid-portion of the left

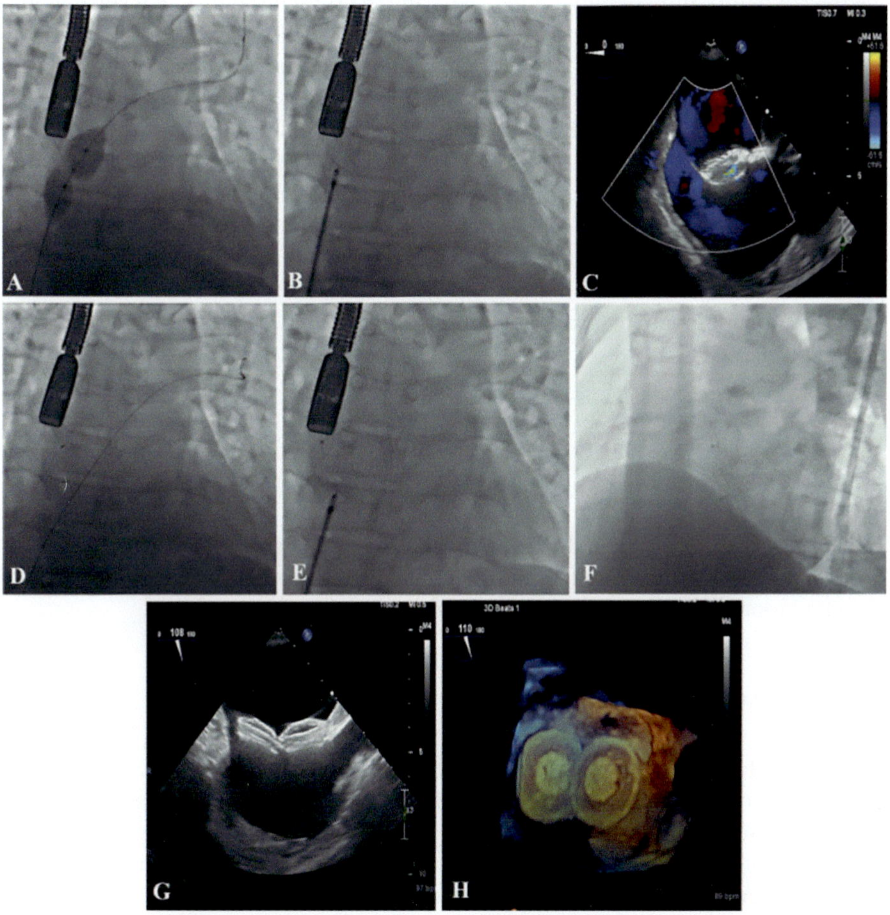

Fig. 7.1 (**a**–**h**) Under TEE guidance (**a**), the SFD was confirmed, and an Occlutech Figulla Flex II device (12 mm) was successfully deployed (**b**, **c**). The inferior defect was wired (**d**) and subsequently closed with another Occlutech Figulla Flex II device (12 mm) (**e**, **f**). The final result demonstrates the proper positioning and configuration of both devices, with no residual shunting observed (**g**, **h**)

atrium. The dilator was then removed to allow blood backflow and prevent air embolism. Subsequently, the delivery sheath was advanced further, positioning it just outside the left upper pulmonary vein.

Next, the delivery cable was loaded, and the cable was connected to the hub of the right atrial disk. The entire device was retracted into the loader (slenderized) and thoroughly flushed with heparinized saline.

Subsequently, the loader was connected to the delivery sheath, and the device was advanced through the sheath. The left atrial disk of the device was deployed in the left atrium. Once the left atrial disk was fully extruded, the entire system was retracted together under TEE guidance to assess the alignment of the device with the interatrial septum. At this stage, the operator confirmed the optimal positioning of the device before fully deploying the waist and the right atrial disk. All surrounding rims were thoroughly evaluated to ensure the device did not encroach on nearby cardiac structures. TEE and Minnesota maneuvers were employed to verify the stable position of the device before its final release (Fig. 7.1b, c).

Afterward, the smaller defect, located inferiorly and anteriorly to the larger defect, was wired (Fig. 7.1d). The SFD measured by the sizing balloon was 10 mm, and an Occlutech Figulla Flex II device (12 mm) was selected. The second device was deployed using a procedure similar to that of the first device (Fig. 7.1e, f). All surrounding rims and structures were thoroughly evaluated to confirm the stable positioning of the device and to ensure no interaction with the first device. Based on 3D TEE imaging, no residual shunting was observed. Finally, the device was released (Fig. 7.1g, h).

Case 2

The patient was a 38-year-old woman presenting with atypical chest pain and exertional dyspnea lasting 6 months. She had a medical history of hyperthyroidism and chronic kidney disease, both of which were under close monitoring.

Diagnostic Workup

During the physical examination, the patient exhibited stable vital signs. Auscultation revealed a normal S1 and a loud P2 in the left second intercostal space, with no murmurs detected. ECG showed right axis deviation and incomplete RBBB. TTE and TEE demonstrated normal left ventricular size and function, moderate right ventricular enlargement with mild-to-moderate dysfunction, and mild right atrial enlargement. The estimated systolic PAP was 40 mm Hg. TEE identified two separate ASDs. The non-stretched major axis diameters of the larger and smaller defects were approximately 14 mm and 11 mm, respectively. The smaller defect was situated more anteroinferiorly, with an IDD of about 9 mm. The surrounding rims of both defects were adequate, and no fenestrations, PFO, or PAPVC were observed.

Given the patient's symptoms, right ventricular volume overload, and the suitability of the defects for transcatheter closure, she was scheduled for percutaneous device closure.

RHC Data

- Systolic blood pressure: 120/70 mm Hg;
- Mean PAP: 18 mm Hg;
- Right ventricular pressure: 40/5 mm Hg;
- Mean right atrial pressure: 4 mm Hg;
- Qp/Qs ratio: 2.2;
- Pulmonary vascular resistance: 1.5 WU; and
- Oxygen saturation: 98%.

Procedural Technique

Under general anesthesia and continuous hemodynamic monitoring, short sheaths were inserted into the right common femoral artery, as well as the right and left common femoral veins. Intravenous heparin was administered to achieve an activated clotting time >200 s. Following this, RHC was performed.

The initial plan was to close both defects using a single device. The larger defect was wired, and the tip of the Amplatz Super Stiff Guidewire was positioned in the left upper pulmonary vein. The sizing balloon (NuMED 35) was then advanced to the mid-portion of the left atrium. Under TEE and fluoroscopic guidance, the balloon was inflated with diluted contrast before being retracted into the right atrium. Still, despite multiple attempts, the balloon remained completely unstable and prolapsed into the right atrium, preventing the measurement of the SFD (Fig. 7.2a).

As a result, using the left femoral vein access, both defects were wired, and two sizing balloons (NuMED 25) were inflated separately to measure the SFD of each defect. The SFD of the larger defect was 16 mm, while the SFD of the smaller defect, located superiorly and posteriorly to the larger one, was 12 mm (Fig. 7.2b).

Next, both balloons were simultaneously inflated to measure the combined SFD of the defects. Since the IDD was >7 mm and the SFD of each defect was sufficiently large, it was decided to close the defects with two separate devices. Based on the SFD and the surrounding rims, an Occlutech Figulla Flex II device (13.5 mm) and an Occlutech Figulla Flex II device (18 mm) were selected. After flushing and de-airing all segments of the delivery sheaths (Occlutech 8 Fr and 10 Fr), the sheaths were advanced over the guidewires to the mid-portion of the left atrium. The dilators were then removed to allow blood backflow and prevent air embolism. Subsequently, the delivery sheaths were advanced further, positioning them just outside the left upper pulmonary vein. The delivery cables were loaded and connected to the hub of the right atrial disk. Both devices were fully retracted into their respective loaders and thoroughly flushed with heparinized saline.

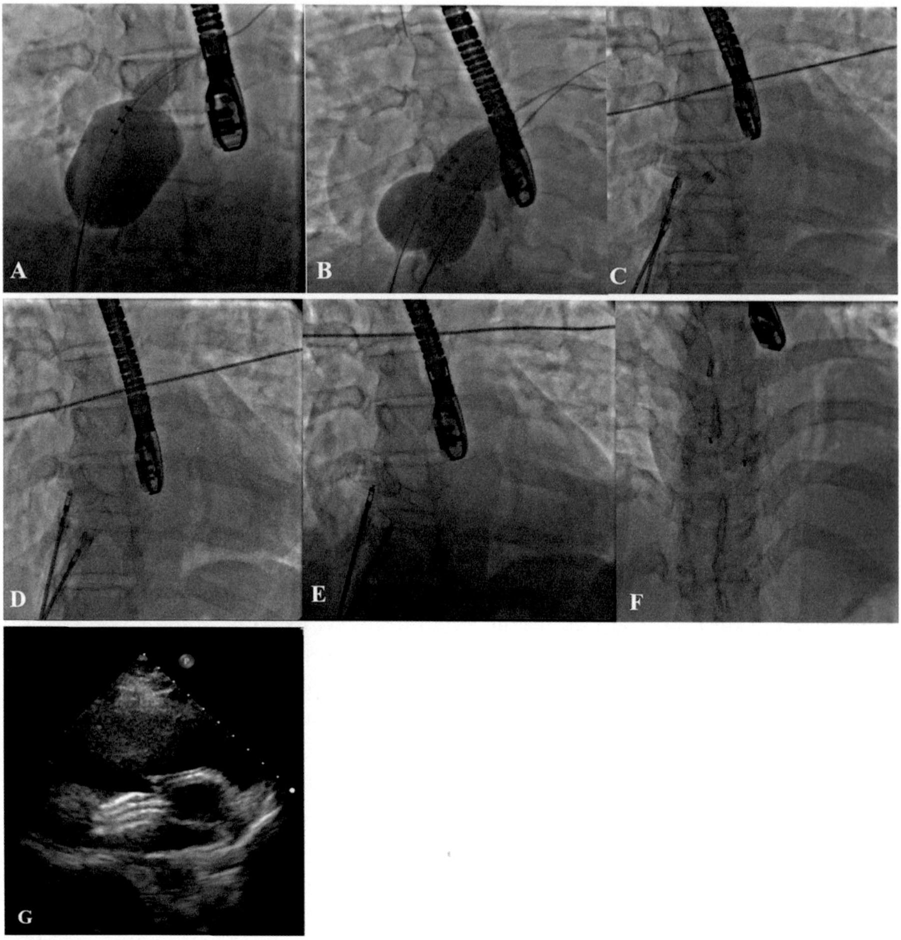

Fig. 7.2 (**a**–**f**) The initial large sizing balloon was unstable (**a**); consequently, two separate smaller sizing balloons were inflated simultaneously (**b**). The smaller defect was then closed using an Occlutech Figulla Flex II device (13.5 mm), and its stability was assessed via TEE using the pull-and-push maneuver, ensuring there was no impingement on the surrounding structures (**c**). The left atrial disk of the second device was then extruded into the left atrium. (**g**) The entire system was maneuvered toward the septum, with half of its waist deployed into the left atrium. The system was then fully retracted, allowing the remainder of the waist and the right atrial disk to be deployed by pulling and pushing the device. Subsequently, under TEE and fluoroscopic guidance, the stability of the device was confirmed. The device closely conformed to the smaller device (**c**–**e**). The correct configuration of the devices is depicted in (**f**) and (**g**)

Next, the loaders were connected to the delivery sheaths, and the devices were advanced through the sheaths. The sheath for the smaller defect was retracted, and the left atrial disk was extruded. The entire system was then retracted toward the septum.

In the next stage, under TEE and fluoroscopic guidance, the remainder of the devices were deployed in the proper position using the pull-and-push maneuver (Fig. 7.2c).

Subsequently, the sheath for the second defect was retracted, and the left atrial disk was released. The entire device was slowly pulled out, allowing part of the waist to deploy in the left atrium. Finally, the sheath was fully retracted into the right atrium, and the right atrial disk was deployed (Fig. 7.2c, d).

During the deployment of the right disk in the right atrium, the entire system was advanced toward the interatrial septum, ensuring the device securely hugged the smaller defect (Fig. 7.2e).

Following confirmation by TEE that both devices were properly positioned and not interfering with adjacent structures, the devices were successfully released (Fig. 7.2f, g).

Case 3

The patient was a 60-year-old woman with a medical history of diabetes mellitus and hypertension. She presented to our clinic with symptoms of exertional dyspnea (classified as New York Heart Association functional class III) and palpitations, which had persisted for approximately 2 years and had significantly worsened in the preceding months.

Diagnostic Workup

During the physical examination, the patient exhibited hypertension with a wide pulse pressure and elevated jugular venous pressure. Auscultation revealed mid-systolic murmurs at the left sternal border and a loud P2 at the apex. ECG showed right axis deviation, RBBB, and left ventricular hypertrophy with a strain pattern.

TTE and TEE revealed mild left ventricular enlargement with preserved systolic function, mild left ventricular hypertrophy accompanied by moderate diastolic dysfunction (Grade II), moderate right ventricular enlargement with severe systolic dysfunction, and moderate biatrial enlargement. Additionally, rheumatic changes were noted in the mitral valve, with no mitral stenosis, mild mitral regurgitation, no aortic stenosis, mild aortic insufficiency, and moderate-to-severe tricuspid regurgitation. The systolic PAP was estimated at 55 mm Hg.

A moderate-sized secundum-type ASD was also identified, with a non-stretched major axis diameter of approximately 15 mm, a significant left-to-right shunt, and adequate surrounding rims (Fig. 7.3a). Moreover, mild pericardial effusion was observed.

Given the patient's symptoms and echocardiographic findings, she was scheduled for coronary angiography and RHC.

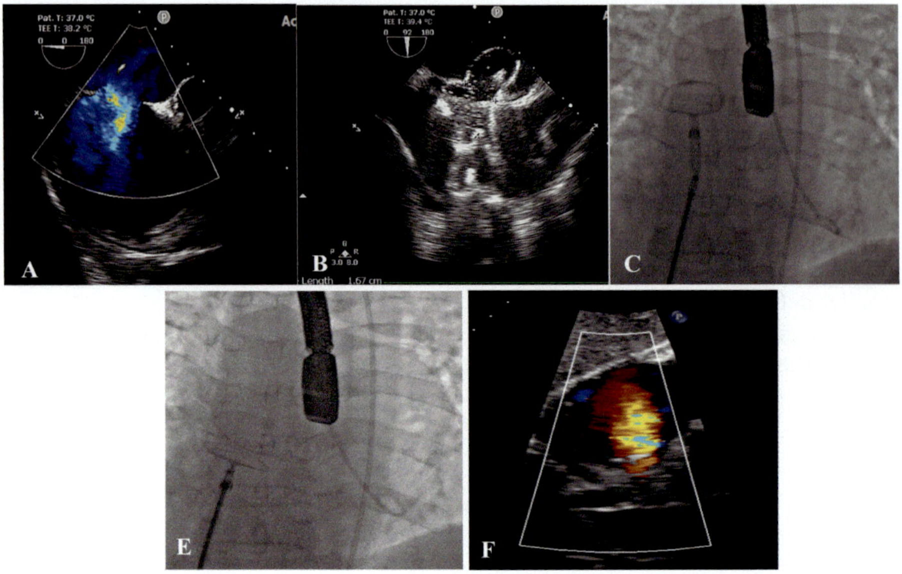

Fig. 7.3 (**a**–**e**) A moderate-sized secundum ASD is illustrated herein. The non-stretched major axis diameter of the defect measured approximately 15 mm. Also observable were a significant net left-to-right shunt and sufficient rims (**a**). Initially, a sizing-balloon occlusion test was performed (**b**). Given the hemodynamic changes after long-term balloon inflation, an Occlutech Fenestrated ASD Device (ASD 21 with 5 mm fenestration, compatible with a 12 Fr sheath) was chosen for closure. During the monitoring of the left ventricular end-diastolic pressure, the left disk was deployed (**c**). Thereafter, under TEE guidance, the waist and the right disk were deployed with the appropriate alignment of the device and the interatrial septum (**d**). The device's stability and configuration, along with the amount of residual shunting, were comprehensively assessed, and any impingement on the surrounding structures was ruled out. Ultimately, the device was released. The final echo image illustrated a minimal left-to-right shunt through the fenestration of the device (**e**)

The patient underwent selective coronary angiography, and the procedure revealed normal epicardial coronary arteries with no evidence of significant stenosis or abnormalities.

RHC Data

- Blood pressure: 160/70 mm Hg;
- Left ventricular pressure: 150/18 mm Hg;
- Right ventricular pressure: 60/20 mm Hg;
- Mean PAP: 35 mm Hg;
- Pulmonary vascular resistance: 4 WU;
- Mean right atrial pressure: 15 mm Hg;
- Qp/Qs ratio: 3.2; and
- Oxygen saturation: 95%.

Procedural Technique

Under general anesthesia and continuous hemodynamic monitoring, short sheaths were inserted into the right femoral artery and vein. Intravenous heparin was administered to achieve an activated clotting time >200 s. Following this, RHC was performed.

Based on the hemodynamic data, a balloon-sizing occlusion test was conducted. The defect was wired, and the sizing balloon was inflated for 15 min. During this period, a Pigtail Catheter was positioned in the left ventricle, and a Multipurpose Catheter was placed in the left pulmonary artery to measure the left ventricular end-diastolic pressure and PAP.

After 15 min, the following measurements were recorded:

- Left ventricular end-diastolic pressure: 25 mm Hg;
- PAP: 60/30 mm Hg; and
- Systemic blood pressure: 140/60 mm Hg.

Given the persistent elevation of the left ventricular end-diastolic pressure (>20 mm Hg) and the reduction in systemic blood pressure, the decision was made to use a fenestrated ASD device. TEE revealed an SFD of 17 mm (Fig. 7.3b) and adequate surrounding rims. Consequently, an Occlutech Fenestrated ASD Device (ASD 21 with 5 mm fenestration, compatible with a 12 Fr sheath) was selected (Fig. 7.3).

The implantation procedure for the fenestrated ASD device was technically similar to that of standard ASD closure devices. After flushing and de-airing all segments of the delivery sheath (Occlutech 12 Fr), the sheath was advanced over the guidewire to the mid-portion of the left atrium. The dilator was then removed to allow blood backflow and prevent air embolism.

Subsequently, the delivery sheath was advanced into the left atrium, positioning it just outside the left upper pulmonary vein. The delivery cable was then loaded and connected to the hub of the right atrial disk. The entire device was retracted into the loader and thoroughly flushed with saline.

The loader was connected to the delivery sheath, and the device was advanced through the sheath. At the level of the pulmonary vein, the sheath was retracted to deploy the left atrial disk (Fig. 7.3c). The entire system was then pulled toward the interatrial septum as a single unit under TEE guidance to ensure proper alignment of the device with the septum. The waist and right atrial disk were subsequently deployed (Fig. 7.3d).

Following deployment, the stability and configuration of the device, as well as the degree of residual shunting, were thoroughly evaluated. Any potential impingement on surrounding structures was ruled out. Finally, the device was released.

Case 4

The patient was a 36-year-old man presenting with exertional dyspnea (New York Heart Association functional class II) lasting 6 months. He also reported one episode of a transient ischemic attack approximately 3 months prior, though no definitive cause was identified.

Diagnostic Workup

During the physical examination, the patient exhibited elevated jugular venous pressure, a normal S1, and a loud P2 at the left sternal border, along with mid-systolic murmurs (Grade II) in the same area. Palpation revealed prominent right ventricular heaves. ECG demonstrated right axis deviation, an rSr' pattern in V_1, and an incomplete RBBB pattern.

TTE and TEE showed normal left ventricular size and systolic function, moderate-to-severe right ventricular enlargement with moderate systolic dysfunction, and moderate right atrial enlargement. There was no mitral stenosis, mild-to-moderate mitral regurgitation, no aortic stenosis, no aortic insufficiency, and moderate-to-severe tricuspid regurgitation (gradient: 35 mm Hg).

Additionally, a large secundum-type ASD was identified, with a non-stretched major axis diameter of 26 mm and a significant left-to-right shunt. The surrounding rims were generally adequate, except for the anterosuperior rim (2 mm) and the posterior rim (3 mm), which were deficient. No other anomalies were observed.

Based on these findings, the patient was scheduled for RHC and percutaneous ASD closure.

RHC Data

- Blood pressure: 130/80 mm Hg;
- Right ventricular pressure: 45/12 mm Hg;
- Mean PAP: 26 mm Hg;
- Pulmonary vascular resistance: 2.5 WU;
- Mean right atrial pressure: 12 mm Hg;
- Qp/Qs ratio: 4; and
- Oxygen saturation: 98%.

Procedural Technique

Under general anesthesia and continuous hemodynamic monitoring, arterial and venous short sheaths (6 Fr) were inserted, and RHC was performed. A comprehensive evaluation of the defect revealed a round secundum-type ASD with the following rim measurements:

- Anterosuperior rim: 2 mm (floppy);
- Posterior rim: 3 mm;
- Inferior rim: 5 mm; and
- Anteroinferior rim: 4 mm.

No additional anomalies, such as fenestrations, PFO, or PAPVC, were observed. The net shunt direction was left to right.

Intravenous heparin was administered to achieve an activated clotting time >200 s. The defect was crossed using a Multipurpose Catheter, with its tip positioned in the left upper pulmonary vein. A floppy-tip Amplatz Super Stiff Guidewire was then placed in the left upper pulmonary vein.

After preparing and de-airing the sizing balloon (NuMED 35)—a critical step in the procedure—the balloon was slowly inflated with diluted contrast in the left atrium under TEE guidance. It was then gently pulled back toward the right atrium to visualize the central waist on fluoroscopy.

The balloon was continuously inflated until the SFD was reported by the echocardiographer. Stabilizing the sizing balloon proved challenging due to melon seeding. Nonetheless, after several attempts, the central waist was successfully visualized (Fig. 7.4a). The SFD measured 30 mm.

Given the large size of the defect and the deficiency in the anterosuperior and posterior rims, an Occlutech Figulla Flex II device (33 mm) was selected. After flushing and de-airing the delivery sheath (COOK 14 Fr), the sheath was positioned at the inferior vena cava–right atrium junction. The dilator was then removed to allow blood backflow and prevent air embolism.

Subsequently, the delivery sheath was advanced into the left atrium, positioning it just outside the left upper pulmonary vein. The delivery cable was loaded and connected to the hub of the right atrial disk.

Next, the entire device was retracted into the loader and thoroughly flushed with heparinized saline. The loader was then connected to the delivery sheath, and the device was advanced through the sheath. The left atrial disk was fully extruded into the left atrium, and the entire system was gently pulled back as a unit under TEE guidance to assess alignment with the interatrial septum.

In the subsequent stage, the waist and right atrial disk were fully deployed. Nevertheless, TEE revealed that the entire device had embolized into the right atrium. As a result, the sheath was advanced toward the interatrial septum, and the right atrial disk and waist were retracted. Various techniques were employed at this stage to address the issue.

Using the left atrial roof technique (Fig. 7.4b) and the right atrial roof technique (Fig. 7.4c), the waist was partially deployed in the left atrium, followed by full deployment of the device in the left atrium. The device was then pulled back toward the interatrial septum while the right atrial disk was retracted.

With the left upper pulmonary vein technique (Fig. 7.4d), the device was deployed correctly. However, none of these techniques achieved proper alignment of the device with the interatrial septum.

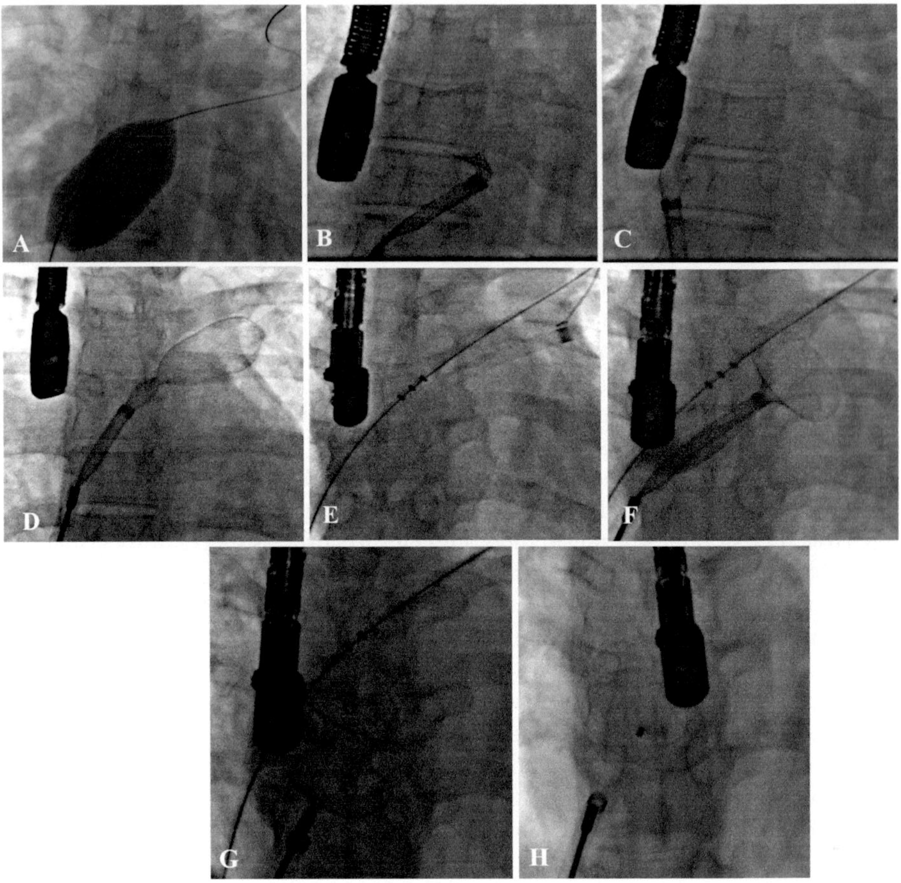

Fig. 7.4 (**a**–**h**) The images depict sizing-balloon inflation. The waist was not visible (**a**). The left atrial roof technique (**b**), the right atrial roof technique (**c**), and the left upper pulmonary vein approach (**d**) are shown. None of the techniques contributed to the appropriate device alignment in the images. As a final technique, the sizing balloon was advanced toward the left atrium (**e**), while it was partially inflated in the left atrium. Next, the delivery sheath was pulled back, and the left disk was deployed (**f**). Then, the waist and half of the right disk were extruded, and the entire system was pulled back toward the interatrial septum. Thereafter, the right disk was completely deployed (**g**), and the balloon was deflated and retrieved slowly, causing the left disk to engage the septum. After a complete assessment of the surrounding structures and confirmation of the device's stability, it was released (**h**)

As a final attempt, the sizing balloon was inserted into the left atrium via the left femoral vein and partially inflated. The left atrial disk was deployed while the balloon anchored it in the left atrium. The sheath was then pulled back, allowing the waist and right atrial disk to deploy and advance toward the interatrial septum (Fig. 7.4e–g).

The balloon was slowly deflated and removed, enabling the left atrial disk to securely engage the left side of the septum. Finally, TEE was used to confirm proper device deployment by thoroughly evaluating the device configuration, residual shunting, and surrounding structures (Fig. 7.4h).

Discussion

Percutaneous closure of secundum-type ASDs is a safe and effective procedure and is regarded as the preferred treatment for patients with appropriate morphological and clinical conditions. Echocardiography, whether TEE or intracardiac, along with sizing balloons, plays a critical role in percutaneous closure.

Under echocardiography guidance, the following can be accurately assessed, as follows:

- The location and morphology of the defect;
- The number of defects and any fenestrations;
- The IDD;
- The dimensions of the surrounding rims and adjacent structures;
- The total length of the interatrial septum;
- The alignment with the interatrial septum; and
- Possible aneurysmal variations.

Further, sizing balloons, used under echocardiography guidance, aid in selecting the appropriate device size and type for optimal outcomes.

The closure of most ASDs is typically straightforward and free of serious complications. Nonetheless, in rare instances, the procedure can be challenging, requiring greater expertise and alternative techniques. Even in cases of large ASDs, a high procedural success rate can generally be achieved.

Multiple secundum-type ASDs or secundum-type ASDs with surrounding fenestrations represent one of the less common (approximately 10%) but more complex scenarios for transcatheter septal defect closure [4]. Various classifications and closure techniques have been developed to address these cases.

With the assistance of echocardiographic findings, most secundum-type ASDs can be closed using a single large device, depending on the IDD. The IDD is defined as the distance between the central defect and any surrounding holes that need to be covered. If the IDD is <7 mm, it is often possible to close the defect with a single device. (In such cases, the larger defect is treated as the central hole, and a single device is implanted to occlude the larger hole while compressing the smaller ones.)

For patients with a larger ASD accompanied by multiple small ASDs or fenestrations, a viable alternative is the implantation of a large Amplatzer Cribriform Septal Occluder, a PFO Occluder, or an Occlutech Uniform Occluder, particularly in those with an aneurysmal interatrial septum.

When the IDD exceeds 7 mm, double-device deployment is typically necessary. Nevertheless, there is ongoing debate regarding the optimal strategy for multiple-device implantation. While some advocate closing the larger defect first, others recommend prioritizing the smaller defect [5, 6]. Another perspective suggests closing the inferior defect (often the smaller one) first.

Small-hole devices tend to have stronger anchors, allowing them to securely attach to the other device and reduce the exposed device area. Still, if large holes are located inferiorly, their closure should take precedence. Superior small holes can be addressed with smaller devices, leveraging the rigid support provided by the lower large device [7].

In section "Case 1", we initially closed the larger defect, which was located superiorly and posteriorly to the smaller one. The smaller, inferior defect was then occluded separately using a larger device to overlap the rim of the first device, despite its smaller SFD. This approach resulted in no residual shunting or interference with surrounding structures.

In section "Case 2", we first measured the SFD of both defects under TEE guidance. The smaller, superior defect was closed first, followed by the larger defect using a larger device. This device securely hugged the smaller superior device without causing impingement on surrounding structures or leaving residual shunting.

There is currently no consensus on the optimal strategy for closing multiple ASDs. Therefore, the interventionist must determine the most effective technique with the fewest side effects, taking into account factors such as the location, size, and orientation of the defects, the IDD, the type of interatrial septum, the dimensions of the rims, the surrounding structures, and the types of devices available in the catheterization laboratory.

Another significant challenge for interventionists during percutaneous secundum-type ASD closure is managing older patients and those with pulmonary hypertension. Aging is often associated with comorbidities such as systemic hypertension, coronary heart disease, left and right ventricular systolic and diastolic dysfunction, and valvular heart disease. These conditions can reduce left ventricular compliance, leading to increased left atrial pressure, atrial arrhythmias, pulmonary hypertension, and ultimately, congestive heart failure.

Percutaneous ASD closure in these patients carries high risks, making careful case selection essential. The indications for ASD closure in older patients are similar to those in younger individuals. For elderly patients with congestive heart failure or pulmonary hypertension, the balloon occlusive test is strongly recommended.

During the balloon occlusion of the ASD, abrupt closure may cause pulmonary edema if any of the following conditions persist for at least 15 min:

- An increase in left ventricular end-diastolic pressure to >50% or a persistent elevation >20 mm Hg;
- A rise in pulmonary capillary wedge pressure by >10 mm Hg or a persistent elevation >20 mm Hg; or
- A decrease in systemic blood pressure by >20% [8].

In patients with ASDs, the likelihood of pulmonary hypertension is relatively low (approximately 8%–10%), making it advisable to investigate other potential causes. However, there is no clear linear relationship between the severity of pulmonary hypertension and the size of the ASD or the duration of shunting. Furthermore, the underlying pathogenic mechanisms remain poorly understood.

The complete closure of ASDs in patients with pulmonary hypertension is a subject of debate. This is due to the potential risk of worsening pulmonary vascular disease, including increased pulmonary vascular resistance and right heart failure, without the compensatory benefits of right-to-left shunting.

In symptomatic patients with pulmonary hypertension, medical therapy is initiated using diuretics, endothelin receptor inhibitors, prostaglandins, or phosphodiesterase inhibitors. Nonetheless, for patients with refractory symptoms, an alternative strategy is partial defect closure using fenestrated devices. This approach aims to reduce the left-to-right shunt while permitting a right-to-left shunt to act as a "pop-off valve."

Fenestrated closure allows blood to shunt from right to left if pulmonary vascular resistance increases, helping to maintain cardiac output. As a result, fenestrated ASD closure is the sole interventional treatment option for patients with ASDs and moderate-to-severe pulmonary hypertension. The implantation technique for fenestrated devices is identical to that used for standard ASD devices.

In section "Case 3", during the balloon occlusion test, both the left ventricular end-diastolic pressure and the PAP increased. Considering the patient's clinical condition and the ASD morphology, a fenestrated ASD Occlutech Occluder was implanted.

Another challenging scenario in the transcatheter closure of secundum-type ASDs involves large defects, which often coexist with deficient rims, mal-aligned interatrial septa, or aneurysmal septa. While several closure techniques have been developed, they require the expertise of highly skilled interventionists.

In section "Case 4", a 33 mm device was selected. Initial attempts using various techniques were unsuccessful. Nevertheless, an unconventional technique was employed at the operator's discretion, resulting in the successful closure of the defect.

Predictors of difficult ASD closure include the following:

- The diameter and location of the defect;
- The dimensions of the surrounding rims;
- The presence of interatrial septum aneurysms; and
- The alignment with the interatrial septum.

The following are some recommended maneuvers for challenging ASD closure cases:

1. Adjusting the curvature of the delivery sheath to improve alignment between the device and the interatrial septum;
2. Rotating the sheath or device (clockwise or counterclockwise) within the left atrium;

3. Clockwise rotation of the left atrial disk in the left atrial roof;
4. Deploying the left atrial disk and half of the waist in the mid-portion of the left atrium, followed by a rapid pullback toward the septum;
5. Deploying the entire device in the left atrium and then simultaneously pulling back the device while pushing the delivery sheath forward;
6. Deploying the left atrial disk in the left or right upper pulmonary veins and subsequently pulling the sheath into the right atrium to enhance alignment with the interatrial septum;
7. Using a straight, side-hole delivery sheath (e.g., a modified Mullins sheath); and
8. Balloon-, dilator-, or catheter-assisted approaches (as demonstrated in section "Case 4").

Conclusions

Today, even complex secundum-type ASDs can be successfully closed using the percutaneous transcatheter approach when performed by experienced interventionists. These specialists have access to a variety of device designs, including self-centering and non-self-centering options, to address numerous anatomical, device-related, and physiological challenges.

Although rare, serious complications can occur unexpectedly during the procedure or manifest later, typically within 1 month to 1 year. Device embolization, for instance, is an uncommon event (occurring in up to 1% of cases) and most often happens within the first 24 h after implantation.

Predictors of device dislodgement include the following:

- Undersized ASD devices;
- A deficient posteroinferior rim (rather than the posterior rim);
- A left atrium too small to accommodate the device; and
- Technical issues during implantation.

The most common sites of embolization are the left atrium (24.6%), the aorta (18.4%), and the right ventricle (16.7%). While most devices can be retrieved percutaneously if embolization occurs early, late embolization often requires surgical retrieval due to epithelialization of the device.

Another rare complication is complete atrioventricular block, which is usually transient and responds to steroid therapy. Still, some case reports suggest that it can become permanent, necessitating either surgical removal of the device or pacemaker implantation.

The most serious complication is cardiac erosion (occurring in 0.1%–0.3% of cases), which typically arises within the first year post-implantation [9]. Predictors of this rare but feared complication include the following:

- Deficient anterior and/or superior rims;
- Device movement within the heart;

- Oversized devices;
- Adult age;
- Use of Amplatzer Septal Occluder devices;
- Mal-aligned interatrial septum; and
- Multiple deployment attempts before final device placement.

Wall perforation most commonly affects the roof of the right or left atrium or the atrial-aortic junction. Some cases of delayed erosion have also been reported, highlighting the importance of long-term follow-up.

Atrial arrhythmias may occur during the procedure, though they are usually transient.

Learning from reported experiences, the judicious use of imaging modalities, and the appropriate selection of equipment are crucial for successfully managing complex ASD closure cases.

References

1. Wiktor DM, Carroll JD. ASD closure in structural heart disease. Curr Cardiol Rep. 2018;20:37. https://doi.org/10.1007/s11886-018-0984-z. PMID: 29705879
2. Yong G, Khairy P, De Guise P, Dore A, Marcotte F, Mercier LA, et al. Pulmonary arterial hypertension in patients with transcatheter closure of secundum atrial septal defects: a longitudinal study. Circ Cardiovasc Interv. 2009;2:455–62. https://doi.org/10.1161/CIRCINTERVENTIONS.109.868945. PMID: 20031755
3. Firouzi A, Hosseini Z. Transcatheter secundum ASD device closure. In: Maleki M, Alizadehasl A, editors. Case-based clinical cardiology. London: Springer; 2021. p. 585–92. https://doi.org/10.1007/978-1-4471-7496-7_63.
4. Diab KA, Cao QL, Bacha EA, Hijazi ZM. Device closure of atrial septal defects with the Amplatzer septal occluder: safety and outcome in infants. J Thorac Cardiovasc Surg. 2007;134:960–6. https://doi.org/10.1016/j.jtcvs.2007.06.018. PMID: 17903513
5. Mahadevan VS, Gomperts N, Haberer K, McKenzie J, Morgan G, Nair P, et al. Transcatheter closure of atrial septal defects with multiple devices in adults: procedural and clinical outcomes. Int J Cardiol. 2009;133:359–63. https://doi.org/10.1016/j.ijcard.2007.12.019. PMID: 18281119
6. Yankun Y, Zhongying X, Shiliang J, Shihua Z, Gejun Z, Jinglin J, et al. Simultaneous transcatheter closure of multiple atrial septal defects using dual Amplatzer septal occluder devices. Am J Med Sci. 2016;352:245–51. https://doi.org/10.1016/j.amjms.2016.05.027. PMID: 27650227
7. Farhaj Z, Hongxin L, Wenbin G, Zhang WL, Liang F, Zhang HZ, et al. Device closure of diverse layout of multi-hole secundum atrial septal defect: different techniques and long-term follow-up. J Cardiothorac Surg. 2019;14:130. https://doi.org/10.1186/s13019-019-0952-3. PMID: 31288830
8. Swan L, Varma C, Yip J, Warr M, Webb G, Benson L, et al. Transcatheter device closure of atrial septal defects in the elderly: technical considerations and short-term outcomes. Int J Cardiol. 2006;107:207–10. https://doi.org/10.1016/j.ijcard.2005.03.009. PMID: 16503257
9. Khajali Z, Firouzi A, Hosseini Z, Pouraliakbar H, Noohi Bezanjani F, Alemzadeh-Ansari MJ. Case report of worsening of preexisting pericardial effusion after ASD device closure: is it calamitous? Egypt Heart J. 2021;73:50. https://doi.org/10.1186/s43044-021-00177-2. PMID: 34097147

Chapter 8
Patent Foramen Ovale Closure

Ata Firouzi, Zahra Hosseini, and Ehsan Khalilipur

Abstract A PFO is a frequently observed benign condition in healthy individuals; however, its prevalence is elevated among patients experiencing strokes of unknown origin (cryptogenic stroke). PFO is thought to contribute to stroke risk via paradoxical embolism. Recent studies indicate that closing a PFO—particularly when it is linked to an ASA and/or a significant interatrial shunt—may lower the likelihood of recurrent strokes. This chapter summarizes the latest advancements in PFO device closure.

Keywords PFO · Device closure · Stroke

Abbreviations

ASA	Atrial septal aneurysm
DVT	Deep vein thrombosis
ESUS	Embolic stroke of undetermined source
PFO	Patent foramen ovale
RoPE	Risk of paradoxical embolism
TEE	Transesophageal echocardiography
TIA	Transient ischemic attack
TTE	Transthoracic echocardiography

A. Firouzi · Z. Hosseini · E. Khalilipur (✉)
Cardiovascular Intervention Research Center, Rajaie Cardiovascular Institute, Iran University of Medical Sciences, Tehran, Iran

A. Firouzi et al. (eds.), *Mastering Skills in Complex Cardiovascular Interventions*, https://doi.org/10.1007/978-3-031-97321-5_8

Introduction

A PFO is a remnant of fetal circulation that persists into adulthood. It plays a role in several complex medical conditions, including cryptogenic stroke, migraine, platypnea-orthodeoxia syndrome, and decompression sickness [1, 2]. By the age of two, the foramen ovale closes completely in 75% of individuals. Nonetheless, it remains open in 25% of cases, making it the most common congenital heart defect [3]. The PFO typically has a tunnel-like structure, with a slit-like width averaging 4.9 mm (range: 1–19 mm) and a length (or depth) of 1–6 mm. The morphology of PFOs varies significantly among individuals and can be categorized as either simple or complex.

Criteria for a complex PFO include:

- Long tunnel length: Maximum overlap between the septum primum and septum secundum (≥8 mm);
- PFO height: Separation between the septum primum and septum secundum >2.8 mm during the Valsalva maneuver;
- ASA: Presence of an ASA;
- Larger septal excursion distance: Increased movement of the atrial septum;
- Thick septum secundum: Thickness ≥10 mm;
- Thin septum primum: Thickness <1.6 mm;
- Very short septum secundum flap: Short or absent aortic rim;
- Prominent Eustachian ridge or valve: Visible prominence of these structures;
- Angle between the inferior vena cava and the PFO flap: ≤10 [4]; and
- Large right-to-left shunting: >20 microbubbles.

A complex PFO is more frequently associated with cryptogenic stroke and may necessitate special considerations when selecting the type and size of the occluder device [5]. Postmortem studies indicate that the typical size of a PFO increases with age, suggesting that larger defects are more likely to remain patent into adulthood, while smaller defects may close over time [5].

Diagnosis of PFO

Detecting a PFO often involves identifying a right-to-left shunt rather than directly visualizing the PFO itself. Commonly used imaging techniques include echocardiography (TTE or TEE) and transcranial Doppler. Transcranial Doppler imaging diagnoses PFOs by detecting air microemboli in the middle cerebral artery after the peripheral injection of agitated saline. TEE remains the gold standard for both detecting and visualizing a PFO, with a prevalence of 39% using this method, compared with 18% with TTE and 27% with transcranial Doppler [6]. TEE is used to identify PFOs by injecting agitated saline into the cephalic vein in the antecubital fossa. This fills the right atrium with microbubbles. If a PFO is present, these

Table 8.1 The RoPE score calculator [12]

Characteristic	Points
No history of hypertension	1 point
No history of diabetes	1 point
No history of stroke or TIA	1 point
Non-smoking status	1 point
Cortical infarct on imaging	1 point
Age, years	
18–29	5 points
30–39	4 points
40–49	3 points
50–59	2 points
60–69	1 point
≥70	0 points
Total score (sum of individual points)	
Maximum score (a patient <30 yr with no hypertension, diabetes, stroke or TIA, non-smoking status, and cortical infarct)	**RoPE score: 10**
Minimum score (a patient ≥70 yr with hypertension, diabetes, stroke or TIA, current smoking status, and no cortical infarct)	**RoPE score: 0**

microbubbles can enter the left atrium within a few heartbeats (approximately 3 s) after appearing in the right atrium. While most shunting occurs from right to left, some reverse shunting may occur when relative atrial pressures shift during the cardiac cycle or with breathing. The Valsalva maneuver can increase right atrial pressure during TTE, improving the sensitivity of PFO detection [6, 7]. Nevertheless, the amount of contrast visible in the left atrium does not allow for a precise estimation of the PFO size. Imaging, on the other hand, allows for risk classification by identifying anatomical abnormalities, such as the presence of an ASA (Table 8.1). TEE is recommended for a comprehensive evaluation of the atrial septal architecture before PFO closure [8]. TEE demonstrates a sensitivity of 90% and a specificity of over 95% in diagnosing PFO, whereas TTE has a sensitivity of 50%–60% and a specificity of over 90%.

ESUS and PFO

ESUS accounts for approximately 25% of stroke cases, and PFO is considered one of the leading causes of ESUS. ESUS is defined as symptomatic cerebral infarcts without an identifiable cause after a thorough investigation, ruling out lacunar infarction, major vessel stenosis >50%, or cardiac etiologies such as atrial fibrillation. Determining the optimal duration of extended cardiac monitoring remains particularly challenging. Atrial fibrillation detection rates have been reported as 3.7% after 1 month, 8.9% after 6 months, 12.4% after 12 months, 21.1% after 2 years, and 30.0% after 3 years [9, 10].

Recurrent cerebrovascular events are common in medically treated patients with a history of paradoxical embolism. Still, in over one-third of these patients, other pathologies—beyond the mere presence of a PFO—are identified. On the other hand, PFOs are found in nearly half of individuals with ESUS. Yet, approximately one-third of PFOs detected in patients with cryptogenic stroke are likely incidental and unrelated to the stroke. As a result, it remains unclear whether a PFO is causally linked to cerebral ischemia or simply an incidental finding during diagnostic workup [11]. Therefore, selecting patients for whom PFO closure is justified remains an unresolved issue.

The RoPE score was developed to assist in decision-making by estimating the likelihood that a PFO is responsible for a cryptogenic stroke [12] (Table 8.1).

PFO and DVT

DVT is more prevalent in individuals with a cryptogenic stroke (20%) than in those with a confirmed cause of stroke (4%). Detecting a thrombus transiting through a PFO, indicating an imminent paradoxical embolism, is rare; most such cases are identified postmortem. For patients diagnosed with DVT, the standard initial treatment is anticoagulation, typically for at least 3 months, before considering PFO closure. If long-term anticoagulation is required for other reasons (e.g., recurrent DVTs), there is insufficient evidence to recommend PFO closure definitively. In such cases, a multidisciplinary team decision involving neurology and cardiology is advised. PFO closure may be considered in patients at high risk of recurrent DVT [13, 14].

A Challenging Case

The patient was a 30-year-old woman who was referred to another center due to recurrent left arm paresis. Following a comprehensive neurological assessment that ruled out any abnormal findings, a TEE examination was performed, revealing a high-risk PFO characterized by a thick septum secundum with a short length, a PFO height of 3 mm, and an aneurysmal septum with a large right-to-left shunt, without any other abnormalities. Based on the patient's symptoms and these findings, she was scheduled for percutaneous PFO closure. The procedure was completed; however, 3 months later, she was referred to our center with complaints of recurrent amaurosis fugax. TEE showed the previous occluder device positioned in the interatrial septum, 12 mm away from the PFO (4 × 6 mm), with significant bubble passage. Another fenestration (3 mm) was observed at the anterosuperior rim, resulting in a left-to-right shunt (Fig. 8.1a, b). Given her symptoms and the echocardiographic findings, she was considered a candidate for percutaneous device closure.

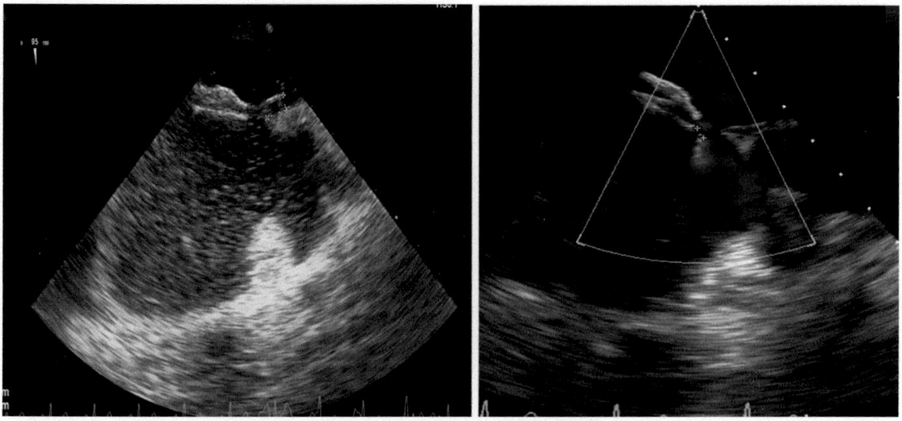

Fig. 8.1 The previous occluder device is shown in the interatrial septum, located 12 mm away from the PFO (4 × 6 mm), with significant bubble passage. Additionally, another fenestration (3 mm) is observed at the anterosuperior rim, resulting in a left-to-right shunt

Procedural Technique

Under general anesthesia, 6 Fr sheaths were inserted into the right common femoral artery and vein. Heparin was administered to achieve an activated clotting time >200 s. Using a multipurpose catheter, a 0.035-in. J-tip wire was advanced through the PFO under TEE guidance. (At our center, TEE is not routinely used during PFO closure except in special cases, such as those with suspected concomitant fenestrations or a large aneurysmal interatrial septum. In addition, a sizing balloon is not employed.) The catheter was then advanced over the wire, with its tip positioned in the right upper pulmonary vein, and the wire was exchanged for a 260 cm super-stiff guidewire. A 10 Fr Occlutech delivery sheath was advanced over the wire into the mid-left atrium. The dilator was subsequently removed to allow back-bleeding. The 10 Fr Occlutech delivery sheath was advanced over the wire into the mid-left atrium. The dilator was then removed to allow back-bleeding. Based on prior TEE findings (a large PFO of 6 mm with an ASA and concomitant fenestration), a Figulla Flex II UNI Occluder (24 mm) was selected. After loading the device, connecting it to the delivery cable, and ensuring it was fully de-aired, the device was advanced through the delivery sheath to the tip of the sheath. By retracting the sheath, the left disk was deployed. The entire system was then pulled back against the septum, and the right disk was deployed (Fig. 8.2). The final agitated saline infusion revealed no evidence of a residual shunt.

Aspirin and clopidogrel were continued for 6 months. During follow-up, the patient experienced no new events.

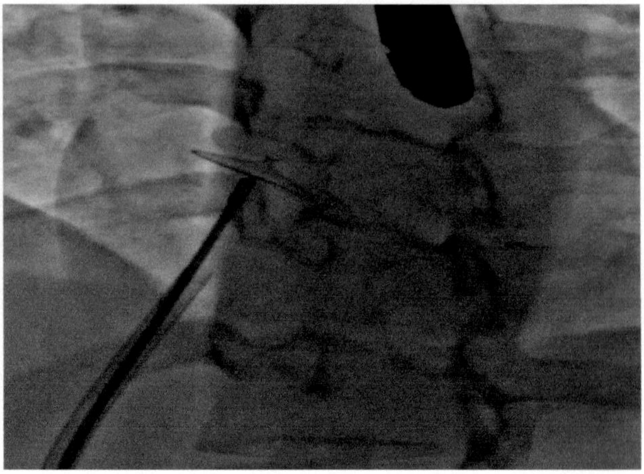

Fig. 8.2 The Figulla Flex II UNI Occluder (24 mm) was deployed correctly, which hinged the previous device

Randomized Trials

Initial randomized controlled trials failed to demonstrate the superiority of PFO closure over conservative treatment. Nonetheless, pre-specified "per-protocol" and "as-treated" analyses in the PC and RESPECT trials showed that PFO closure using the Amplatzer PFO Occluder device was more effective than medical treatment alone, with minimal associated complications [15–17].

Furthermore, extended follow-up in the RESPECT trial, spanning nearly 6 years compared with the initial study, revealed 10 additional strokes in the medical treatment arm versus the closure arm (hazard ratio [HR]: 0.55). Stroke recurrence was significantly reduced by half in the closure arm ($P = 0.046$).

Moreover, subgroup analysis revealed that PFO closure benefited patients with a significant right-to-left shunt, an ASA, or those whose planned medical regimen included antiplatelet therapy rather than anticoagulants (if assigned to the medical treatment group) [17].

Across all six trials, PFO closure was combined with antiplatelet therapy to mitigate risk. Based on randomized studies, an evidence-based protocol for PFO closure in ischemic stroke patients was developed to optimize clinical outcomes. Nevertheless, the duration of therapy varied significantly among trials, ranging from at least 6 months to the duration of the study (Table 8.2).

Table 8.2 Early studies of PFO device closure vs. medical treatment

Study	Number of patients	Criteria	Device	Follow-up, years	Antithrombotic therapy	Primary endpoint	Result
CLOSURE I	909	Patients 16–60 yr old with cryptogenic stroke or TIA and PFO	STARFlex septal closure system	2	Device arm: Aspirin & warfarin (1 mo), followed by aspirin (2 yr). Medical treatment arm: Aspirin or warfarin or aspirin & warfarin.	Early all-cause death, late death due to neurologic cause, stroke, and TIA	Percutaneous PFO closure did not significantly reduce recurrent stroke or TIA compared with medical treatment alone.
PC	414	Patients <60 yr old with cryptogenic stroke, TIA, or systemic embolism and PFO	Amplatzer PFO Occluder	4	Device arm: Aspirin (5–6 mo) & ticlopidine or clopidogrel (1–6 mo). Medical treatment arm: Antiplatelet therapy or anticoagulation therapy.	Death, nonfatal stroke, TIA, or peripheral embolism	Percutaneous PFO closure did not significantly reduce death or recurrent embolism compared with medical treatment alone.
RESPECT	980	Patients 18–60 yr old with cryptogenic stroke and PFO	Amplatzer PFO Occluder	5.9 (median)	Device arm: Aspirin plus clopidogrel (1 mo), followed by aspirin (5 mo). Medical treatment arm: Aspirin or warfarin or clopidogrel or aspirin and extended-release dipyridamole.	Recurrent fatal and nonfatal stroke and early death	Percutaneous PFO closure significantly reduced recurrent stroke rates compared with medical treatment alone.

Meta-Analyses of All Six Randomized Trials

Subsequently, more than a dozen meta-analyses were conducted to pool data from the randomized trials, aiming to increase sample size and reduce the risk of type II errors. All meta-analyses reached a consistent conclusion: percutaneous PFO closure significantly reduces the risk of recurrent stroke in patients with PFO-related stroke compared to medical therapy. The pooled data also demonstrated that this clinical benefit of PFO closure is not associated with an increased risk of major adverse events or a significant impact on major bleeding when compared to standard-of-care blood thinners [15, 18–21].

One notable caveat is that PFO closure was associated with a fourfold increase in the incidence of atrial fibrillation or flutter. However, this difference was less pronounced in the subgroup of patients treated with the Amplatzer PFO Occluder (OR: 2.29, 95% CI: 0.91–5.76) [22].

In summary, the decision to close a PFO in a patient with a cryptogenic stroke should be made by a multidisciplinary team, including a neurologist or stroke specialist and a cardiologist. Individualized decision-making is essential to identify suitable candidates, particularly younger patients (up to 60 years old) who have experienced a stroke with no other identifiable cause despite a comprehensive evaluation. Key factors favoring PFO closure include brain MRI findings suggestive of embolic infarcts, large right-to-left shunts, and/or the presence of an ASA. Conversely, there is limited evidence supporting PFO closure for preventing recurrent stroke in older patients or those with small right-to-left shunts, no ASA, or lacunar infarcts on imaging [23–26].

Device-Associated Atrial Fibrillation

Percutaneous PFO closure has been linked to a fourfold increase in the incidence of new-onset atrial fibrillation, raising valid concerns that, while reducing the risk of PFO-mediated stroke recurrence, the procedure may introduce another potential risk factor for thromboembolic stroke. Still, it is encouraging that the majority of these new atrial arrhythmias occurred during or shortly after the procedure and rarely led to recurrent cerebrovascular events or required long-term anticoagulation [22]. Out of 1889 patients across all six trials, only five (0.1%) experienced recurrent stroke due to device-associated atrial fibrillation, with three of these cases involving the discontinued STARFlex device [27]. As a result, the stroke risk associated with this type of atrial fibrillation appears to be low, and the majority of patients do not require long-term anticoagulation. By way of example, 70% of patients in the CLOSE trial who developed device-related atrial fibrillation eventually discontinued anticoagulation [28].

Technical Considerations

Before planning the closure procedure, it is essential to review contraindications.

Examples of Contraindications to PFO Closure

- *Presence of an intracardiac mass, vegetation, tumor, or thrombus at the intended implant site, or documented venous thrombus in the vessels used to access the PFO.*
- *Inadequate vasculature (used for access to the PFO) that cannot accommodate the appropriate sheath size.*
- *Anatomical constraints where the required device size would interfere with other intracardiac or intravascular structures, such as valves or pulmonary veins.*
- *Active endocarditis or other untreated infections.*

Some operators use balloon sizing instead of the method outlined in the instructions. In this approach, an inflatable sizing balloon is advanced across the PFO and gradually inflated until a waist becomes visible. Gentle retraction of the inflated sizing balloon can evert or expand the tunnel, optimizing the final orientation of the implanted device. A device size at least twice the diameter of the sizing balloon's waist is typically selected. A larger device may be considered if the septum secundum is thick or if a significant ASA is present.

During fluoroscopy, the "Pacman sign" (where the septum secundum is positioned between the left and right atrial disks, and the cranial halves of the disks resemble open jaws biting into the thick septum secundum) confirms the correct device position before release (Fig. 8.3).

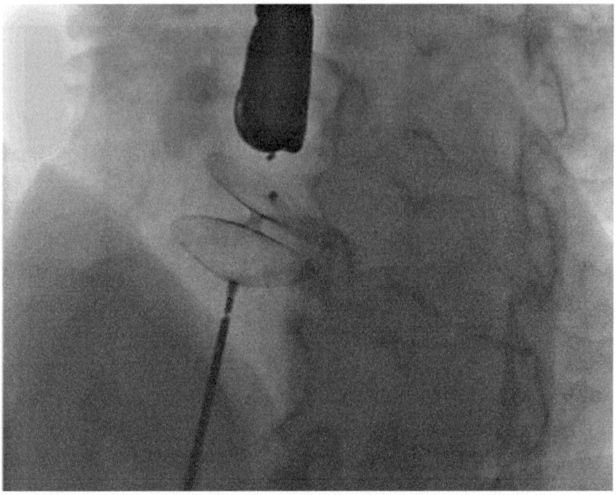

Fig. 8.3 The correct device position is illustrated by the "Pacman sign"

Procedural and Device-Related Complications

Transcatheter PFO closure is associated with a very low complication rate. In the RESPECT trial, 22 out of 499 patients (4.2%) experienced device- or procedure-related adverse events. The rates of atrial fibrillation, tamponade, and cardiac thrombus were extremely low (0.2%, 0.4%, and 0.4%, respectively). Additionally, three patients (0.6%) experienced vascular complications. No cases of device embolization or intraprocedural stroke were reported. After a follow-up period of 6.3 years (3141 patient-years), no patients experienced device-related perforation (e.g., erosion).

In the nonrandomized pivotal trial of the CARDIOFORM Septal Occluder for transcatheter ASD closure, no significant procedure- or device-related adverse events were observed in 50 patients at the 6-month follow-up (CARDIOFORM package insert). Among 60 patients who underwent transcatheter PFO closure with the CARDIOFORM Septal Occluder, stroke, thrombus formation, and atrial fibrillation/flutter occurred in one (1.7%), one (1.7%), and five (8.3%) patients, respectively.

To minimize complications, the following techniques are recommended:

- Use ultrasound guidance for femoral vein cannulation.
- Perform all catheter exchanges within the left atrium over a stiff wire placed in one of the pulmonary veins or over a pigtail-shaped wire (e.g., an Inouye-type wire).
- Thoroughly de-air and flush the delivery sheath and occluder to prevent air emboli during deployment, which can lead to stroke, chest pain, ST-segment elevation, or myocardial infarction [29, 30].

References

1. Hara H, Virmani R, Ladich E, Mackey-Bojack S, Titus J, Reisman M, et al. Patent foramen ovale: current pathology, pathophysiology, and clinical status. J Am Coll Cardiol. 2005;46:1768–76. https://doi.org/10.1016/j.jacc.2005.08.038. PMID: 16256881
2. Wilmshurst PT, Ellis BG, Jenkins BS. Paradoxical gas embolism in a scuba diver with an atrial septal defect. Br Med J (Clin Res Ed). 1986;293:1277. https://doi.org/10.1136/bmj.293.6557.1277. PMID: 3096459
3. Kerut EK, Norfleet WT, Plotnick GD, Giles TD. Patent foramen ovale: a review of associated conditions and the impact of physiological size. J Am Coll Cardiol. 2001;38:613–23. https://doi.org/10.1016/S0735-1097(01)01427-9. PMID: 11527606
4. Nakayama R, Takaya Y, Akagi T, Watanabe N, Ikeda M, Nakagawa K, et al. Low-angle patent foramen ovale (PFO): high-risk PFO morphology associated with paradoxical embolism. JACC Cardiovasc Imaging. 2021;14:645–7. https://doi.org/10.1016/j.jcmg.2020.08.017. PMID: 33221214
5. Hagen PT, Scholz DG, Edwards WD. Incidence and size of patent foramen ovale during the first 10 decades of life: an autopsy study of 965 normal hearts. Mayo Clin Proc. 1984;59:17–20. https://doi.org/10.1016/S0025-6196(12)60231-X. PMID: 6694427

6. Di Tullio M, Sacco RL, Venketasubramanian N, Sherman D, Mohr JP, Homma S. Comparison of diagnostic techniques for the detection of a patent foramen ovale in stroke patients. Stroke. 1993;24:1020–4. https://doi.org/10.1161/01.STR.24.7.1020. PMID: 8322375
7. Takaya Y, Watanabe N, Ikeda M, Akagi T, Nakayama R, Nakagawa K, et al. Importance of abdominal compression Valsalva maneuver and microbubble grading in contrast transthoracic echocardiography for detecting patent foramen ovale. J Am Soc Echocardiogr. 2020;33:201–6. https://doi.org/10.1016/j.echo.2019.09.008. PMID: 31759799
8. Silvestry FE, Cohen MS, Armsby LB, Burkule NJ, Fleishman CE, Hijazi ZM, et al. Guidelines for the echocardiographic assessment of atrial septal defect and patent foramen ovale: from the American Society of Echocardiography and Society for Cardiac Angiography and Interventions. J Am Soc Echocardiogr. 2015;28:910–58. https://doi.org/10.1016/j.echo.2015.05.015. PMID: 26239900
9. Best JG, Bell R, Haque M, Chandratheva A, Werring DJ. Atrial fibrillation and stroke: a practical guide. Pract Neurol. 2019;19:208–24. https://doi.org/10.1136/practneurol-2018-002089. PMID: 30612135
10. Saver JL. Cryptogenic stroke in patients with patent foramen ovale. Curr Atheroscler Rep. 2007;9:319–25. https://doi.org/10.1007/s11883-007-0038-z. PMID: 18377787
11. Alsheikh-Ali AA, Thaler DE, Kent DM. Patent foramen ovale in cryptogenic stroke: incidental or pathogenic? Stroke. 2009;40:2349–55. https://doi.org/10.1161/STROKEAHA.109.547828. PMID: 19443798
12. Kent DM, Saver JL, Ruthazer R, Furlan AJ, Reisman M, Carroll JD, et al. Risk of paradoxical embolism (RoPE)-estimated attributable fraction correlates with the benefit of patent foramen ovale closure: an analysis of 3 trials. Stroke. 2020;51:3119–23. https://doi.org/10.1161/STROKEAHA.120.029350. PMID: 32912092
13. Hart RG, Diener HC, Coutts SB, Easton JD, Granger CB, O'Donnell MJ, et al. Embolic strokes of undetermined source: the case for a new clinical construct. Lancet Neurol. 2014;13:429–38. https://doi.org/10.1016/S1474-4422(13)70310-7. PMID: 24646875
14. Kent DM, Ruthazer R, Weimar C, Mas JL, Serena J, Homma S, et al. An index to identify stroke-related vs incidental patent foramen ovale in cryptogenic stroke. Neurology. 2013;81:619–25. https://doi.org/10.1212/WNL.0b013e3182a08d59. PMID: 23864311
15. Meier B, Kalesan B, Mattle HP, Khattab AA, Hildick-Smith D, Dudek D, et al. Percutaneous closure of patent foramen ovale in cryptogenic embolism. N Engl J Med. 2013;368:1083–91. https://doi.org/10.1056/NEJMoa1211716. PMID: 23514285
16. Carroll JD, Saver JL, Thaler DE, Smalling RW, Berry S, MacDonald LA, et al. Closure of patent foramen ovale versus medical therapy after cryptogenic stroke. N Engl J Med. 2013;368:1092–100. https://doi.org/10.1056/NEJMoa1301440. PMID: 23514286
17. Saver JL, Carroll JD, Thaler DE, Smalling RW, MacDonald LA, Marks DS, et al. Long-term outcomes of patent foramen ovale closure or medical therapy after stroke. N Engl J Med. 2017;377:1022–32. https://doi.org/10.1056/NEJMoa1610057. PMID: 28902580
18. Arfaras-Melainis A, Palaiodimos L, Mojadidi MK. Transcatheter closure of patent foramen ovale: randomized trial update. Interv Cardiol Clin. 2019;8:341–56. https://doi.org/10.1016/j.iccl.2019.05.002. PMID: 31445719
19. Furlan AJ, Reisman M, Massaro J, Mauri L, Adams H, Albers GW, et al. Closure or medical therapy for cryptogenic stroke with patent foramen ovale. N Engl J Med. 2012;366:991–9. https://doi.org/10.1056/NEJMoa1009639. PMID: 22417252
20. Palaiodimos L, Kokkinidis DG, Faillace RT, Foley TR, Saw J, Marin ML, et al. Percutaneous closure of patent foramen ovale vs. medical treatment for patients with history of cryptogenic stroke: a systematic review and meta-analysis of randomized controlled trials. Cardiovasc Revasc Med. 2018;19(7 Pt B):852–8. https://doi.org/10.1016/j.carrev.2018.03.008. PMID: 29685696
21. Kokkinidis DG, Palaiodimos L, Faillace RT. Percutaneous closure of patent foramen ovale vs. medical treatment for patients with history of cryptogenic stroke—a systematic review

and meta-analysis of randomised controlled trials. Cerebrovasc Dis. 2018;45(Suppl 1):1–68. https://doi.org/10.1159/000489693. (abstract only; full text unavailable).
22. Mojadidi MK, Elgendy AY, Elgendy IY, Mahmoud AN, Elbadawi A, Eshtehardi P, et al. Atrial fibrillation after percutaneous patent foramen ovale closure. Am J Cardiol. 2018;122:915. https://doi.org/10.1016/j.amjcard.2018.05.021. PMID: 30055752
23. Mojadidi MK, Elgendy AY, Elgendy IY, Mahmoud AN, Elbadawi A, Eshtehardi P, et al. Transcatheter patent foramen ovale closure after cryptogenic stroke: an updated meta-analysis of randomized trials. JACC Cardiovasc Interv. 2017;10:2228–30. https://doi.org/10.1016/j.jcin.2017.08.029. PMID: 29122138
24. Kernan WN, Ovbiagele B, Black HR, Bravata DM, Chimowitz MI, Ezekowitz MD, et al. Guidelines for the prevention of stroke in patients with stroke and transient ischemic attack: a guideline for healthcare professionals from the American Heart Association/American Stroke Association. Stroke. 2014;45:2160–236. https://doi.org/10.1161/STR.0000000000000024. PMID: 24788967
25. Kokkinidis DG, Palaiodimos L, Mastoris I, Papanastasiou CA, Sarafidis PA, Faillace RT, et al. The best DEFENSE for high-risk patent foramen ovale: an updated meta-analysis of six randomized trials. Arch Cardiovasc Dis. 2019;112:150–2. https://doi.org/10.1016/j.acvd.2018.10.005. PMID: 30655161
26. Alakbarzade V, Keteepe-Arachi T, Karsan N, Ray R, Pereira AC. Patent foramen ovale. Pract Neurol. 2020;20:225–33. https://doi.org/10.1136/practneurol-2019-002450. PMID: 32238508
27. Elgendy AY, Elgendy IY, Mojadidi MK, Mahmoud AN, Barry IJ, Jneid H, et al. New-onset atrial fibrillation following percutaneous patent foramen ovale closure: a systematic review and meta-analysis of randomised trials. EuroIntervention. 2019;14:1788–90. https://doi.org/10.4244/EIJ-D-18-00767. PMID: 30418167
28. Staubach S, Steinberg DH, Zimmermann W, Wawra N, Wilson N, Wunderlich N, et al. New onset atrial fibrillation after patent foramen ovale closure. Catheter Cardiovasc Interv. 2009;74:889–95. https://doi.org/10.1002/ccd.22110. PMID: 19496129
29. Firouzi A, Hosseini Z. Transcatheter PFO device closure. In: Maleki M, Alizadehasl A, editors. Case-based clinical cardiology. London: Springer; 2021. p. 593–9. https://doi.org/10.1007/978-1-4471-7496-7_64.
30. Turc G, Lee JY, Broderick JP, Kim JS, Albers GW, Choi YC, et al. Patent foramen ovale closure in stroke patients with migraine: a pooled analysis of randomized trials. JACC Cardiovasc Interv. 2024;17:632–41. https://doi.org/10.1016/j.jcin.2023.12.030. PMID: 38385912

Chapter 9
Percutaneous Ventricular Septal Defect Closure in Adults

Ata Firouzi, Zahra Hosseini, and Mohammadreza Baay

Abstract VSD is one of the most common congenital heart defects, with a high likelihood of spontaneous closure. Thanks to rapid advancements in technology, transcatheter VSD closure has become a routine procedure, offering high success rates and low complication rates when appropriate cases and devices are selected.

In addition to congenital VSDs, acquired forms such as post-MI VSR and post-traumatic VSDs can occur. Given the high mortality rate associated with surgical repair of post-MI VSR, there has been a growing preference for transcatheter septal closure in these patients.

Keywords Ventricular septal defect · Spontaneous healing · Appropriate case and device selection · Post-MI VSR · Transcatheter septal closure

Abbreviations

ADO	Amplatzer duct occluder
AI	Aortic insufficiency
AV	Atrioventricular valve
LV	Left ventricle
LVP	Left ventricular pressure
MI	Myocardial infarction
PAP	Pulmonary artery pressure
PMVSD	Perimembranous ventricular septal defect
RAP	Right atrial pressure
RV	Right ventricle
RVP	Right ventricular pressure
TEE	Transesophageal echocardiography

A. Firouzi · Z. Hosseini (✉) · M. Baay
Cardiovascular Intervention Research Center, Rajaie Cardiovascular Institute, Iran University of Medical Sciences, Tehran, Iran

A. Firouzi et al. (eds.), *Mastering Skills in Complex Cardiovascular Interventions*, https://doi.org/10.1007/978-3-031-97321-5_9

TR	Tricuspid regurgitation
TTE	Transthoracic echocardiography
VSD	Ventricular septal defect
VSR	Ventricular septal rupture

Introduction

Isolated VSDs represent approximately 37% of all congenital heart disease cases in children. However, since up to 90% of these defects close spontaneously over time, their prevalence is notably lower in adults. Congenital VSDs have been classified and categorized into four primary groups based on their location within the RV.

Type I (Infundibular, Outlet, Doubly Committed Juxta-Arterial)
This VSD is situated below the semilunar valves (aortic and pulmonary) in the outlet septum of the RV, above the crista supraventricularis. As a result, it is sometimes referred to as supracristal. It is the rarest type, comprising only 6% of all VSDs, except in the Asian population, where it accounts for approximately 30%. AV prolapse and progressive regurgitation are common due to the loss of support for the right and/or non-coronary cusps of the AV. The spontaneous closure of these defects is uncommon.

Type II (Membranous, Perimembranous)
This is the most prevalent type, representing 80% of all VSDs. It is located in the membranous septum, inferior to the crista supraventricularis. The septal leaflet of the tricuspid valve may form a "pouch," which can reduce the shunt and lead to spontaneous closure. Over time, this type can result in the development of a double-chambered RV.

Type III (Inlet or Atrioventricular Canal, AVSD Type)
This VSD is located just below the inlet valves (tricuspid and mitral) within the inlet portion of the RV septum. It accounts for only 8% of all defects and is often associated with a common AV valve. It may also involve AV septal malalignment and a straddling tricuspid valve, frequently occurring in individuals with Down syndrome.

Type IV (Muscular, Trabecular)
This VSD is found in the muscular septum, surrounded by muscle, and typically located in the apical, central, or outlet regions of the interventricular septum. These defects can be multiple, creating a "Swiss cheese" appearance. They represent up to 20% of VSDs in infants, but their incidence decreases in adults due to the high likelihood of spontaneous closure [1].

The clinical presentation of unrepaired VSDs depends on the presence of a hemodynamically significant shunt. This is directly influenced by the size of the defect, pulmonary vascular resistance, systemic vascular resistance, LV and RV systolic and diastolic function, and the presence of RV or LV outflow tract obstruction.

Small VSDs typically cause only a minimal left-to-right shunt, with no hemodynamic consequences or symptoms. Medium-sized VSDs result in moderate LV volume overload and mild pulmonary arterial hypertension, often presenting later in childhood with mild congestive symptoms. In contrast, patients with large VSDs develop congestive heart failure early in childhood due to severe LV overload and significant pulmonary arterial hypertension. Over time, approximately 10%–15% of these patients progress to Eisenmenger syndrome. These individuals may present with dyspnea, cyanosis, syncope, palpitations, and even sudden cardiac death.

Spontaneous closure of VSDs most commonly occurs during the first year of life, with up to 90% closing before the age of 8. This phenomenon is more frequent in small defects (<6 mm) and in muscular VSDs (70%–75% closure rate), compared with PMVSDs, which have a lower closure rate of 20%–30%. Spontaneous closure is particularly common in defects located in the anterior, apical, and mid-ventricular septum [2].

Acquired VSDs are rare and typically result from MI, traumatic events, or surgical procedures.

Diagnostic Workup

In small VSDs, a loud, harsh pansystolic murmur is typically heard at the left sternal border. Nonetheless, as the defect size increases, the murmur becomes softer and may eventually disappear in large VSDs, where an accentuated P2 sound is often present. On ECG, the criteria for LV hypertrophy are prominent in moderate-sized defects, while right axis deviation and RV hypertrophy criteria are observed in patients with Eisenmenger syndrome.

TTE is the primary diagnostic tool for detecting VSDs. It provides critical information, including defect size, number, location, relationship to adjacent structures, hemodynamic effects, Qp/Qs ratio, LV and RV size and function, associated anomalies, severity of AI and TR, and PAP. TTE also helps evaluate the indications for and suitability of surgical or percutaneous closure.

Treatment Approach

While surgical closure has traditionally been the standard treatment for congenital VSDs, recent advancements in transcatheter equipment, knowledge, techniques, and skills have made percutaneous closure a more common option for suitable cases. This approach is now frequently used for muscular VSDs, PMVSDs, iatrogenic VSDs, and post-MI VSDs.

Given the high success rates and low risk of major complications associated with percutaneous closure, it has become the preferred method for selected adult patients with muscular or PMVSDs, as well as for post-MI and post-traumatic VSDs.

Case 1

The patient was a 32-year-old man referred for evaluation of exertional dyspnea (New York Heart Association functional class II) and orthopnea, which had persisted for 2 years. He also reported recent episodes of nocturnal palpitations. Physical examination revealed no evidence of Eisenmenger syndrome. ECG findings indicated pronounced LV hypertrophy criteria.

Diagnostic Workup

TTE demonstrated moderate LV enlargement with mild systolic dysfunction. A large PMVSD measuring 8 mm was identified, with a significant net left-to-right shunt (Qp/Qs: 1.8). The defect was partially restricted by the septal leaflet of the tricuspid valve. The distance between the defect and the aortic cusp was 4 mm, and mild AI was noted. RV size and function were normal, with moderate TR and an estimated RV systolic pressure of 45 mm Hg. No other anomalies were observed. Given the patient's symptoms, LV volume overload, and the suitability of the defect for transcatheter closure, he was scheduled for percutaneous device closure.

Right Heart Catheterization Data

- Oxygen saturation: 95%;
- RV systolic pressure: 55/12 mm Hg;
- mRAP: 12 mm Hg;
- PAP: 55/20 mm Hg;
- Pulmonary vascular resistance: 2 WU;
- Systemic vascular resistance: 120 WU; and
- Qp/Qs ratio: 2.5.

Procedural Technique (Venous Approach) [3]

The right common femoral artery and vein were accessed, and 6 Fr sheaths were inserted. Intravenous heparin was administered to achieve an activated clotting time >200 s. Following LV injection in the left anterior oblique 60°-cranial 20° projection to visualize the defect (Fig. 9.1a), the procedure was continued under general anesthesia.

Using an antegrade approach, the VSD was crossed from the LV side using a JR catheter and a 260 cm, 0.035-in. hydrophilic guidewire. In some cases, depending

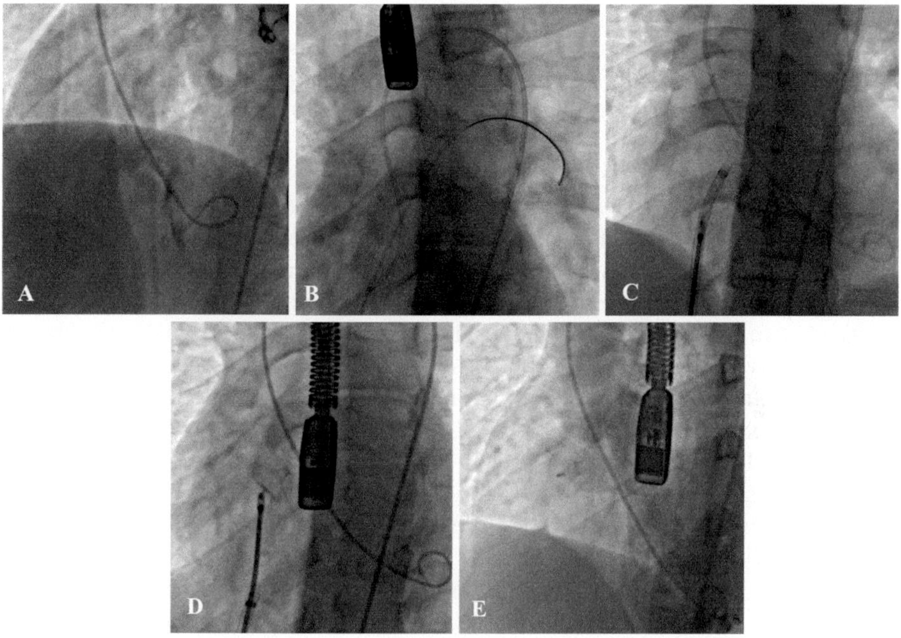

Fig. 9.1 (**a–e**) LV injection in the left anterior oblique 60°/right anterior oblique 20° projection revealed a large PMVSD with a significant left-to-right shunt (**a**). After antegrade crossing of the defect using a long hydrophilic guidewire, the wire was snared in the left pulmonary artery and exteriorized through the venous system, creating an arteriovenous loop (**b**). A long delivery sheath was advanced through the venous access using the kissing catheter technique and positioned in the ascending aorta. The distal disk was then slowly deployed in the aorta, gently passed through the aortic valve, and, with slight clockwise rotation, its proper position was confirmed by LV injection (**c**). Repeated LV injections confirmed the deployment of the waist and the RV disk (**d**). The final LV injection demonstrated the device's optimal positioning with mild residual shunting. TEE confirmed no interaction with adjacent structures

on the defect's location, alternative catheters, such as an internal mammary artery catheter or a cutoff pigtail catheter, were utilized. Once the defect was crossed with the guidewire, the catheter was advanced through it and positioned in the left pulmonary artery.

Through the venous access, a pigtail catheter without a guidewire was inserted into the same pulmonary artery where the guidewire was located (to minimize interference with the tricuspid valve chordae). The pigtail catheter was then exchanged for a JR guiding catheter. To create an arteriovenous loop, the guidewire was snared and exteriorized through the venous access (Fig. 9.1b).

A long delivery sheath (Occlutech, Germany, 10 Fr) was carefully advanced toward the RV, LV, and aortic arch using the kissing catheter technique. The dilator was then removed from the venous access, and the guidewire and end-hole catheter were withdrawn from the arterial line.

Under TEE guidance by an experienced echocardiographer, a Muscular VSD Occluder (Occlutech, Germany) measuring 12 mm was selected based on the defect size observed during ventriculography (8 mm—the largest diameter at the LV end of the defect during the diastolic phase) and the acceptable distance between the defect and the aortic cusps.

The delivery cable was loaded, and the cable was connected to the hub of the RA disk. The entire device was then retracted into the loader (slenderizing the device) and thoroughly flushed with heparinized saline. The loader was connected to the delivery sheath, and the device was advanced through the sheath.

Under simultaneous TEE and fluoroscopic guidance, the left disk was fully unsheathed in the ascending aorta. The entire system was gently pulled back through the aortic valve into the LV (Fig. 9.1c). The system was then further retracted toward the defect. After confirming the appropriate positioning of the LV disk via LV injection, the sheath was withdrawn to deploy the waist and the proximal disk (right disk) (Fig. 9.1d).

The operator must thoroughly assess the device's position and stability to ensure no compromise of surrounding structures. If there is any concern about impingement on the valves or other adjacent structures, the device should be recaptured and repositioned. Once the device's position is confirmed as correct by TEE and LV injection (Fig. 9.1e), it can be safely released.

In this case, the severity of AI and TR did not increase, and no conduction abnormalities were observed. A follow-up LV injection performed 10 min after deployment revealed only minimal residual shunting.

Twenty-four hours after the procedure, the patient reported a change in urine color (brownish) and noted icteric sclera. Follow-up laboratory findings are detailed in Table 9.1.

The patient exhibited signs of clinical intravascular hemolysis [4]. Despite 5 days of conservative management—including aggressive hydration, high-dose hydrocortisone, beta-blocker therapy, and blood transfusions—the patient's hemolysis progressed and remained refractory. This prompted the heart team to schedule him for device removal. Fortunately, by the seventh day, the patient's condition began to improve, and his urine color gradually changed from Coca-Cola-like to normal. After 10 days, he was discharged home in good condition, with no evidence of residual shunting on TTE.

Case 2

The patient was a 45-year-old woman with a known history of tetralogy of Fallot. She had undergone total surgical correction 30 years prior and biologic pulmonary valve replacement 5 years earlier. She presented with frequent, recently onset palpitations and presyncope symptoms. Holter ECG monitoring confirmed sustained monomorphic ventricular arrhythmia and episodes of complete heart block. As a

Table 9.1 Hospitalization course

Variables	Before the procedure	1st Day	2st Day	4st Day	5st Day	7st Day	9st Day	10st Day
Hemoglobin	14	12	10	7.5	9	11	11.5	11.8
Lactate dehydrogenase	100	1100	5850	6700	7300	5550	2100	1000
Indirect Bilirubin	0.8	4.5	6	9	11	8	6	4
Blood Urea Nitrogen	16	22	32	56	68	50	45	30
Creatinine	0.8	1	1.8	3	3.8	2.5	1.5	1
Reticulocyte count	-	-	%2.2	-	%3.8	-	-	%3

result, the patient was deemed a candidate for implantable cardioverter-defibrillator insertion.

ECG findings revealed right axis deviation, complete right bundle branch block with a QRS duration of 180 ms, and a strain pattern.

Diagnostic Workup

TTE revealed mild to moderate LV enlargement with moderate systolic dysfunction (ejection fraction: 40%) and moderate RV enlargement with moderate systolic dysfunction. The tricuspid aortic valve showed no evidence of stenosis, with mild AI and a normal aortic root size. The biologic pulmonary valve demonstrated normal motion with acceptable hemodynamic function, no leakage, and no subvalvular or supravalvular gradient.

A large residual VSD was identified at the site of the previous surgical patch, with a significant left-to-right shunt. The distance between the defect and the aortic valve was 3 mm. No other anomalies were observed.

Given the plan for transvenous implantable cardioverter-defibrillator insertion and the presence of an intracardiac shunt with a risk of systemic embolization, the patient was scheduled for percutaneous VSD closure.

Procedural Technique (Venous Approach)

The right common femoral artery and vein were accessed, and 6 Fr sheaths were inserted. Intravenous heparin was administered to achieve an activated clotting time >200 s. LV injection in the left anterior oblique 60°-cranial 20° projection revealed a conical-shaped PMVSD at the site of the previous surgical patch (Fig. 9.2a).

Under general anesthesia, the VSD was crossed from the LV side using an antegrade approach with a JR catheter and a 260 cm, 0.035-in. hydrophilic guidewire. The distal end of the guidewire was positioned in the superior vena cava. To create an arteriovenous circuit, the guidewire was snared and exteriorized through the venous access.

A long delivery sheath (Occlutech, Germany, 9 Fr) was carefully advanced toward the RV, LV, and aortic arch using the kissing catheter technique. Under TEE guidance by an experienced echocardiographer, an ADO type 1 measuring 12–10 mm was selected based on the defect's shape and size (8 mm—the largest diameter at the LV end of the defect during the diastolic phase) and the acceptable distance between the defect and the aortic cusps.

Under simultaneous TEE and fluoroscopic guidance, the left retention disk was fully unsheathed in the ascending aorta. The entire system was gently pulled back through the aortic valve into the LV and positioned at the defect. After confirming the appropriate placement of the LV disk within the defect's pouch via LV

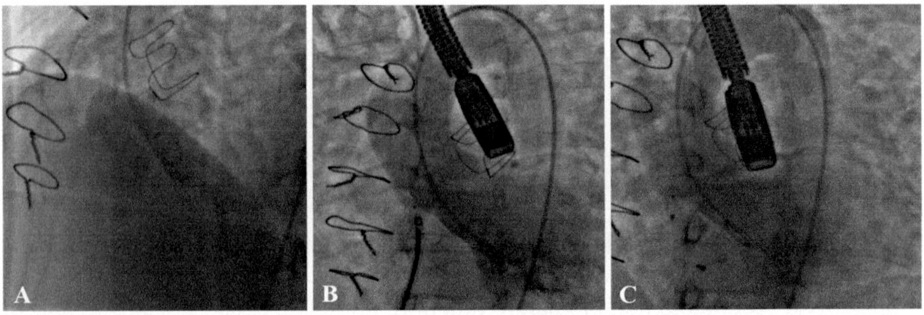

Fig. 9.2 (**a**–**c**) LV injection revealed a large, conical-shaped PMVSD at the site of the previous surgical patch (**a**). After antegrade wiring of the defect and creating an arteriovenous loop, a long delivery sheath was advanced through the venous access toward the ascending aorta. Given the VSD's shape, an ADO type 1 device (12–10 mm) was selected. The retention disk was deployed in the pouch neck, followed by the gentle deployment of the proximal disk (**b**). After confirming the device's proper position via LV injection and TEE guidance, the device was released with no residual shunting (**c**)

angiography (Fig. 9.2b), the sheath was retracted to deploy the waist and the proximal disk. After confirming the appropriate position and orientation of the device, with no interaction with adjacent structures (aortic valve and tricuspid valve) as assessed by TEE and LV injection, the device was released. No residual shunting was observed (Fig. 9.2c).

Case 3

A 49-year-old woman with a recent anterior ST-segment elevation MI was referred to the emergency department with pulmonary edema. Seven days prior, she had undergone primary percutaneous coronary intervention on the left anterior descending artery at another center.

On physical examination, she presented in cardiogenic shock (blood pressure: 70/30 mm Hg, pulse rate: 120 bpm, respiratory rate: 22 bpm, and oxygen saturation: 78%). A harsh, holosystolic murmur with a thrill was audible at the left sternal border, accompanied by an S3 gallop. ECG findings showed prominent Q waves in leads V_1–V_3 with T-wave inversion.

Diagnostic Workup

TTE revealed moderate LV enlargement with severe systolic dysfunction (ejection fraction: 15%). The apex was aneurysmal, with akinesia of the anterior, anterolateral, and anteroseptal segments. Moderate LV diastolic dysfunction was also noted. The RV was mildly enlarged with moderate systolic dysfunction.

A large, slit-like apical VSR measuring 15 mm was observed, with a significant left-to-right shunt. Additional findings included moderate functional mitral regurgitation, moderate TR with a peak gradient of 75 mm Hg, severe pulmonary hypertension (mPAP: 45 mm Hg), and no pericardial effusion (Fig. 9.3).

Bedside pulmonary capillary wedge pressure catheter measurements revealed the following:

- PAP: 80/30 mm Hg;
- Pulmonary capillary wedge pressure: 35 mm Hg;
- RVP: 80/20 mm Hg; and
- mRAP: 18 mm Hg.

Given the patient's critical condition, an intra-aortic balloon pump was emergently implanted at the bedside. High-dose diuretics, vasopressors, and inodilators were initiated.

The patient was in a low cardiac output state with multi-organ dysfunction, including a decreased level of consciousness, acute renal failure, and acute liver failure. Although emergent surgical repair of the VSR was considered, the heart

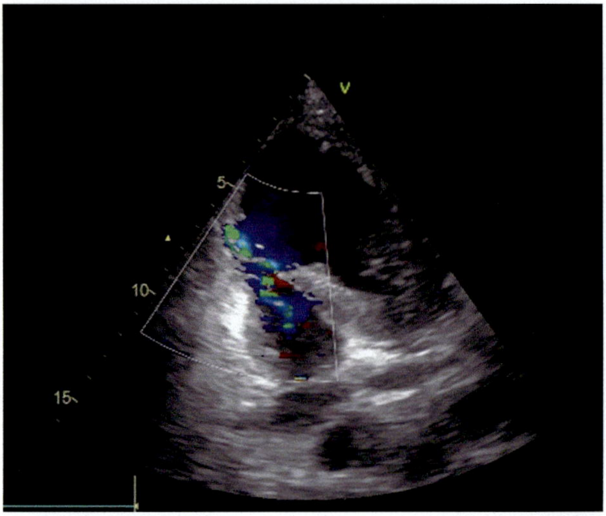

Fig. 9.3 The image shows the patient's aneurysmal apex and a large slit-like post-MI VSR with a significant left-to-right shunt

team opted for transcatheter septal closure due to the high surgical mortality risk and the patient's relatively young age.

Within 48 h, the patient's hemodynamic status improved, and she was deemed a candidate for transcatheter septal closure.

Right Heart Catheterization Data

- Blood pressure: 85/40 mm Hg;
- LVP: 80/30 mm Hg;
- RVP: 80/20 mm Hg;
- PAP: 80/30 mm Hg;
- mRAP: 20 mm Hg;
- Qp/Qs Ratio: 2.7;
- Mixed venous saturation: 45%; and
- Cardiac index: 1.7 L/min/m^2.

Procedural Technique (Trans-Septal Closure)

The right femoral artery and right internal jugular vein were accessed, and 6 Fr and 8 Fr sheaths were inserted, respectively. Intravenous heparin was administered to achieve an activated clotting time >200 s. Under general anesthesia, the VSR was thoroughly assessed using TEE and LV angiography, which revealed a large apical VSR (Fig. 9.4a).

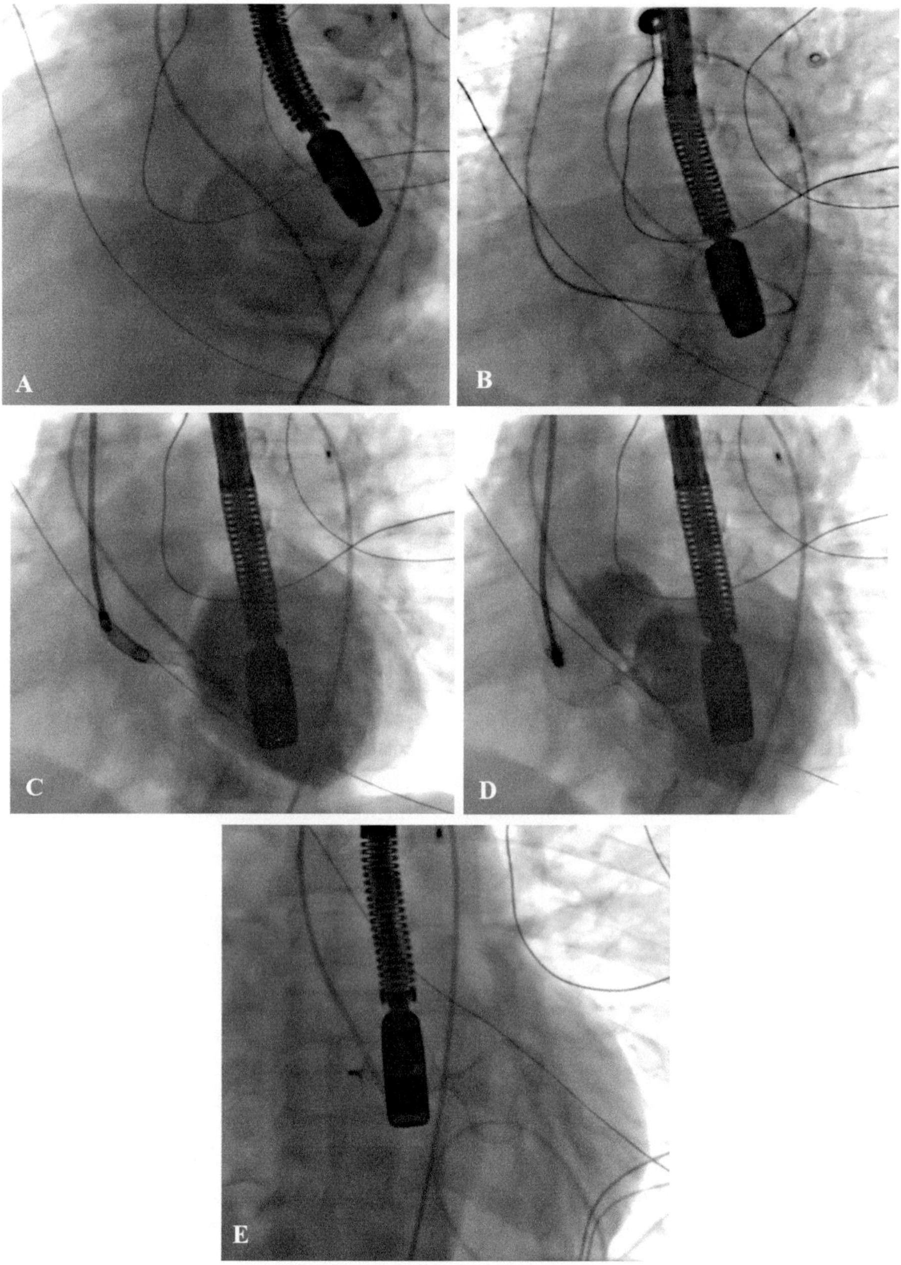

Fig. 9.4 (**a**–**e**) LV injection revealed a significant post-MI VSR with considerable left-to-right shunting (**a**). Following the wiring of the defect and its exteriorization through the venous access to create an arteriovenous circuit (**b**), a Figulla II Occlutech device measuring 30 mm was selected based on the defect's size. By retracting the delivery sheath, the LV disk was positioned, and with additional traction on the device, the RV disk was deployed (**c**, **d**). Both disks were well-expanded, demonstrating proper device positioning with only mild residual shunting at the conclusion (E).

The defect was crossed from the LV side using a JR catheter and a 0.014-in. coronary guidewire (BMW, 300 cm). After crossing the defect, the distal tip of the guidewire was snared in the superior vena cava to create an arteriovenous circuit. The guidewire was then exteriorized through the venous access, and the JR catheter was also exteriorized simultaneously. The guidewire was subsequently exchanged for a long 0.035-in. hydrophilic guidewire (Fig. 9.4b).

A kink-resistant delivery sheath (Lifetech, 12 Fr) was carefully advanced from the venous access toward the RV and LV using the kissing catheter technique. The dilator was then removed, and the guidewire and end-hole catheter were withdrawn from the arterial line.

Under TEE guidance by an experienced echocardiographer, an atrial septal occluder (Figulla II Occlutech) measuring 30 mm was selected based on the defect size (20 mm—the largest diameter at the LV end during the diastolic phase). The device was attached to the delivery cable, de-aired, and advanced through the long sheath under simultaneous TEE and fluoroscopic guidance.

The distal disk (left disk) was gently deployed on the LV side of the VSR by retracting the delivery sheath (Fig. 9.4c). The entire system was then pulled back toward the defect, and the sheath was further retracted to deploy the waist. After confirming the correct position of the device using TEE and LV injection, the proximal disk (right disk) was deployed (Fig. 9.4d).

The operator meticulously assessed the device's position and stability to ensure no compromise of surrounding structures. Once the device's proper placement was confirmed by TEE and LV injection, it was released. Mild residual shunting was noted at the end of the procedure (Fig. 9.4e).

Fortunately, the patient was discharged 5 days after the procedure without complications. She remained symptom-free during 6-month and 2-year clinical follow-ups.

Case 4

The patient was a 27-year-old man who presented with recent penetrating trauma (a stabbing wound) to his anterior chest wall. Due to his cardiogenic shock status, he underwent an emergent sternotomy, during which a rupture of the RV free wall and tamponade were diagnosed. Successful RV repair and pericardial drainage were performed, and he was discharged in stable condition. During clinical follow-up, a harsh pansystolic murmur (grade III) was noted at the left sternal border, and TTE confirmed the presence of a moderate-sized muscular VSD (Fig. 9.5). The ECG was normal.

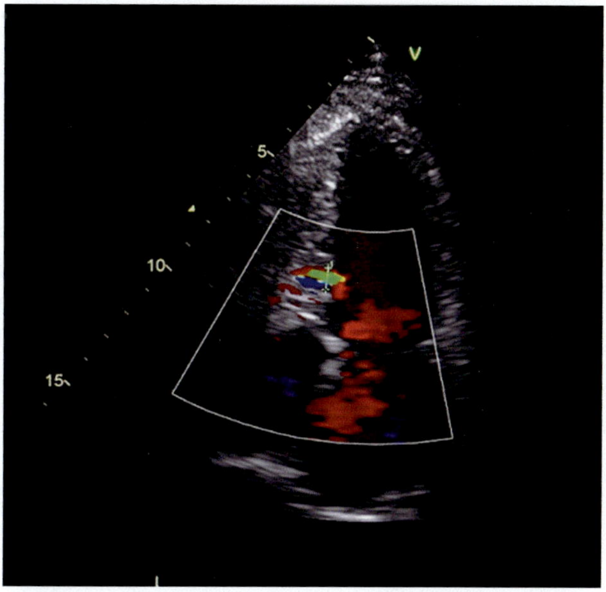

Fig. 9.5 TTE illustrated a small to moderately-sized muscular VSD in the posterobasal part of the septum

Diagnostic Workup

TEE indicated mild LV enlargement with normal systolic function (ejection fraction: 55%). The RV size was normal, but there was mild-to-moderate systolic dysfunction. The mitral valve apparatus appeared normal, with mild mitral regurgitation noted. The aortic valve was normal, showing no visible dissection flap. The tricuspid valve apparatus was also normal, with mild TR (TR gradient: 30 mm Hg). A small to moderate-sized VSD measuring 6 mm was identified in the posterobasal region of the muscular interventricular septum, accompanied by significant left-to-right shunting. The distance from the defect to the mitral valve and the coronary sinus was deemed acceptable. No pericardial effusion was present. Cardiac computed tomography angiography confirmed that the defect was suitable for percutaneous device closure. In this case, although the patient was asymptomatic, the LV size was enlarged, and the net left-to-right shunt was significant. Given the high risk associated with reoperation, the heart team decided to proceed with percutaneous device closure.

Right Heart Catheterization Data

- Blood Pressure 120/70 mm Hg;
- LVP: 120/15 mm Hg;
- RVP: 40/10 mm Hg;

- PAP: 40/25 mm Hg;
- mRAP: 10 mm Hg; and
- Qp/Qs ratio: 1.9.

Procedural Technique (Venous Approach)

The right common femoral artery and vein were accessed, and 6 Fr sheaths were inserted. Intravenous heparin was administered to achieve an activated clotting time >200 s. LV injection in the left anterior oblique 60°-cranial 20° projection was performed to characterize the defect's shape and size (Fig. 9.6a).

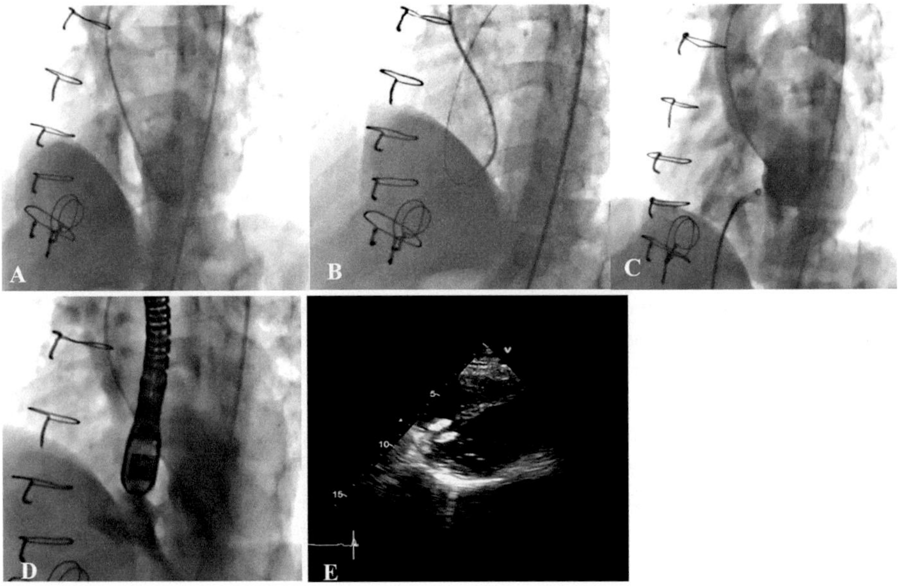

Fig. 9.6 LV injection in the left anterior oblique projection revealed a moderate-sized muscular VSD (**a**) with significant left-to-right shunting (**a**). Using an antegrade approach, the defect was wired with a hydrophilic wire, which was then snared and exteriorized through the jugular vein to create an arteriovenous loop (**b**). A long delivery sheath (Occlutech, Germany, 10 Fr) was carefully advanced toward the RV, LV, and aortic arch using the kissing catheter technique. Under TEE guidance, a muscular VSD occluder (Occlutech, Germany) measuring 10 mm was selected based on the defect size (6 mm—the largest measurement at the LV end of the defect during diastole) and the acceptable distance between the defect and the aortic cusps, mitral valve, and coronary sinus. Under TEE and fluoroscopic guidance, the left disk was fully unsheathed in the ascending aorta and gently pulled back through the arteriovenous loop toward the LV. The entire system was then retracted toward the defect, and after confirming the appropriate position of the LV disk via LV injection (**c**), the sheath was retracted to deploy the waist and proximal disk (**d**). Follow-up TTE demonstrated the proper positioning of the device with no residual shunt (**e**)

Under general anesthesia, the VSD was crossed from the LV side using an antegrade approach with a JR catheter and a 260 cm, 0.035-in. hydrophilic guidewire. After crossing the defect, the catheter was advanced through it, and the wire tip was positioned in the inferior vena cava. To create an arteriovenous loop, the guidewire was snared and exteriorized through the venous access (Fig. 9.6b).

A long delivery sheath (Occlutech, Germany, 10 Fr) was carefully advanced toward the RV, LV, and aortic arch using the kissing catheter technique. Under TEE guidance by an experienced echocardiographer, a muscular VSD Occluder (Occlutech, Germany) measuring 10 mm was selected based on the defect size (6 mm—the largest diameter at the LV end during the diastolic phase) and the acceptable distance between the defect and the aortic cusps, mitral valve, and coronary sinus.

Under simultaneous TEE and fluoroscopic guidance, the left disk was fully unsheathed in the ascending aorta. The entire system was gently pulled back through the aortic valve into the LV and positioned at the defect. After confirming the appropriate placement of the LV disk via LV injection (Fig. 9.6c), the sheath was retracted to deploy the waist and the proximal disk (right disk) (Fig. 9.6d).

Finally, after confirming the device's proper position and orientation, with no interaction with surrounding structures and stable rhythm, the device was released. A tiny residual shunt was noted at the end of the procedure (Fig. 9.6e).

Discussion

VSDs are the most common congenital heart disease, with PMVSDs being the most prevalent subtype. Although spontaneous closure rates are high, surgical repair may be necessary in some adult patients (Table 9.1).

With advancements in cardiac imaging modalities and techniques, interventional PMVSD closure has gained increasing acceptance. The availability of various occlusive device designs, coupled with the high success rates of percutaneous closure for muscular VSDs over the past decade, has further supported this approach. However, percutaneous closure of PMVSDs remains a challenging and high-risk procedure due to the variable anatomical morphology, proximity to valves and conduction tissues, and the complexity of the manipulation process.

The key to improving outcomes and minimizing complications lies in careful patient and device selection, as well as an accurate defect sizing strategy.

Compared with surgical repair, the percutaneous approach avoids sternotomy and offers potential advantages such as lower morbidity, faster recovery, shorter hospital stays, and reduced costs. Ideally, device closure should be straightforward, with low rates of residual shunting. Special attention must be given to preserving the function of the aortic and tricuspid valves and avoiding damage to the conduction system.

Left ventriculography is performed at a 55°–60° left anterior oblique with 20° cranial projection to define the defect. This is combined with intraoperative TEE to accurately determine the VSD's location, shape, depth, and its relationship with adjacent structures such as the aortic and tricuspid valves. Additionally, the following measurements are obtained:

- VSD size: The smallest diameter of the defect, typically located on the RV side;
- VSD ampulla size: The largest diameter of the defect, usually located on the LV side; and
- VSD length.

The defect's entry diameter is measured on angiography during the largest diastolic phase on the LV side.

The most preferred closure technique is the antegrade approach (venous approach), which involves creating an arteriovenous circuit. In this method, the distal disk is initially deployed on the LV side of the VSD. Caution is essential when passing the wire and catheter through the tricuspid valve to establish the arteriovenous circuit, as an aneurysm is often adjacent to or part of the tricuspid valve apparatus.

In contrast, the retrograde approach (arterial approach) involves closing the RV side of the defect first without creating an arteriovenous circuit.

One of the most challenging aspects of percutaneous VSD closure is selecting the appropriate device, particularly for PMVSDs. Undersized devices increase the risk of device embolization and residual shunting, while oversized devices may cause damage to adjacent structures, leading to complications such as complete heart block or injury to the aortic or tricuspid valves.

Today, a variety of devices are used (both on-label and off-label) depending on the defect's location, size, orientation, presence of an aneurysm, and proximity to surrounding structures and the conduction system. These include the following:

- Muscular septal occluder devices;
- Symmetrical and asymmetrical double-disk PMVSD occluder devices;
- Duct occluder devices (types I and II); and
- Post-MI VSD occluder devices.

Although muscular septal occluder devices (symmetric double-disk with a 7 mm central waist length) are primarily used for muscular VSDs (via the trans-femoral vein for more basal defects or the trans-jugular vein for more apical defects), they are also frequently chosen for closing large PMVSDs with at least a 4 mm subaortic rim (section "Case 1"). The device waist is selected based on the LV entry diameter without oversizing. During deployment, muscular septal occluder devices are more controllable and easier to maneuver compared to PMVSD-specific devices. Still, careful attention must be paid to the aortic valve and the atrioventricular structures to avoid complications.

For PMVSDs with an aneurysm or those with a ductal appearance, the conventional double-disk device must be large enough to cover the entire aneurysm. Still, this approach may increase the risk of complications.

Nonetheless, duct occluder devices (type I) are often more suitable for PMVSDs with an aneurysm. The retention disk can be positioned entirely within the aneurysm, while the cylindrical portion of the device secures the opening of the aneurysm on the RV side. This design minimizes contact with the aortic valve and reduces pressure on the ventricular septum (section "Case 2").

The device size is selected such that the right disk diameter (pulmonary end) is 2 mm (±0.5 mm) larger than the smallest VSD diameter. This ensures that the left retention skirt (aortic end of the device) is fully accommodated within the aneurysm, particularly when the defect depth allows the retention disk to reach the RV exit.

In some PMVSD cases where the distance between the upper border of the defect and the aortic cusp is <3 mm, ADO type II devices may be more suitable due to their softer profile. Recent advancements in the design of dedicated PMVSD devices and coils have facilitated percutaneous closure of even those PMVSDs located 1–2 mm from the aortic valve, achieving high success rates and low complication rates.

One of the rare complications of transcatheter device closure is clinical intravascular hemolysis, which occurs in 1%–3% of cases. The exact mechanisms underlying this phenomenon remain unclear, but potential causes include incomplete closure with residual shunting, abnormal device configuration, and mechanical exposure of the device to blood flow.

The optimal treatment strategy for clinical hemolysis following device closure depends on several factors, including the following:

- The degree of hemolysis;
- The magnitude of the hemoglobin drop;
- The duration of gross hematuria;
- The number of packed red blood cell units required for transfusion;
- The degree of elevation in creatinine and bilirubin levels; and
- The severity of clinical symptoms.

In most cases, severe hemolysis is self-limiting, with symptoms and signs resolving within 5–7 days with conservative management. Medical therapies include the following:

- Massive hydration;
- Beta-blockers (to reduce shear stress);
- Iron supplementation;
- Blood transfusion;
- Pentoxifylline (to improve blood viscosity and erythrocyte deformability); and
- Corticosteroids.

The role of these medications in treating hemolysis remains unclear; they may have a placebo effect, as hemolysis often resolves spontaneously within several days due to the reduction of residual shunts and thrombosis of the device mesh.

However, in patients with refractory hemolysis, invasive approaches—such as transcatheter closure of the residual shunt with another device or surgical device removal—should be considered if:

- Two weeks of comprehensive medical therapy fails to resolve elevated renal function markers (which may require hemodialysis) or bilirubin levels, or
- There is ongoing hemoglobin drop requiring recurrent blood transfusions or persistent hematuria.

In section "Case 1", severe clinical hemolysis improved by the seventh day with conservative management.

One of the uncommon causes of acquired VSDs is post-MI VSR. Post-MI VSR is one of the most catastrophic mechanical complications associated with high morbidity and mortality. The incidence of VSR has declined from 1%–3% in the pre-reperfusion era to 0.17%–0.31% following primary percutaneous coronary intervention for ST-segment elevation MI.

Based on the Becker classification (types I–III), VSR can present within hours, 3–5 days, or even later after the index MI. Risk factors include female gender, hypertension, no history of previous MI, and delayed reperfusion. VSR occurs with equal frequency in both anterior and inferior MIs.

In anterior MI, the VSR typically occurs in the apical segment, presenting as a simple, slit-like defect at the same level on both sides of the septum, with thinner rims. In contrast, VSR in inferior MI is more complex, often following a serpiginous path.

Depending on the degree of left-to-right shunt, LV and RV volume overload and dysfunction, and PAP, patients may present with a range of symptoms—from relatively stable hemodynamics to pulmonary edema and frank cardiogenic shock.

If pharmacologic therapy is not tolerated or fails to achieve hemodynamic stability, mechanical circulatory support devices—such as an intra-aortic balloon pump, as used in section "Case 3"—should be rapidly initiated as a bridge to definitive repair. In most cases, surgery should not be delayed for patients with a correctable lesion who agree to an aggressive management strategy and require both pharmacologic and mechanical support.

Be that as it may, in a subset of patients with stable hemodynamics, surgery may be postponed for 2–4 weeks to allow partial healing of the infarcted tissue. Determining the optimal timing for surgery requires a multidisciplinary "heart team" approach, integrating multiple aspects of the clinical course and the anatomy of the mechanical complication.

Conservative therapy invariably leads to death, with a 90–95% mortality rate within 2 months of diagnosis without intervention. Recently, transcatheter septal closure has emerged as a viable option, particularly for patients deemed inoperable and whose anatomy is suitable for closure device application.

Although dedicated post-MI VSR devices have been developed in recent decades, their limited available sizes and the reduced thickness of the perforation site and surrounding rims pose challenges. Based on our experience, the most appropriate occlusive devices—especially in the acute phase—are the largest available atrial septal occluder devices, regardless of the defect size observed on echocardiography. Device selection should consider the LV and RV sizes and the distance between the defect and the aortic valve.

If a percutaneous technique is being considered, it is essential to thoroughly evaluate the anatomy of the lesion, the surrounding tissue rims, and the defect's location relative to potential percutaneous access routes, as these factors can significantly influence device delivery.

In CSE III, the patient was discussed during a multidisciplinary team meeting. Given the high surgical risk, transcatheter closure of the VSR was performed using a 30 mm atrial septal occluder (Figulla II Occlutech) via the jugular vein, resulting in mild residual shunting.

Another cause of acquired VSDs is chest trauma, whether penetrating or blunt. Despite significant advancements in trauma care, over 90% of patients may die before reaching the emergency room [5]. The most vulnerable parts of the heart following trauma include the RV, LV, interventricular septum, tricuspid valve, aortic valve, conduction tissue, and pericardium [6].

VSDs occur in 1%–5% of penetrating chest trauma cases. Apical VSDs with irregular, ragged walls are the most common type of traumatic VSDs, although VSDs can also result from penetrating injuries to any part of the septum. Clinical presentation varies widely, ranging from stable conditions to critical hypotension, shock, and cardiac tamponade.

The diagnosis of VSDs resulting from stab-wound trauma may not be identified during the initial evaluation in the emergency department. Closure of VSDs in this context is recommended for patients with a large defect (>10 mm) or a significant left-to-right shunt.

In our case, the unusual location of the VSD initially led us to consider a congenital VSD diagnosis. The percutaneous closure procedure in this scenario follows the same principles as in other cases.

Conclusions

Percutaneous closure of a VSD is a challenging procedure due to the variability in anatomical morphology, proximity to valves and conduction systems, and the complexity of the manipulation process. The key to improving the outcomes of this treatment while minimizing complications—such as complete heart block, device embolization, valvular malfunction, and hemolysis—lies in careful case and device selection, as well as an accurate defect sizing strategy.

References

1. Lopez L, Houyel L, Colan SD, Anderson RH, Béland MJ, Aiello VD, et al. Classification of ventricular septal defects for the eleventh iteration of the International Classification of Diseases—striving for consensus: a report from the International Society for Nomenclature

of Paediatric and Congenital Heart Disease. Ann Thorac Surg. 2018;106:1578–89. https://doi.org/10.1016/j.athoracsur.2018.06.020. PMID: 30055140
2. Zhang J, Ko JM, Guileyardo JM, Roberts WC. A review of spontaneous closure of ventricular septal defect. Proc (Baylor Univ Med Cent). 2015;28:516–20. https://doi.org/10.1080/08998280.2015.11929324. PMID: 26424959
3. Firouzi A, Hosseini Z. Transcatheter perimembranous VSD device closure. In: Maleki M, Alizadehasl A, editors. Case-based clinical cardiology. London: Springer; 2021. p. 607–14. https://doi.org/10.1007/978-1-4471-7496-7_66.
4. Firouzi A, Alemzadeh-Ansari MJ, Hosseini Z, Shafiei A, Gholoobi A, Shakerian F, et al. Management of severe clinical intravascular hemolysis following percutaneous transcatheter closure in adulthood. J Clin Exp Cardiolog. 2019;11:684. https://doi.org/10.35248/2155-9880.19.11.684.
5. Carr JA, Buterakos R, Bowling WM, Janson L, Kralovich KA, Copeland C, et al. Long-term functional and echocardiographic assessment after penetrating cardiac injury: 5-year follow-up results. J Trauma Acute Care Surg. 2011;70:701–4. https://doi.org/10.1097/TA.0b013e3181f762d9. PMID: 21610363
6. Wani ML, Ahangar AG, Lone GN, Singh S, Dar AM, Bhat MA, et al. Penetrating cardiac injury: a review. Trauma Mon. 2012;17:230–2. https://doi.org/10.5812/traumamon.3461. PMID: 24350102

Chapter 10
Percutaneous Transcatheter Closure of Patent Ductus Arteriosus

Zahra Khajali, Ata Firouzi, and Zahra Hosseini

Abstract A PDA is a congenital anomaly resulting from the persistence of the distal portion of the left sixth aortic arch. The persistence of the duct after birth can have detrimental effects on the cardiovascular and pulmonary systems, depending on the size of the duct and the volume of blood flow through it. The resulting volume overload can lead to enlargement and failure of the left cardiac chambers, as well as pulmonary artery hypertension.

In adults, percutaneous intervention is often preferred over surgical closure due to factors such as ductal calcification, aortic fragility, and elevated PAP. Approved devices for PDA closure include ADO types I and II, as well as certain coils. However, the variability in the shape and size of the duct in some cases makes closure with standard devices challenging. As a result, the off-label use of other devices, such as vascular plugs, muscular ventricular septal occluders, and atrial septal occluders, may be necessary.

Keywords Patent ductus arteriosus · Left cardiac chamber enlargement
Pulmonary hypertension · Percutaneous closure · Amplatzer duct occluder device (types I and II) · Coils · Muscular ventricular septal occluder device

Abbreviations

ADO	Amplatzer duct occluder
PAP	Pulmonary artery pressure
PDA	Patent ductus arteriosus
VSD	Ventricular septal defect

Z. Khajali · A. Firouzi · Z. Hosseini (✉)
Cardiovascular Intervention Research Center, Rajaie Cardiovascular Institute, Iran University of Medical Sciences, Tehran, Iran

A. Firouzi et al. (eds.), *Mastering Skills in Complex Cardiovascular Interventions*, https://doi.org/10.1007/978-3-031-97321-5_10

Introduction

A PDA is a congenital structure resulting from the persistence of the distal portion of the left sixth aortic arch. This anomaly creates a communication between the proximal left pulmonary artery and the descending thoracic aorta, just distal to the origin of the left subclavian artery. The duct plays a crucial role in fetal circulation but typically closes within the first 48–72 h after birth.

PDAs account for 6%–11% of all congenital heart defects. While most PDAs are diagnosed and treated in infancy or early childhood, their prevalence in adulthood is approximately 0.05%. Persistent patency of the duct after birth can have adverse effects on the cardiovascular and pulmonary systems, depending on the size of the duct and the volume of blood flow through it. Small, silent PDAs are often discovered incidentally during routine evaluations [1].

A continuous murmur with a thrill is typically heard over the left upper sternal border. Larger shunts may present with clinical manifestations such as exertional dyspnea, orthopnea, palpitations, syncope, signs and symptoms of pulmonary hypertension, right ventricular failure, and differential hypoxemia or cyanosis in cases of irreversible pulmonary vascular disease with reverse shunting [2].

In infants and children, PDAs are often associated with other congenital heart defects, whereas in adults, they are usually an isolated finding. The volume overload caused by a PDA can lead to left ventricular and left atrial enlargement, heart failure, and pulmonary artery hypertension. Additionally, small PDAs may increase the risk of endarteritis.

PDAs should be closed in all cases except for silent PDAs and large PDAs that have led to Eisenmenger syndrome [3]. Percutaneous catheter-based interventions are usually the preferred treatment, except in cases of complex PDA anatomy or active endarteritis. In adults, percutaneous interventions are particularly favored over surgery due to factors such as ductal calcification, aortic fragility, and elevated PAP.

Treatment Approach

In all patients, baseline transthoracic echocardiography should be performed to thoroughly evaluate the size and function of the left and right ventricles, the mean PAP, the shape and size of the PDA, the length of the ductal tunnel, the dimensions of the aortic and pulmonic ends, and the Qp/Qs ratio. This imaging modality is also essential for identifying any associated congenital anomalies.

During angiography, hemodynamic parameters are carefully assessed. Diagnostic angiography is performed using left and right anterior oblique as well as lateral projections to evaluate shunt severity, the mean PAP, pulmonary vascular resistance, systemic vascular resistance, cardiac output, the Qp/Qs ratio, and PDA type (to guide device selection).

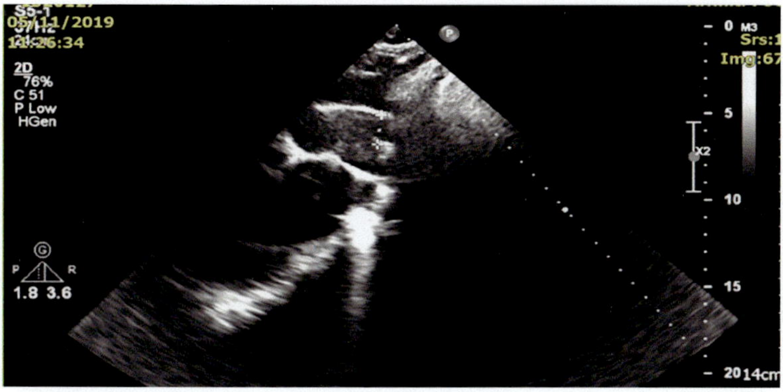

Fig. 10.1 The suprasternal view shows a large PDA

The default strategy is the antegrade approach, which involves accessing the PDA via the right femoral vein, inferior vena cava, right atrium, left pulmonary artery, and descending aorta, using a straight-tipped 0.035-in. guidewire. In rare cases where the antegrade approach is unsuccessful, the retrograde technique is employed.

The choice of device type and size is based on the operator's judgment and experience, the morphology of the duct, device accessibility, and cost considerations.

The devices approved for PDA closure include ADO I, ADO II, and certain types of coils. Nonetheless, the variability in the shape and size of the duct in some cases reduces the feasibility of closure with standard devices (Fig. 10.1). As a result, alternative devices such as vascular plugs, muscular VSD occluders, and atrial septal defect occluders may be used off-label.

Muscular VSD occluders, with their higher waist-to-disc ratios and larger maximum waist diameters, are particularly suitable for larger PDAs with diameters exceeding 14 mm. Key factors in preventing device embolization include the position of the waist on the duct, the relationship between the retention disc and the duct/aorta, and the appropriate sizing of the device.

Case 1

A 29-year-old woman presented to our clinic for evaluation of valvular heart disease. She reported mild dyspnea, which she had not previously noticed. On physical examination, continuous machinery-like murmurs were detected. An ECG revealed signs of biventricular hypertrophy, particularly left ventricular volume overload.

Echocardiography identified a large PDA (Fig. 10.1), moderate-to-severe mitral regurgitation due to mitral valve prolapse, and severe pulmonary arterial hypertension. Cardiac catheterization confirmed a significant left-to-right shunt through the

PDA and demonstrated pulmonary artery vasoreactivity. Based on these findings, the decision was made to close the PDA using a device to improve PAP and reduce the severity of mitral regurgitation.

Cardiac Catheterization and Procedure

Hemodynamic and pressure studies confirmed a left-to-right shunt through the PDA, along with elevated pulmonary pressure and right ventricular systolic pressure (Table 10.1).

Left ventricular injection revealed severe left ventricular enlargement, mild systolic dysfunction, and moderate mitral regurgitation. Aortic isthmus injection demonstrated a large PDA (Fig. 10.2a), while aortic root injection showed normal coronary arteries and no evidence of aortic insufficiency. A 20-min vasoreactivity test using 100% oxygen resulted in a significant reduction in the mean PAP.

The presence of a significant left-to-right shunt, pulmonary vasoreactivity, and moderate mitral regurgitation led to the decision to close the PDA with a device. This approach aimed to delay the need for surgical mitral valve repair and reduce PAP, thereby lowering the surgical risk for future interventions. Cefazolin (1000 mg) was administered intravenously, and a guidewire was advanced through the catheter from the femoral vein to the inferior vena cava, then to the right atrium, right ventricle, and left pulmonary artery, before crossing the PDA into the descending aorta. The stiff wire was threaded through the catheter, after which both the venous sheath and the catheter were removed. Subsequently, an introducer sheath was advanced over the stiff guidewire into the descending aorta, and the guidewire was extracted. A muscular VSD occluder (18 mm) (Occlutech) was then mounted on a delivery cable and advanced through the introducer sheath across the PDA into the aorta.

Table 10.1 The patient's hemodynamic data during right heart catheterization

	Pressure (mm Hg)	Saturation
Left ventricle	120/0-10	95
Right ventricle	95/0-7	70
Pulmonary artery Pressure	95/40	85
Ascending aora	120/80	95
Descending aora	120/80	95

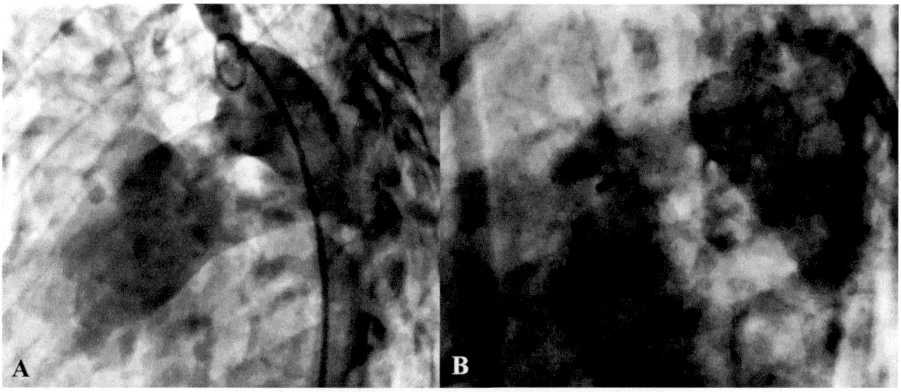

Fig. 10.2 (**a**, **b**) The aortogram shows a large PDA. The post-PDA-closure aortogram demonstrates a mild residual flow and the position of the device in the duct

Following this, the first aortic disc was opened, the waist of the device was gently deployed on the duct, and the pulmonic disc was opened. The deployment site was verified using a gentle push-pull maneuver, and the device was released. Angiography of the descending aorta confirmed the proper positioning of the device (Fig. 10.2b). The defect closure was performed using the largest available device (an 18 mm muscular VSD).

Case 2

A 19-year-old female patient with a known history of congenital heart disease and irregular follow-ups presented to our clinic with ascites and severe dyspnea. She was subsequently admitted to our hospital. The patient had not received any prior medication or management. Physical examination revealed differential cyanosis, ascites, and very weak pulses in the lower limbs. An ECG indicated atrial fibrillation and showed evidence of biventricular hypertrophy in the precordial leads. A chest X-ray demonstrated severe cardiomegaly. Echocardiography revealed significant biventricular dysfunction, a large VSD with a left-to-right shunt, a large PDA with a right-to-left shunt, and severe narrowing of the aorta distal to the PDA. Computed tomography angiography subsequently confirmed the echocardiographic findings and identified significant preductal coarctation of the aorta (Fig. 10.3a). Catheterization was performed for hemodynamic evaluation and potential interventions. The hemodynamic data are presented in Table 10.2. Based on these findings and the patient's pulmonary vasoreactivity following oxygen therapy, we scheduled her for PDA closure and coarctation stenting during the same session.

The favorable anatomy of the coarctation allowed for a simultaneous procedure using a covered stent to address both the coarctation and the PDA (Fig. 10.3b). A

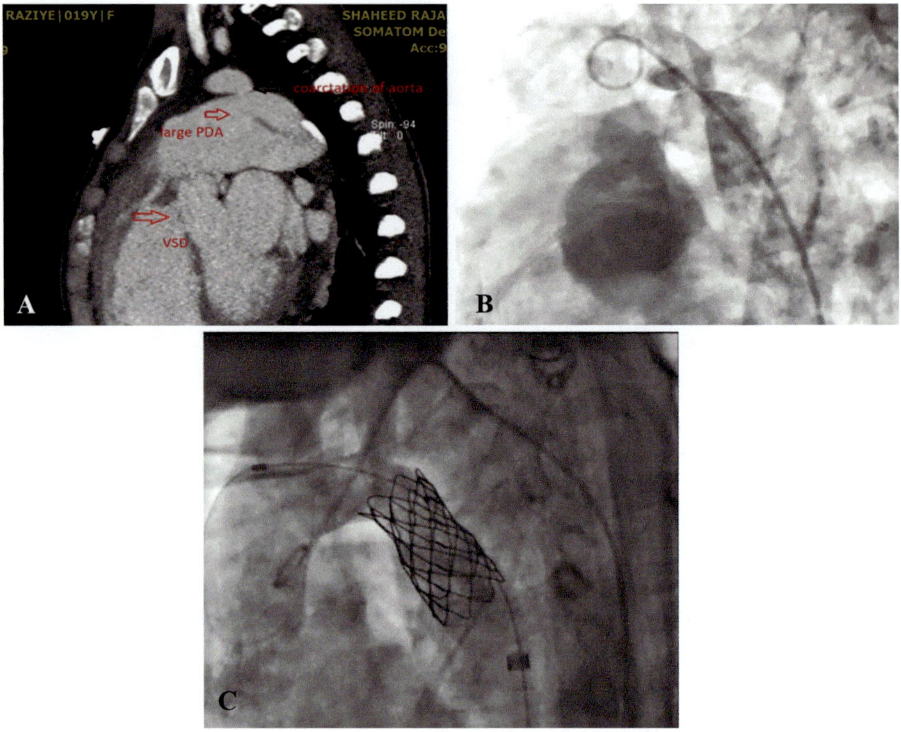

Fig. 10.3 (**a**–**c**) Computed tomography angiography confirmed the echocardiographic data and the existence of significant preductal coarctation of the aorta (**a**). Aortic arc injection in deep left anterior oblique projection demonstrated a large PDA with concomitant coarctation (**b**). Aortic arch injection after stenting confirmed that the position of the stent was appropriate with mild residual PDA shunt (**c**)

Table 10.2 The patient's hemodynamic data during right heart catheterization

		Pressure	Saturation
1	Left ventricle	140/0-14	93
2	Right ventricle	95/0-10	62
3	PAP	95/65	73
4	Ascending aorta	140/70	93
5	Descending aora	95/60	88

covered CP stent (45 mm) in an 8-zig configuration was selected. This stent was mounted on a balloon-in-balloon catheter (22 × 45 mm) and deployed at the coarctation site, effectively opening the stenosis and covering most of the PDA. A 14 Fr Cook introducer sheath was used for this procedure. The gradient across the coarctation improved significantly, and PAP decreased to 75/35 mm Hg. Postprocedural echocardiography indicated a marked improvement in left ventricular ejection fraction, with only a small residual flow through the PDA. Additionally, the stent was positioned appropriately, with no significant gradient observed (Fig. 10.3c).

Discussion

Since the first successful PDA device closure by Porstmann in 1967, numerous devices have been developed and introduced. Today, PDA device closure is typically completed in about 30 min, with very few major complications reported. The effectiveness of this approach in children is well-established, and there is no debate regarding its long-term patency and outcomes.

Still, in adults, there is a lack of long-term follow-up data and large-scale registries [4, 5]. The most time-consuming aspect of the PDA closure procedure is successfully wiring the duct, which is usually attempted via the antegrade venous approach. If this fails, the retrograde approach is used. In most reported cases, the ADO has been the preferred device, with successful deployment and minimal complications [6].

The choice of device for PDA closure depends on factors such as the size and shape of the PDA, the length of the duct, the presence of associated anomalies, and the operator's experience. While numerous devices are commercially available for transcatheter PDA closure, only five have received approval from the Food and Drug Administration (FDA): the Gianturco coil (1992), the Cook detachable coil (1994), the Flipper detachable coil (1996), the Gianturco-Grifka vascular occlusion device (1996), and ADO I (1996).

Another option, although used off-label, is ADO II (2009), which features two equal retention discs on either side to reduce the risk of device embolization. Additionally, it has a flexible waist that not only allows for a good fit within the lumen but also aligns with the duct, preventing disc protrusion and obstruction of the pulmonary artery or aorta. Nevertheless, the large diameter (12 mm) of ADO II limits its use for closing a large PDA. The most commonly used devices are duct occluders, with septal occluders being utilized in selected cases. Immediate postprocedural angiography can confirm whether the device is correctly positioned and whether any residual shunting is present. In certain situations, off-label Amplatzer vascular plugs and septal occluders have been used for PDA closure due to their varying sizes and shapes [7].

For larger ducts, the Amplatzer VSD occluder is a more suitable option due to its unique structure, which allows it to accommodate a thicker portion of the septum. Further, it provides a better approximation to the PDA wall because of its larger waist and two concentric discs. The higher waist-to-disc ratio (0.44 for the smallest size to 0.69 for the largest) compared with ADO II (0.33 for the smallest to 0.5 for the largest) and the larger maximum waist diameter (22 mm for the muscular VSD occluder vs. 12 mm for ADO II) make this device appropriate for closing larger PDAs (>14 mm). When used with short lengths and large diameters, this device can cause residual shunting.

Studies have indicated that the presence of double discs decreases the likelihood of complications such as device migration, making this design particularly effective for patients with high pulmonary pressure. Research has demonstrated that 50% of the compressive effects of the device are attributed to the waist, while only 25% are due to each of the discs. A crucial factor for the successful implantation of a muscular VSD device for PDA closure is the positioning of the aortic retention disc within the duct. Dislodgment of the retention disc in the duct significantly elevates the risk of embolization.

The association between coarctation of the aorta and PDA is uncommon, particularly in adults. There are two types of aortic coarctation based on the position of the duct: preductal and postductal. The preductal, or infantile type, is typically observed in neonates and infants. In this type, the coarctation occurs before the duct, causing blood flow in the descending aorta to originate from the pulmonary artery via the duct. Consequently, closing the duct could exacerbate the patient's condition. This anomaly is characterized by differential cyanosis, and patients rarely survive into adulthood. In contrast, the postductal or adult type of coarctation occurs after the duct, meaning that patients are not reliant on ductal flow. This type is primarily observed in adults. The management of the preductal type is more complicated, particularly in adults, due to the size of the duct and the presence of associated pulmonary artery hypertension. Surgical repair is the preferred method for neonates, as the incidence of re-coarctation following balloon angioplasty is higher in infants. In adults, however, catheter-based interventions are preferred due to the increased risks associated with surgery. The percutaneous procedure can be conducted in a single stage or sequentially. A simultaneous approach helps to minimize the duration of the intervention, reduce radiation exposure, and lower overall costs.

Conclusions

Currently, transcatheter closure of suitable PDA cases is considered the gold standard treatment for these patients, offering minimal complications and favorable long-term outcomes. While optimal results, such as cardiac reverse remodeling, are typically achieved in children, this procedure also yields acceptable results in adult patients. Notably, within 24 h post-procedure, there is a significant reduction in left ventricular size and PAP. Although some patients may exhibit left ventricular systolic dysfunction, sustained improvements in left ventricular ejection fraction, residual shunting, and the severity of mitral regurgitation are observed after 6 months and during long-term follow-up, with minimal adverse events reported [5].

Therefore, we recommend that, although reverse remodeling following transcatheter closure of the PDA in adults is a time-consuming process, PDA closure remains advisable for these patients. Additionally, there is a clear need for long-term outcome studies to further explore this issue.

References

1. Cassidy HD, Cassidy LA, Blackshear JL. Incidental discovery of a patent ductus arteriosus in adults. J Am Board Fam Med. 2009;22:214–8. https://doi.org/10.3122/jabfm.2009.02.070257. PMID: 19264946
2. Krichenko A, Benson LN, Burrows P, Moes CA, McLaughlin P, Freedom RM. Angiographic classification of the isolated, persistently patent ductus arteriosus and implications for percutaneous catheter occlusion. Am J Cardiol. 1989;63:877–80. https://doi.org/10.1016/0002-9149(89)90064-7. PMID: 2929462
3. Harrison DA, Benson LN, Lazzam C, Walters JE, Siu S, McLaughlin PR. Percutaneous catheter closure of the persistently patent ductus arteriosus in the adult. Am J Cardiol. 1996;77:1094–7. https://doi.org/10.1016/S0002-9149(96)00145-9. PMID: 8644634
4. Sudhakar P, Jose J, George OK. Contemporary outcomes of percutaneous closure of patent ductus arteriosus in adolescents and adults. Indian Heart J. 2018;70:308–15. https://doi.org/10.1016/j.ihj.2017.08.018. PMID: 29455770
5. Khajali Z, Firouzi A, Hosseini Z, Pouraliakbar H, Noohi Bezanjani F, Alemzadeh-Ansari MJ, et al. Cardiac reverse remodeling after transcatheter patent ductus arteriosus closure in adults. Curr Probl Cardiol. 2021;46:100938. https://doi.org/10.1016/j.cpcardiol.2021.100938. PMID: 34456082
6. Hong TE, Hellenbrand WE, Hijazi ZM. Transcatheter closure of patent ductus arteriosus in adults using the Amplatzer duct occluder: initial results and follow-up. Indian Heart J. 2002;54:384–9. PMID: 12457126
7. Firouzi A, Hosseini Z. Transcatheter patent ductus arteriosus (PDA) device closure. In: Maleki M, Alizadehasl A, editors. Case-based clinical cardiology. London: Springer; 2021. p. 601–6. https://doi.org/10.1007/978-1-4471-7496-7_65.

Chapter 11
Percutaneous Transcatheter Closure of Ruptured Sinus of Valsalva Aneurysms

Zahra Khajali, Ata Firouzi, and Zahra Hosseini

Abstract Rupture of SVAs is a rare congenital anomaly that can lead to acute symptoms, including chest pain, shortness of breath, and fatigue. The most common site of rupture is the right sinus, with the rupture typically occurring into the right-sided heart chambers. The primary diagnostic modality is TTE, which is confirmed by TEE or CT. Surgical repair is the standard treatment; however, transcutaneous closure of the defect has become an appropriate alternative. The devices that can be used include septal and ductal occluders.

Keywords Sinus of Valsalva aneurysm · Rupture of Valsalva sinus · Transcutaneous closure

Abbreviations

ADO	Amplatzer duct occluder
CT	Computed tomography
SVA	Sinus of Valsalva aneurysm
TEE	Transesophageal echocardiography
TTE	Transthoracic echocardiography
VSD	Ventricular septal defect

Introduction

Rupture of SVAs is a rare congenital anomaly that can lead to acute symptoms, including chest pain, shortness of breath, and fatigue. The most common site of rupture is the right sinus, with the rupture typically occurring into the right-sided

Z. Khajali · A. Firouzi · Z. Hosseini (✉)
Cardiovascular Intervention Research Center, Rajaie Cardiovascular Institute, Iran University of Medical Sciences, Tehran, Iran

A. Firouzi et al. (eds.), *Mastering Skills in Complex Cardiovascular Interventions*, https://doi.org/10.1007/978-3-031-97321-5_11

heart chambers. The primary diagnostic modality is TTE, confirmed by TEE or CT. Surgical repair is the standard treatment; nonetheless, transcutaneous closure of the defect has become an appropriate alternative. The devices that can be used include septal and ductal occluders.

The most common sites for SVA rupture are the right-sided heart chambers. Nevertheless, such ruptures can also occur, albeit rarely, in the left ventricle, left atrium, pericardium, pulmonary artery, and superior vena cava. The primary and most valuable diagnostic modality is TTE, while TEE provides an accurate assessment of the affected sinus, the extent of left-to-right shunts, chamber enlargement, and any associated cardiac lesions. Magnetic resonance imaging may be useful for a more precise diagnosis of coexisting cardiac conditions.

The treatment for SVA rupture is surgical repair. However, in recent years, percutaneous transcatheter closure of ruptured SVAs has emerged as a less invasive alternative to open surgical repair, yielding favorable results. The devices used for occlusion include ductal occluders, patent foramen ovale, and atrial septal defect occluders, as well as VSD occluders.

Complications associated with non-ruptured SVAs, such as ventricular outflow tract obstruction and arrhythmias, are additional indications for surgical repair. In asymptomatic patients, the decision to perform surgical interventions for non-ruptured aneurysms is controversial and depends on the size of the aneurysm. Generally, large aneurysms should be repaired to prevent complications, while smaller aneurysms may be monitored.

Challenging Cases

Case 1

A 29-year-old man presented to our hospital with a two-week history of chest pain and dyspnea. During the physical examination, findings consistent with aortic regurgitation were observed, including an early diastolic murmur along the left sternal border. TTE and TEE revealed a non-ruptured SVA extending into the right ventricular outflow tract. The imaging also showed enlargement of the left and right ventricles, as well as the left atrium. Mild-to-moderate mitral regurgitation and mild aortic valve regurgitation were also detected. The aneurysm featured a large fenestration measuring 10 mm in diameter, with the main opening measuring 19 mm in diameter (Fig. 11.1a). Additionally, the right coronary artery was noted to be near the aneurysm's opening. Although the aneurysm protruded into the right ventricular outflow tract, it did not cause obstruction. The patient underwent cardiac catheterization guided by TEE to address the ruptured aneurysm. Access was achieved through the right femoral artery and vein using 7 Fr and 6 Fr sheaths, respectively. Coronary angiography confirmed normal coronary arteries, and the Qp:Qs ratio was measured at 3:1. The defect was successfully crossed, and an arteriovenous loop was established (Fig. 11.2a, b). A 12 mm Occlutech muscular VSD occluder (Occlutech GmbH, Jena, Germany) was deployed antegradely via the venous

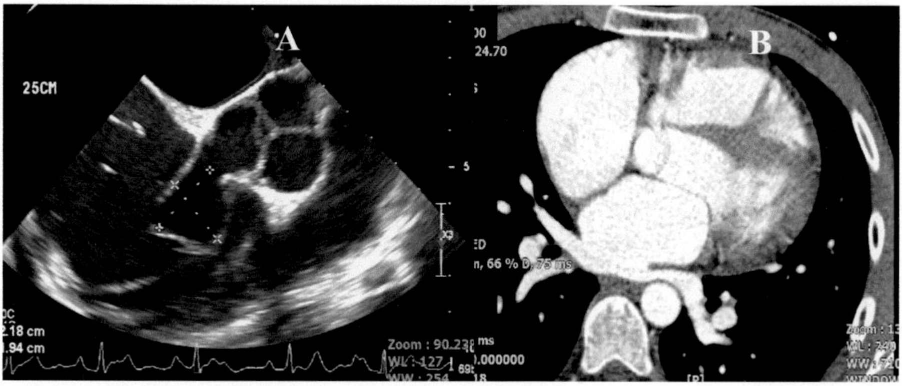

Fig. 11.1 (**a**, **b**) TEE and CT images demonstrate the rupture of the noncoronary cusp aneurysm into the right atrium, exhibiting a windsock appearance

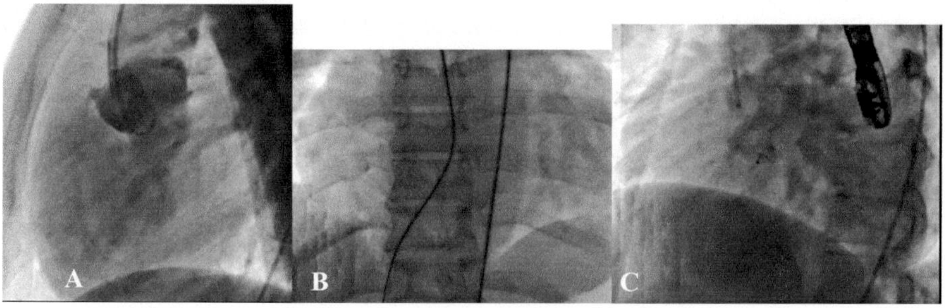

Fig. 11.2 (**a–c**) The aortic angiogram in the lateral view reveals the rupture of the noncoronary cusp aneurysm into the right atrium (**a**). An arteriovenous loop is created for the closure of the rupture (**b**). The defect was successfully closed using a 12 mm VSD occluder device (**c**)

approach to close the defect (Fig. 11.2c). Post-procedure, the Qp:Qs ratio normalized to 1:1, and TEE showed no evidence of aortic regurgitation, with only a mild residual defect related to the device.

Case 2

A 35-year-old woman with no significant medical history presented with a one-month history of chest pain and shortness of breath. She reported no episodes of fever. On admission, her blood pressure was 130/50 mm Hg, and her heart rate was 95 bpm. Physical examination revealed a continuous murmur localized to the third left intercostal space. TTE and TEE identified a right SVA originating from the noncoronary sinus and extending into the right atrium. The aneurysm exhibited a windsock-like morphology, with the aortic end measuring 14 mm and the opening

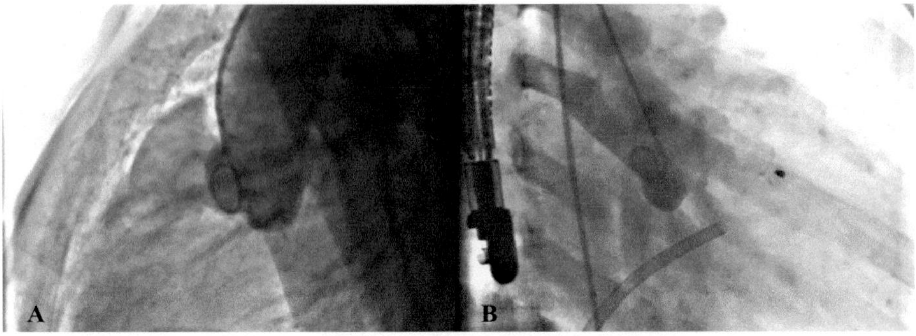

Fig. 11.3 (**a**, **b**) The aortic angiograms illustrate the rupture of the right coronary cusp aneurysm into the right ventricle. An ADO type I device was utilized to close the defect

into the right atrium measuring 10 mm. Imaging also confirmed a trileaflet aortic valve without regurgitation and no associated anomalies. Both the right and left ventricles were dilated but maintained normal systolic function.

The patient underwent cardiac catheterization guided by TEE to address the ruptured aneurysm. Access was obtained through the right femoral artery and vein using 7 Fr and 6 Fr sheaths, respectively. Coronary angiography revealed normal coronary arteries, and the Qp:Qs ratio was measured at 2.5:1. The defect was successfully crossed, and an arteriovenous loop was established (Fig. 11.2). A 12/15 mm Occlutech ductal occluder (Occlutech GmbH, Jena, Germany) was deployed antegradely via the venous approach to close the defect (Fig. 11.3a, b). The device placement did not compromise the right coronary artery. Post-procedure, the Qp:Qs ratio normalized to 1:1, and TEE showed mild aortic regurgitation along with a tiny residual defect related to the device.

Discussion

The first surgical repair of ruptured SVAs was performed by Lillehei on three patients [1]. Surgical repair remains the gold-standard treatment for such aneurysms, offering a low mortality rate and excellent long-term outcomes. Nonetheless, the risks associated with open-heart surgery—such as extended stays in the intensive care unit and hospital, as well as postoperative complications like septicemia—have increased the appeal of percutaneous device closure as a viable alternative. In 1994, Cullen et al. [2] pioneered the use of a Rashkind umbrella device for the transcatheter closure of ruptured SVAs. Today, a variety of devices and coils, deployed via retrograde or antegrade approaches, are utilized for the closure of ruptured SVAs.

Two classification systems—surgical and angiographic—are used to categorize ruptured SVAs based on the site of rupture.

- Surgical classification types include the following:
  - Type I: Rupture into the right atrium;
  - Type II: Rupture into the right atrium or the right ventricle near the tricuspid annulus;
  - Type III: Rupture into the right ventricular outflow tract; and
  - Type IV: Other types of ruptures [2].

In most reported cases, duct occluders and VSD occluders are the most frequently used devices for the transcatheter closure of ruptured SVAs, particularly due to the windsock-shaped morphology of the defect. For defects with a broader aortic end, ADO type I is considered the optimal device. In cases of windsock-like ruptured SVAs, a ductal occluder and/or a muscular VSD occluder is typically preferred. For funnel-shaped ruptures with a long aortic rim, a muscular VSD occluder is a suitable choice. The device size is generally 2–4 mm larger than the aortic end of the rupture. Alternative devices, such as coils, atrial septal occluders, patent foramen ovale occluders, and vascular plugs, can also be utilized in these patients. Across various studies, the success rate of transcatheter closure ranges between 80% and 100% [3].

Follow-Up

Although ruptures can recur after surgical closure of ruptured SVAs, no such cases have been reported to date following transcatheter closure. Still, data on the long-term outcomes of transcatheter closure remain limited. To minimize the risk of dilation or recurrence of a ruptured SVA, the retention disc should be deployed at the aortic orifice of the SVA rather than within the aneurysm itself [4].

Successful transcatheter closure of ruptured SVAs has not been associated with the progression of existing aortic regurgitation or the development of new aortic regurgitation.

The ADO type II device (AGA Medical Corporation) can be easily deployed using a retrograde approach. Nonetheless, its use is limited by its available waist sizes of 3–6 mm, making it suitable only for ruptured SVAs with aortic ends smaller than 4 mm.

Technical Considerations [5]

The technique for closing a ruptured SVA is similar to that used for perimembranous VSD, as the anatomic locations of these two defects are in close proximity: one is situated just above the aortic valve, and the other lies below it. Intraprocedural TEE plays a crucial role in guiding the procedure. It assists operators in measuring the aortic end of the defect for device selection, usually opting for a device that is

2–4 mm larger than the aortic end. TEE also ensures the proper positioning of the device on the aortic side and confirms that it does not interfere with adjacent structures. The length of the ruptured SVA is another critical factor in selecting the appropriate size and type of device.

Color flow Doppler via TEE is instrumental in detecting new or worsening aortic regurgitation and identifying residual shunts. While some interventionists prefer balloon sizing to measure the defect and select the device, TEE remains a valuable tool throughout the procedure for real-time guidance and assessment.

Complications

One of the most significant complications during transcatheter closure of ruptured SVAs is the failure to properly deploy the device on the aortic side. This is particularly challenging when the defect has a short margin for the insertion of the aortic leaflet or lacks an aneurysmal pouch. Another potential complication is residual shunting, although such shunts often resolve spontaneously over time. However, persistent shunts may lead to hemolysis, and severe cases might necessitate repeat interventions or surgical repair.

The development or worsening of aortic regurgitation is another notable concern during transcatheter closure. This can occur due to the device affecting the shape of the aortic leaflet or as a result of a sudden increase in systemic afterload following the closure of large shunts [6]. In many cases, trivial aortic regurgitation observed post-procedure tends to improve during follow-up.

Theoretically, coronary artery impingement by the device could pose a serious complication. Nevertheless, due to the high location of the coronary arteries relative to the site of rupture and aneurysm, this complication has not been reported to date.

Conclusions

The percutaneous closure of ruptured SVAs is a safe and effective treatment method, offering favorable clinical outcomes. In cases where complications such as recurrent shunts persist, they can often be successfully managed through additional transcatheter interventions. Close and long-term follow-up is essential to monitor patient progress and address potential issues. With careful patient selection, the percutaneous approach can be a preferable alternative to surgical repair, providing a less invasive option with excellent results.

References

1. Wang ZJ, Zou CW, Li DC, Li HX, Wang AB, Yuan GD, et al. Surgical repair of sinus of Valsalva aneurysm in Asian patients. Ann Thorac Surg. 2007;84:156–60. https://doi.org/10.1016/j.athoracsur.2007.03.022.
2. Post MC, Braam RL, Groenemeijer BE, Nicastia D, Rensing BJ, Schepens MA. Rupture of right coronary sinus of Valsalva aneurysm into right ventricle. Neth Hear J. 2010;18:209–11. https://doi.org/10.1007/BF03091762.
3. Kerkar PG, Lanjewar CP, Mishra N, Nyayadhish P, Mammen I. Transcatheter closure of ruptured sinus of Valsalva aneurysm using the Amplatzer duct occluder: immediate results and mid-term follow-up. Eur Heart J. 2010;31:2881–7. https://doi.org/10.1093/eurheartj/ehq323.
4. Tong S, Zhong L, Liu J, Yao Q, Guo Y, Shu M, et al. The immediate and follow-up results of transcatheter occlusion of the ruptured sinus of Valsalva aneurysm with duct occluder. J Invasive Cardiol. 2014;26:55–9.
5. Firouzi A, Hosseini Z. Transcatheter device closure of ruptured sinus of Valsalva. In: Maleki M, Alizadehasl A, editors. Case-based clinical cardiology. London: Springer; 2021. p. 641–6. https://doi.org/10.1007/978-1-4471-7496-7_71.
6. Jung SH, Yun TJ, Im YM, Park JJ, Song H, Lee JW, et al. Ruptured sinus of Valsalva aneurysm: transaortic repair may cause sinus of Valsalva distortion and aortic regurgitation. J Thorac Cardiovasc Surg. 2008;135:1153–8. https://doi.org/10.1016/j.jtcvs.2007.11.018.

Chapter 12
Transcatheter Coarctoplasty and Isolated Interrupted Aortic Arch Treatment in Adults

Ata Firouzi, Zahra Hosseini, and Armin Elahifar

Abstract CoA is a congenital heart defect that results in an obstruction of blood flow through the aorta. This condition is characterized by a narrowed segment of the aorta, which features localized thickening of the medial layer, accompanied by some infolding of the media and the presence of neo-intimal tissue. BAV is frequently associated with CoA, occurring in over half of CoA patients, and can lead to AI, AS, or aortic dilation and dissection. Additionally, intracranial aneurysms may develop in 2%–10% of cases. In adults, the most common initial symptom is systemic hypertension, which is often discovered incidentally.

The leading causes of death associated with CoA include congestive heart failure, aortic dissection and rupture, endocarditis, premature coronary artery disease, and intracranial hemorrhage. For neonates and adolescents, the preferred treatment is surgical intervention, although there is a risk of recoarctation and aneurysm formation. In adult patients, the more suitable treatment for those indicated for coarctoplasty is a transcatheter approach, with the primary goal being stenting.

Keywords Coarctation · Bicuspid aortic valve · Systemic hypertension · Aortic dissection · Surgical approach · Recoarctation · Transcatheter coarctoplasty Primary stenting

Abbreviations

AI	Aortic insufficiency
AS	Aortic stenosis
AV	Aortic valve
BAV	Bicuspid aortic valve

A. Firouzi · Z. Hosseini (✉) · A. Elahifar
Cardiovascular Intervention Research Center, Rajaie Cardiovascular Institute, Iran University of Medical Sciences, Tehran, Iran

A. Firouzi et al. (eds.), *Mastering Skills in Complex Cardiovascular Interventions*, https://doi.org/10.1007/978-3-031-97321-5_12

BIB	Balloon-in-balloon
BP	Blood pressure
CoA	Coarctation of the aorta
CP	Cheatham-Platinum
IAA	Interrupted aortic arch
LV	Left ventricle
PDA	Patent ductus arteriosus
RV	Right ventricle
VSD	Ventricular septal defect

Introduction

CoA is a congenital heart defect characterized by a narrowing of the aorta, which obstructs blood flow. This condition involves a constricted segment of the aorta, marked by localized thickening of the medial layer, infolding of the media, and the presence of overlying neo-intimal tissue. The obstruction may manifest as a shelf-like structure along the posterolateral aortic wall or as a membranous, curtain-like formation with either an eccentric or central opening. Typically, CoA occurs at the junction of the ductus arteriosus and the aortic arch, just distal to the left subclavian artery. It accounts for 5%–8% of congenital heart defects in children and is frequently associated with other cardiac anomalies, such as VSD, PDA, hypoplastic aortic arch, Shone complex, and BAV [1].

A BAV is frequently associated with CoA and is found in over half of CoA patients. This association can lead to complications such as AI, AS, aortic dilatation, and dissection. Moreover, intracranial aneurysms may develop in 2%–10% of cases. Timely and accurate diagnosis of CoA is critical for a positive prognosis, as early intervention significantly reduces the risks of long-term morbidity and mortality. Without treatment, the natural course of CoA is poor, with an average life expectancy extending only into the fourth decade. Approximately 75% of patients do not survive beyond their fiftieth birthday [2]. The leading causes of death include congestive heart failure, aortic dissection or rupture, endocarditis, premature coronary artery disease, and intracranial hemorrhage [3].

In adults, systemic hypertension is the most common presenting symptom of CoA, often detected incidentally during routine evaluations. For every new case of hypertension, at least during the initial visit, BP should be measured in both upper limbs, and radial-femoral pulse assessments should be performed. CoA is responsible for approximately 0.2% of all hypertension cases in adults. Additionally, offspring and first-degree relatives of individuals diagnosed with an obstructive left-sided cardiac lesion face a tenfold increased risk of developing coarctation or other cardiac abnormalities. In 1945, Crafoord and Nyhlin [4] achieved the first successful surgical correction of CoA. Later, in 1983, Lock [5] reported the first use of balloon angioplasty for treating the condition. The first stent implantation for CoA was performed by O'Laughlin et al. [6] in 1993, followed by the introduction of the

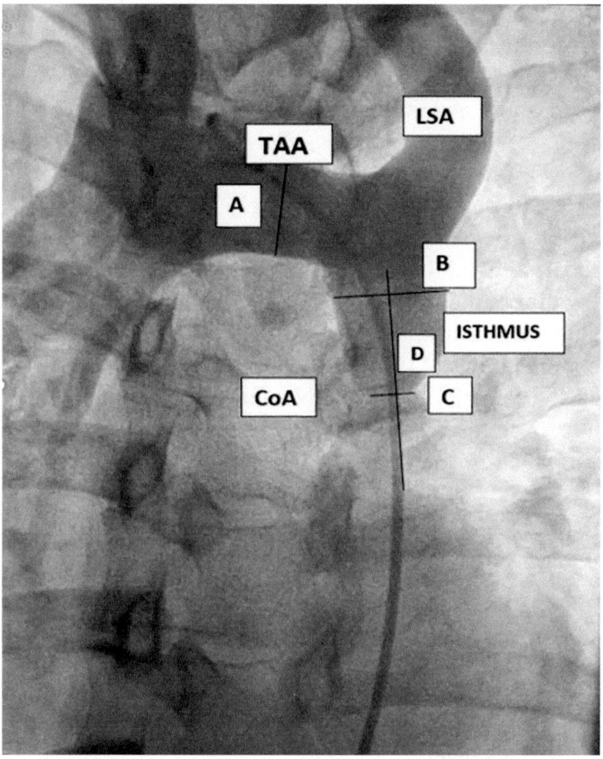

Fig. 12.1 The image shows (**a**) the transverse aortic diameter (TAA), (**b**) the isthmus diameter), (**c**) CoA, and (**d**) the distance between the isthmus and the coarctation segment. LSA left subclavian artery

first covered stent by Gunn et al. [7] in 1999. The covered CP stent received the CE mark in 2003 and gained FDA approval in 2016 (Fig. 12.1).

The transverse aortic arch is the segment of the aorta located between the innominate artery and the left subclavian artery. The isthmus refers to the portion of the aorta just distal to the left subclavian artery. A transverse aortic arch is considered hypoplastic if its diameter is <60% of the ascending aortic diameter. Similarly, the isthmus is deemed hypoplastic when its diameter is <40% of the ascending aortic diameter. Under normal conditions, the aortic isthmus measures 80%–90% of the transverse aortic arch diameter and is nearly equal in size to the aorta at the level of the diaphragm [8].

IAA

An IAA is an anomaly that represents the most severe form of CoA. It is characterized by a complete anatomical and luminal discontinuity between the ascending and descending aorta. IAA occurs in approximately 1.5% of patients with congenital heart defects [9]. Over 98% of individuals with IAA also present with other congenital cardiac anomalies, including PDA, truncus arteriosus, single ventricle, double-outlet RV, VSD, and LV outflow tract obstruction. In 80%–90% of cases, posterior malalignment of the conal septum accompanies IAA, leading to a VSD as an associated lesion. This malalignment can also contribute to LV outflow tract obstruction. Additional anomalies often associated with IAA include BAV, transposition of the great arteries, aberrant innominate arteries, and aortopulmonary windows [10].

If left untreated, approximately 90% of affected infants may die within the first year of life, with most fatalities occurring within the first few days. In the rare cases reported in adults, the presentation can range from being asymptomatic to exhibiting differential BP readings in the extremities, refractory systemic arterial hypertension, headaches, claudication, and congestive heart failure with its associated complications [11]. Survival into adulthood depends on the development of significant collateral circulation. Nonetheless, these collateral vessels are prone to atrophy, atherosclerosis, and even spontaneous rupture, which can lead to secondary complications [12].

IAA Classification [13]

Type A: The disruption occurs distal to the left subclavian artery. This is the second most common type in neonates, accounting for approximately 13% of cases. Nevertheless, Type A is more frequently reported in adults, representing about 79% of cases.

Type B: The disruption is located between the left carotid artery and the left subclavian artery. This is the most common anomaly in neonates, comprising roughly 84% of cases. In up to 50% of Type B interruptions, the right subclavian artery arises aberrantly from the descending aorta below the ductus arteriosus. Type B is the most prevalent form in neonates (53%) and is often associated with DiGeorge syndrome.

Type C: The disruption occurs between the innominate artery and the left carotid artery. This is the rarest type, representing approximately 3% of all cases.

The absence of a significant gradient across a trans-stenotic segment with extensive collateral vessels, narrowing of the abdominal aorta, a slow systolic upstroke, and antegrade diastolic flow on spectral Doppler imaging indicate severe CoA or IAA.

ESC Recommendations for Intervention in Native or Recurrent Coarctation in Adults [14]

Significant native CoA or recoarctation is defined as a mean systolic gradient ≥20 mm Hg on Doppler, or a mean systolic gradient ≥10 mm Hg on Doppler plus evidence of reduced LV systolic function, aortic regurgitation, or collateral circulation.

Class I: For hypertensive patients with a peak-to-peak gradient ≥20 mm Hg, stenting is preferred over surgery when technically feasible.

Class IIa: For hypertensive patients with a peak-to-peak gradient <20 mm Hg but aortic narrowing ≥50% relative to the diameter of the aorta at the diaphragm, catheter-based treatment (stenting) should be considered when technically feasible. For normotensive patients with a peak-to-peak gradient ≥20 mm Hg, stenting should also be considered when technically feasible.

Class IIb: For normotensive patients with a peak-to-peak gradient <20 mm Hg but aortic narrowing ≥50% compared with the diameter of the aorta at the diaphragm, stenting may be considered when technically feasible.

Challenging Cases

Case 1

A 19-year-old postpartum woman was referred to our hospital for further evaluation of resistant systemic hypertension. The patient was G1P1, with a history of pre-eclampsia and premature delivery via cesarean section due to malignant hypertension. Twelve weeks after delivery, she continued to experience refractory hypertension despite being on full guideline-directed medical therapy.

A physical examination revealed the following:

- BP:
  - Right upper extremity: 170/90 mm Hg (SpO_2: 97%);
  - Left upper extremity: 160/90 mm Hg (SpO_2: 97%); and
  - Both lower extremities: Absent common femoral artery, posterior tibial, and dorsalis pedis pulses (SpO_2: 95%).

A cardiac examination revealed a mid-systolic murmur (grade II/VI) audible at the left sternal border. ECG demonstrated LV hypertrophy criteria with a strain pattern in the precordial leads.

Transthoracic echocardiography revealed the following:

- Normal LV size and systolic function (LV ejection fraction: 55%);
- Mild concentric LV hypertrophy and mild LV diastolic dysfunction;
- No VSD;
- No PDA;
- Normal RV size and function;
- BAV without calcification;
- Moderate AS;
- Mitral valve area: 1.7 cm^2;
- Mean pressure gradient: 20 mm Hg;
- Peak pressure gradient: 35 mm Hg;
- Moderate AI;

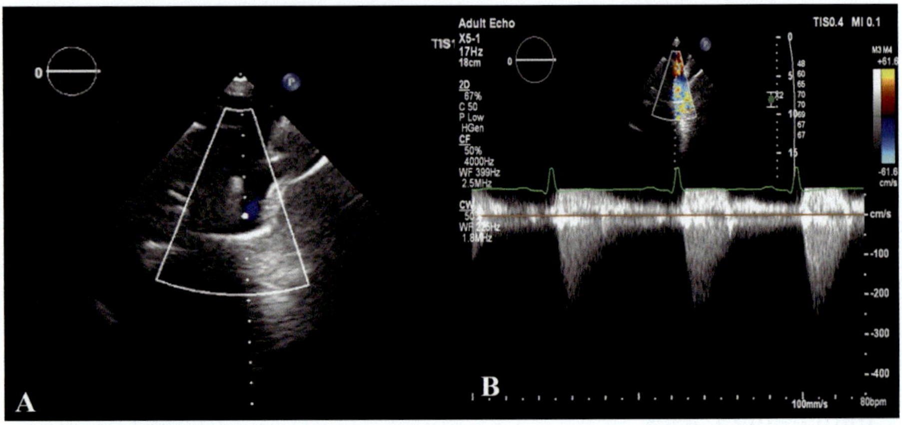

Fig. 12.2 The image shows severe stenosis or interruption immediately following the left subclavian artery (**a**). Turbulent flow is detectable at the site of obstruction, accompanied by a diastolic tail in Doppler imaging (**b**). The exact pressure gradient at the coarctation site could not be determined

- Mild aortic root dilatation (3.8 cm); and
- Severe stenosis or interruption distal to the left subclavian artery.

The exact pressure gradient at the coarctation site could not be measured. Numerous large collateral arteries were observed, with continuous low systolic and antegrade diastolic flow in the abdominal aorta, consistent with CoA (Fig. 12.2a, b).

Given the patient's postpartum status and breastfeeding, aortic computed tomography angiography was avoided to minimize radiation exposure. Instead, the patient was scheduled for aortic angiography with the possibility of transcatheter coarctoplasty.

Procedural Technique

Under conscious sedation, a sonography-guided puncture of the right common femoral artery was performed, and a 6 Fr sheath was inserted. Right radial artery access was also obtained. Intravenous heparin (5000 units) was administered, and the baseline gradient was measured at 50 mm Hg. Angiography demonstrated acceptable diameters of the superficial femoral artery, common femoral artery, external iliac artery, and common iliac artery (Fig. 12.3a). Two Perclose Proglide devices (Abbott Medical, Santa Clara, CA, USA) were deployed in the right common femoral artery.

Simultaneous antegrade and retrograde aortography was performed at the level of the aortic isthmus and descending aorta using a pigtail catheter and a multipurpose catheter, respectively. Imaging was conducted in deep left anterior oblique and anteroposterior projections. The angiography revealed an interrupted aorta just distal to the left subclavian artery. The distance between the interrupted segment and

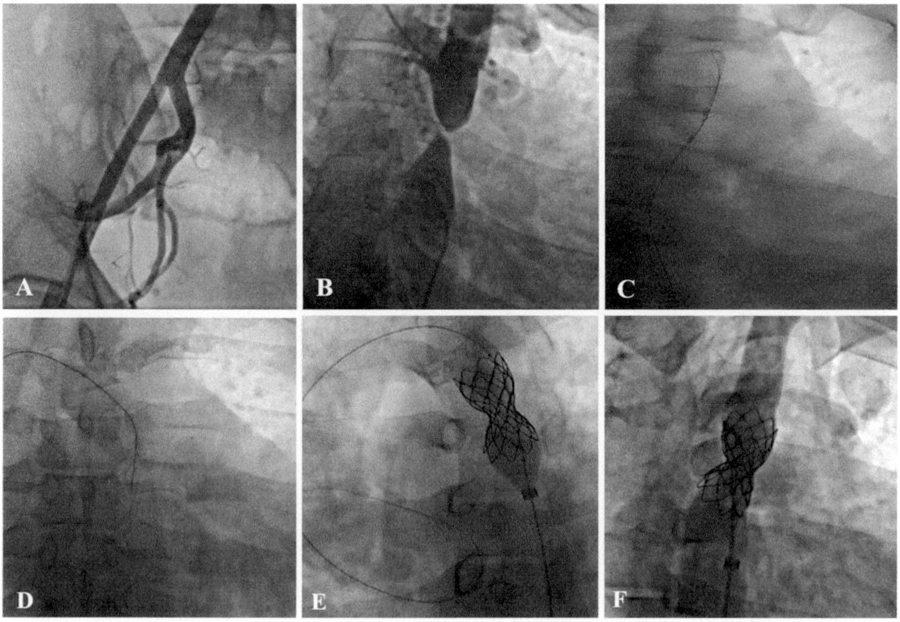

Fig. 12.3 (**a**) Right common femoral artery angiography revealed an acceptable diameter of the superficial femoral artery, common femoral artery, external iliac artery, and common iliac artery. (**b**) Simultaneous antegrade and retrograde aortography showed an interrupted aorta just beyond the left subclavian artery, with a distance of more than 1 cm between the interrupted segment and the left subclavian artery, along with numerous collateral arteries. The length of the interrupted segment was discrete, with both ends tapered, but their alignment was complex. (**c**) In the right anterior oblique caudal projection, the retrograde wire traversed the interrupted segment and was advanced to the aortic arch. (**d**) Using a coronary balloon (2.5 × 15 mm) through a diagnostic multipurpose catheter, the interrupted segment was dilated, and the catheter was advanced to the aortic arch using the balloon swallow technique. (**e**) A CP8Z39 stent was selected; first, the inner balloon was inflated to stabilize the stent, followed by inflation of the outer balloon. (**f**) The final angiography confirmed the proper positioning of the stent without any complications

the left subclavian artery exceeded 1 cm, with numerous collateral arteries visible. The interrupted segment was discrete in length, with tapered ends, but the alignment was complex (Fig. 12.3b).

Under meticulous hemodynamic monitoring, one operator attempted to pass a chronic total occlusion wire (Pilot 150) retrogradely, while a second operator tried to advance a Gaia Second wire antegradely. Despite multiple attempts, the alignment of the wires was not optimal. However, after adjusting the projection to right anterior oblique caudal, the retrograde wire successfully crossed the interrupted segment and was advanced into the aortic arch (Fig. 12.3c).

Once the wire's position in the true lumen was confirmed, a coronary balloon (2.5 × 15 mm) was used to dilate the interrupted segment through the diagnostic multipurpose catheter. Using the balloon swallow technique, the catheter was then advanced into the aortic arch (Fig. 12.3d).

Following this, the wire was exchanged for a 260 cm super-stiff guidewire (Amplatzer). A long delivery sheath (Cook, 14 Fr) was advanced, with its tip positioned in the aortic arch. The dilator was removed, and the sheath was de-aired. A covered CP stent (CP8Z39) was selected, paired with a BIB balloon (20 × 50 mm). The stent was introduced through the sheath, and its optimal position was confirmed using repeated aortography. After retracting the sheath, the inner balloon was inflated first to stabilize the stent, followed by inflation of the outer balloon (Fig. 12.3e).

Once the balloon was deflated, the sheath was advanced over it, and the balloon was removed. The final angiography confirmed the proper positioning of the stent with no complications (Fig. 12.3f). The final gradient across the coarctation site was zero.

Case 2

A 36-year-old man was referred for further evaluation due to resistant systemic hypertension despite optimized guideline-directed medical therapy.

A physical examination revealed the following:

- BP:
  - Right upper extremity: 150/70 mm Hg;
  - Left upper extremity: 150/60 mm Hg; and
  - Both lower extremities: Delayed radiofemoral pulses (100/40 mm Hg).

A cardiac examination revealed a mid-systolic murmur in the interscapular zone and a mid-diastolic murmur at the left sternal border. ECG showed criteria for LV hypertrophy.

Transthoracic echocardiography findings included the following:

- Normal LV size and systolic function (ejection fraction: 55%);
- Mild-to-moderate concentric LV hypertrophy and mild diastolic dysfunction;
- No VSD;
- No PDA
- Normal RV size and systolic function;
- BAV with no AS but moderate-to-severe eccentric AI;
- Mild aortic root dilatation; and
- Significant stenosis just distal to the left subclavian artery with turbulent flow (peak pressure gradient: 50 mm Hg) and a minimal width at the stenotic site.

The patient was deemed a candidate for aortic angiography and transcatheter coarctoplasty.

Procedural Technique

Under conscious sedation, a 6 Fr sheath was inserted into the right common femoral artery, and 5000 units of intravenous heparin were administered (baseline gradient: 50 mm Hg). Angiography of the common femoral artery demonstrated a well-developed common femoral artery, external iliac artery, and common iliac artery (Fig. 12.4a). Injection into the descending thoracic aorta revealed a totally occluded aortic segment with well-developed collateral arteries (Fig. 12.4b). Using a multipurpose catheter, a long hydrophilic wire (260 cm) was advanced through the occluded segment without difficulty. The catheter was then advanced over the wire, and the wire was exchanged for a super-stiff guidewire. Aortic root injection illustrated BAV, severe AI, mild aortic root dilation, and grossly normal coronary arteries (Fig. 12.4c). A temporary pacemaker lead was inserted into the RV apex. A long delivery sheath (Cook, 12 Fr) was advanced toward the aortic arch, and the dilator was removed. The sheath was then de-aired. A bare metal stent (CP8Z39) paired with a BIB balloon (18 × 40 mm) was selected. After crimping the stent onto the balloon and confirming its correct position under fluoroscopy, the stent was

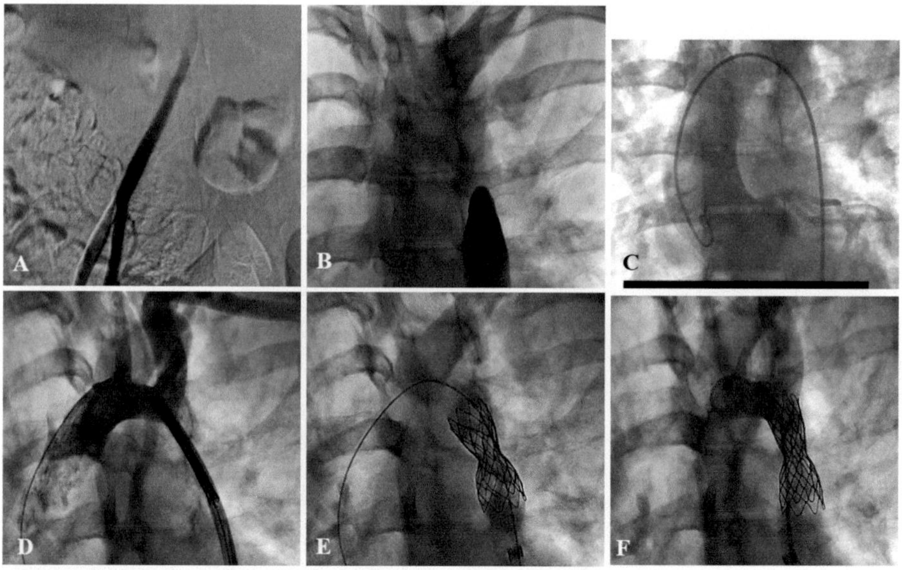

Fig. 12.4 (**a**) Common femoral artery angiography revealed a well-developed common femoral artery, external iliac artery, and common iliac artery. (**b**) Injection into the descending thoracic aorta showed a completely occluded aorta with well-developed collateral arteries. (**c**) Aortic root injection demonstrated a BAV, severe aortic AI, mild dilation of the aortic root, and grossly normal coronary arteries. (**d**) Following repeated aortography, the proper positioning of the stent was confirmed. (**e**) During ventricular over-pacing, the inner balloon was deployed first, followed by the outer balloon; after balloon deflation, pacing was discontinued (**f**). The final aortography confirmed the proper positioning of the stent without any complications

introduced and advanced to the end of the sheath. Repeated aortography confirmed the proper positioning of the stent (Fig. 12.4d). The sheath was retracted, and during ventricular over-pacing, the inner balloon was inflated first, followed by the outer balloon. After deflating the balloon, pacing was discontinued (Fig. 12.4e). Finally, the sheath was advanced over the balloon, and the balloon was removed. Final aortography confirmed that the stent was well-positioned without any complications (Fig. 12.4f). The final gradient across the coarctation site was zero. The patient was scheduled for AV replacement at a later date.

Given the concern for aortic dissection or perforation, a full aortography using digital subtraction angiography was conducted, but no significant abnormalities were noted. Despite this, the patient's hemodynamic instability persisted. At this point, transthoracic echocardiography was performed, revealing signs of physiologic tamponade (Fig. 12.5). Immediate pericardiocentesis was performed, yielding bloody drainage, and the patient's condition improved.

The most likely cause of the tamponade was thought to be the insertion of the pacemaker lead into the RV during the procedure. For further reassurance, a full aortic computed tomography angiography was performed, which showed no abnormalities.

Case 3

A 27-year-old single woman, a known case of Turner syndrome, was referred for transcatheter coarctoplasty due to a diagnosis of CoA. On physical examination, she exhibited typical features of Turner syndrome. BP measurements were as follows:

- Right upper extremity: 120/80 mm Hg;
- Left upper extremity: 120/80 mm Hg; and
- Delayed radiofemoral pulses (lower extremity BP: 80/40 mm Hg).

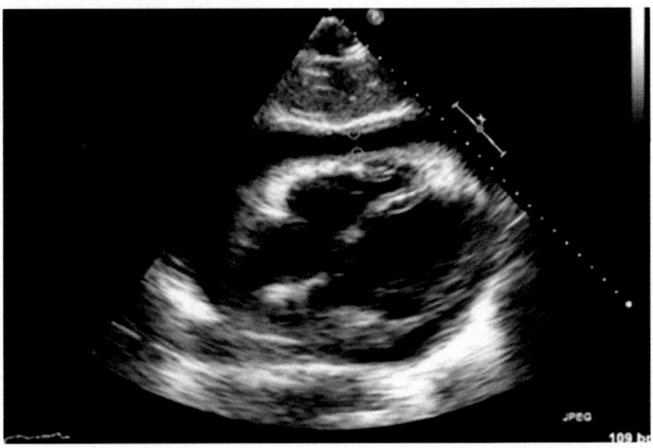

Fig. 12.5 Following hemodynamic deterioration, tamponade resulting from lead perforation was identified

Her body mass index was 18. A cardiac examination revealed no abnormal findings, and ECG was normal.

Transthoracic echocardiography findings included the following:

- Normal LV size and function, with no LV hypertrophy or regional wall motion abnormalities;
- No VSD;
- No PDA;
- Normal RV size and function;
- Tricuspid AV with no AS or AI;
- Normal aortic root size; and
- Significant narrowing just distal to the left subclavian artery with turbulent flow (peak pressure gradient: 40 mm Hg).

The patient was scheduled for aortic angiography and transcatheter coarctoplasty.

Procedural Technique

Under conscious sedation, a 6 Fr sheath was inserted into the right common femoral artery, and right radial artery access was obtained. Intravenous heparin (3000 units) was administered (baseline gradient: 40 mm Hg). Angiography of the right common femoral artery, external iliac artery, and common iliac artery demonstrated acceptable vessel diameters (Fig. 12.6a).

Aortic root injection via radial access revealed a tricuspid AV with no AS or AI, a normal aortic root size, a gothic and hypoplastic aortic arch, and discrete stenosis just distal to the left subclavian artery with post-stenotic dilatation (Fig. 12.6b). Using a retrograde approach, a long 0.035-inch hydrophilic wire (260 cm) was advanced across the stenotic segment (Fig. 12.6c).

The wire was then exchanged for a super-stiff guidewire. A bare-metal stent (CP8Z34) paired with a BIB balloon (16 × 40 mm) was selected. A long delivery sheath (Cook, 10 Fr) was positioned just distal to the left subclavian artery. After crimping the stent onto the balloon and confirming its proper position under fluoroscopy, the stent was introduced and advanced toward the aortic arch. Aortography confirmed the stent's appropriate placement (Fig. 12.6d).

The sheath was retracted, and the inner balloon was inflated first, followed by the outer balloon. Final aortography demonstrated the stent was well-positioned without complications (Fig. 12.6e). The final gradient across the coarctation site was 5 mm Hg.

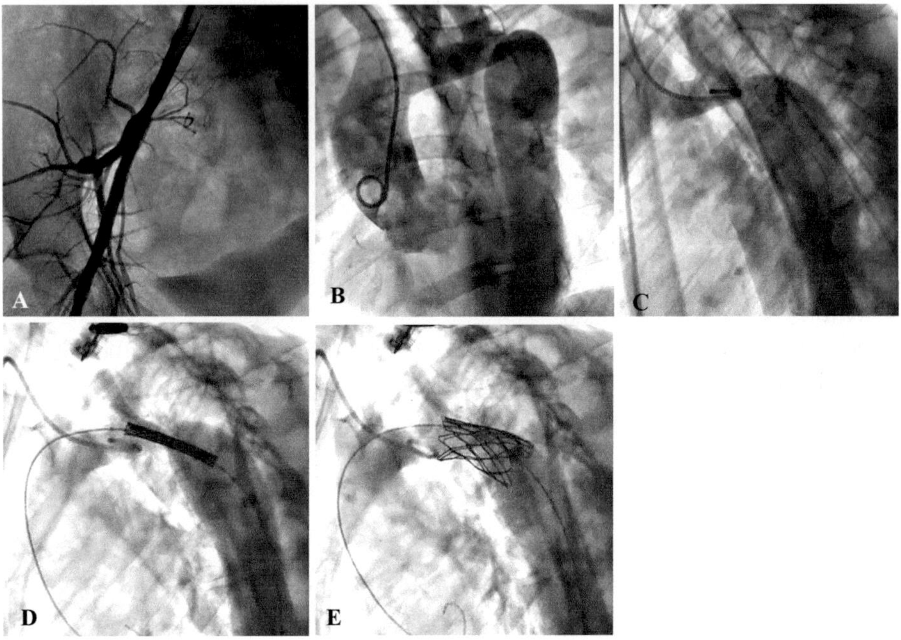

Fig. 12.6 (**a**) Angiography revealed acceptable diameters of the right common femoral artery, external iliac artery, and common iliac artery. (**b**) Aortic root injection via radial access demonstrated a tricuspid AV with no AS, no AI, a normal aortic root size, a gothic and hypoplastic arch, and discrete stenosis just beyond the left subclavian artery accompanied by post-stenotic dilation. (**c**) Using a retrograde approach with a long 0.035-inch hydrophilic wire (260 cm), the stenotic segment was successfully traversed. (**d**) Aortography confirmed the appropriate positioning of the stent. E. The final aortography showed the stent in an optimal position without any complications

The patient was admitted to the CCU and remained stable for approximately 6 hours. At around 10:00 PM, she suddenly developed severe back pain, although her hemodynamics remained stable. An emergent aortic computed tomography angiography was performed, revealing an aortic intramural hematoma at the site of stent placement, along with bilateral hemothorax (Fig. 12.7a, b).

The patient was subsequently deemed a candidate for emergent thoracic endovascular aortic repair (TEVAR) (Fig. 12.8a–c). Through left common femoral artery access, a Cook ZSL 13-24-56-ZT limb stent graft was implanted, extending from just distal to the left subclavian artery to the mid-descending aorta. Repeated contrast injections confirmed no evidence of endoleak (Fig. 12.8a, b).

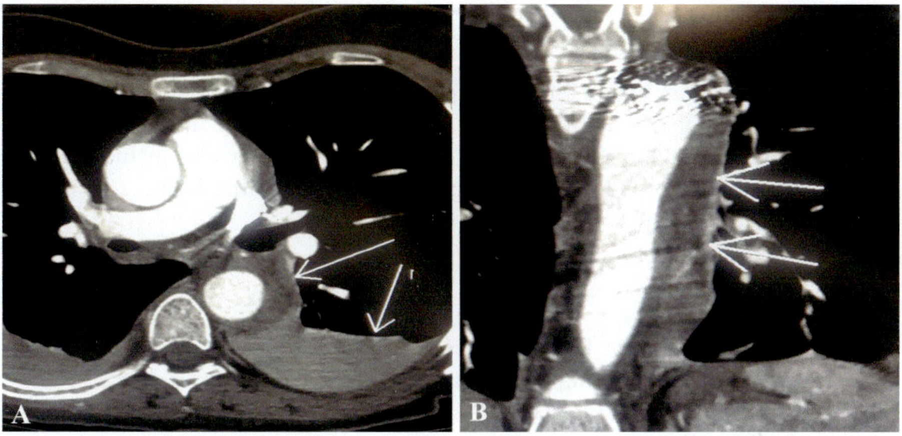

Fig. 12.7 (**a**) Intramural hematoma with bilateral hemothorax is indicated by the white arrows. (**b**) Diffuse IMH extends from the site of stenting to the descending aorta, as shown by the white arrows

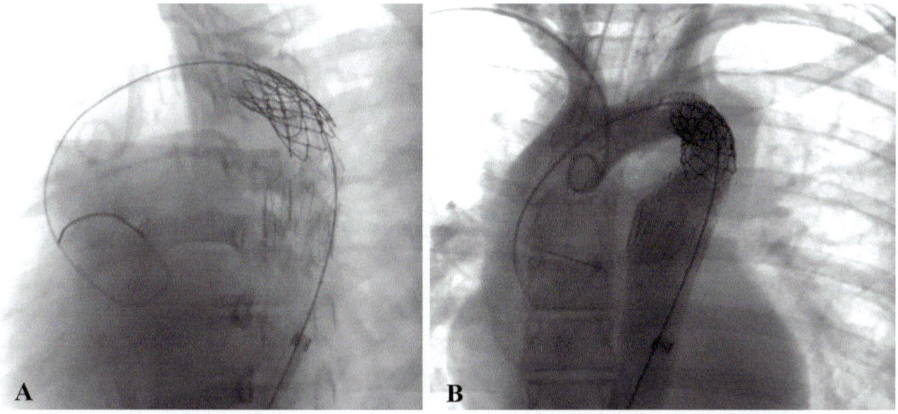

Fig. 12.8 (**a**) A COOK ZSL 13-24-56-ZT limb stent graft was deployed at the origin of the left subclavian artery extending to the mid-portion of the descending aorta. (**b**) The stent graft is shown in position

Case 4

A 45-year-old woman with a known history of thrombophilia, including prior cavernous sinus thrombosis and suspected segmental pulmonary thromboembolism, was referred for systemic BP monitoring. Her medical history also included G1P1, with a history of pre-eclampsia and postpartum deep vein thrombosis.

A physical examination revealed the following:

- BP:
  - Right upper extremity: 150/90 mm Hg;
  - Left upper extremity: 150/80 mm Hg;
  - Delayed brachiofemoral pulse; and
  - Lower extremities: 130/70 mm Hg.

A cardiac examination revealed no significant findings, and ECG was normal.

Transthoracic echocardiography showed the following:

- Normal LV size and function, with no LV hypertrophy;
- Normal RV size and function;
- Tricuspid AV with no AS or AI;
- Normal ascending aorta size;
- Discrete shelf-like stenosis just distal to the left subclavian artery with turbulent flow (peak pressure gradient: 30 mm Hg); and
- No PDA.

The patient was scheduled for aortic angiography and transcatheter coarctoplasty.

Procedural Technique

Under conscious sedation, a 6 Fr sheath was inserted into the right common femoral artery, and 5000 units of intravenous heparin were administered (baseline gradient: 35 mm Hg). Using a multipurpose catheter, the narrowed segment was wired with a 0.035-inch hydrophilic wire (260 cm). The wire was then exchanged for a super-stiff long wire.

Aortic root injection in the left anterior oblique projection revealed the following:

- Tricuspid AV with no AS and mild (1+) AI; and
- Normal aortic root size (Fig. 12.9a).

Subsequent aortic arch injection was performed to assess the residual segment, its length, the distance to the left subclavian artery, and the size of the aorta before and after the coarctation (Fig. 12.9b).

After advancing a long delivery sheath (Cook, 10 Fr) toward the aortic arch, the dilator was removed, and the sheath was de-aired. Based on the aortic arch diameter (20 mm) and distal aortic size (24 mm), a self-expandable stent (20 × 40 mm) was

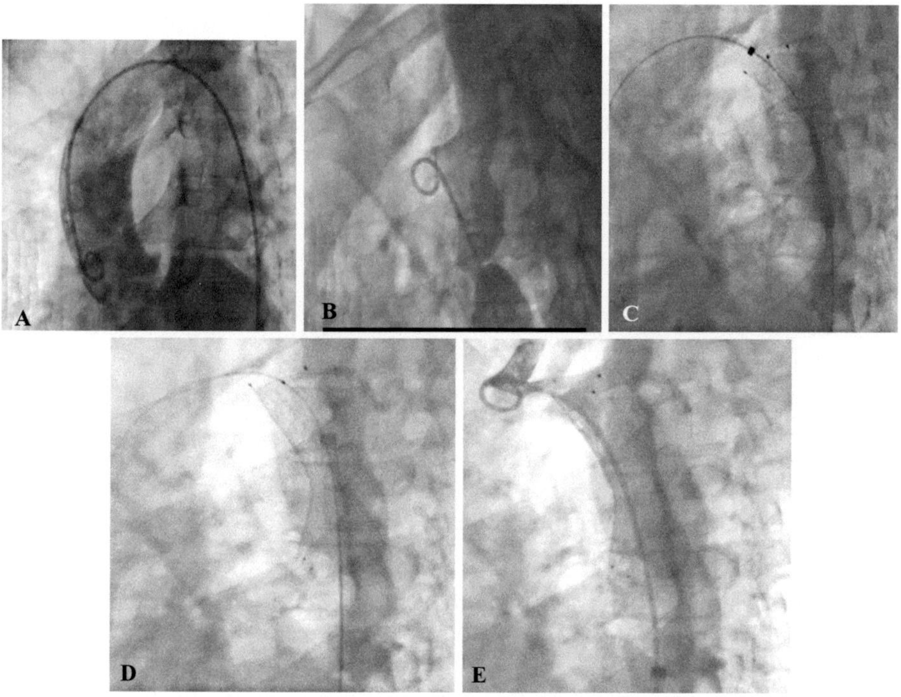

Fig. 12.9 (**a**) Aortic root injection in the left anterior oblique projection revealed a tricuspid AV, no AS, mild (1+) AI, and a normal aortic root size. (**b**) Aortic arch injection was performed to assess the characteristics of the residual segment, including its length, the distance to the left subclavian artery, and the size of the aorta before and after the coarctation. Based on the aortic arch size (20 mm) and the distal aortic size (24 mm), a self-expandable stent measuring 20 × 40 mm was selected. The delivery stent was advanced through the sheath to the tip of the sheath. (**c**) In the left anterior oblique view, angiography through the sheath confirmed the correct positioning of the stent. To ensure optimal stent placement, the proximal stent was covered with the delivery sheath while the distal stent was slowly expanded to its full size. (**d**) By retracting the sheath from the stent catheter, the remainder of the stent was deployed across the coarctation segment. (**e**) The final aortic angiography demonstrated excellent expansion of the device with no residual gradient and no complications

selected. The stent delivery system was passed through the sheath until it reached the sheath's tip.

Using angiography in the left anterior oblique view, the correct position of the stent was confirmed. To ensure optimal stent placement, the proximal portion of the stent was covered with the delivery sheath while the distal part of the stent was slowly expanded to its full size. The sheath was then retracted, allowing the remainder of the stent to deploy across the coarctation segment (Fig. 12.9c, d).

Following stent deployment, a pigtail catheter was advanced to obtain simultaneous pressure measurements across the stent. Multiple angiograms were performed to confirm proper stent placement and rule out complications (Fig. 12.9e). The final gradient across the coarctation site was zero.

Case 5

A 26-year-old woman with a history of CoA and end-arteritis 3 months prior was referred for further evaluation. She had been on antibiotics during this period and was managed with conservative therapy.

Imaging Findings

Transthoracic echocardiography findings included the following:

- No evidence of vegetation or active inflammation;
- Normal LV size with mild systolic dysfunction (ejection fraction: 45%);
- Mild LV hypertrophy;
- Normal RV size and function;
- Tricuspid valve and AV with no abnormalities;
- No AS, mild AI, and no visible vegetation or abscess;
- Normal aortic root size;
  - Significant stenosis just distal to the left subclavian artery (distance: 5 mm; turbulent flow with a peak pressure gradient: 30 mm Hg; and minimal lumen width);
  - Large saccular outpouching (6 × 5 cm) at the coarctation site, distal to the left subclavian artery (to-and-fro flow with no evidence of vegetation);
- Aortic arch: 17 mm (suggesting hypoplasia);
- Aorta distal to the coarctation: 24 mm; and
- No PDA.

The other valves appeared normal.

Positron emission tomography showed no evidence of active inflammation.

A physical examination revealed the following:

- Hemodynamic stability:
  - Right upper arm BP: 150/90 mm Hg;
  - Left upper arm BP: 150/80 mm Hg; and
  - Lower extremities BP: 130/70 mm Hg.
- Cardiac auscultation:
  - Mid-systolic murmur audible at the left sternal border
- No signs of embolic complications related to endocarditis

ECG findings were normal.

Given the above findings, the following points were considered:

- Absence of active endocarditis; and
- Presence of a large pseudoaneurysm at the coarctation site.

Consequently, the patient was deemed a candidate for aortography and likely coarctoplasty.

Procedural Technique

Under conscious sedation, a 6 Fr sheath was introduced into the right common femoral artery. Angiography revealed suitable dimensions of the common femoral artery, external iliac artery, and common iliac artery. Two Perclose Proglid devices (Abbott Medical, Santa Clara, CA, USA) were deployed in the right common femoral artery, after which 5000 units of intravenous heparin were administered. (The initial pressure gradient measured 30 mm Hg.) A 0.035-inch hydrophilic wire was advanced through the stenotic segment using a multipurpose catheter. This wire was subsequently exchanged for a long super-stiff guidewire, with its tip placed in the ascending aorta. An aortic root injection showed a tricuspid aortic valve, no AS, mild AI, and a hypoplastic aortic arch. An aortic arch injection in a deep left anterior oblique projection confirmed the presence of a large saccular pseudoaneurysm just distal to the left subclavian artery at the coarctation site (Fig. 12.10a).

A long delivery sheath (Cook-14 Fr) was advanced over the wire, positioning its distal tip in the aortic arch. The dilator was removed, and the sheath was de-aired. A balloon-expandable covered Optimus stent XL (12Z57) paired with an Andratec BIB 16 × 60 balloon was chosen. The stent was mounted onto the balloon and then introduced, advancing it to the distal tip of the sheath.

After aortography confirmed the stent's appropriate positioning, the proximal portion of the stent was covered with the delivery sheath to ensure optimal placement. The distal part of the stent was gradually dilated to its full size. The sheath was then withdrawn from the stent catheter, allowing the remainder of the stent to be expanded across the coarctation segment (Fig. 12.10b). The final aortography confirmed correct stent placement with no signs of pseudoaneurysm or endoleak. The final pressure gradient was zero (Fig. 12.10c, d).

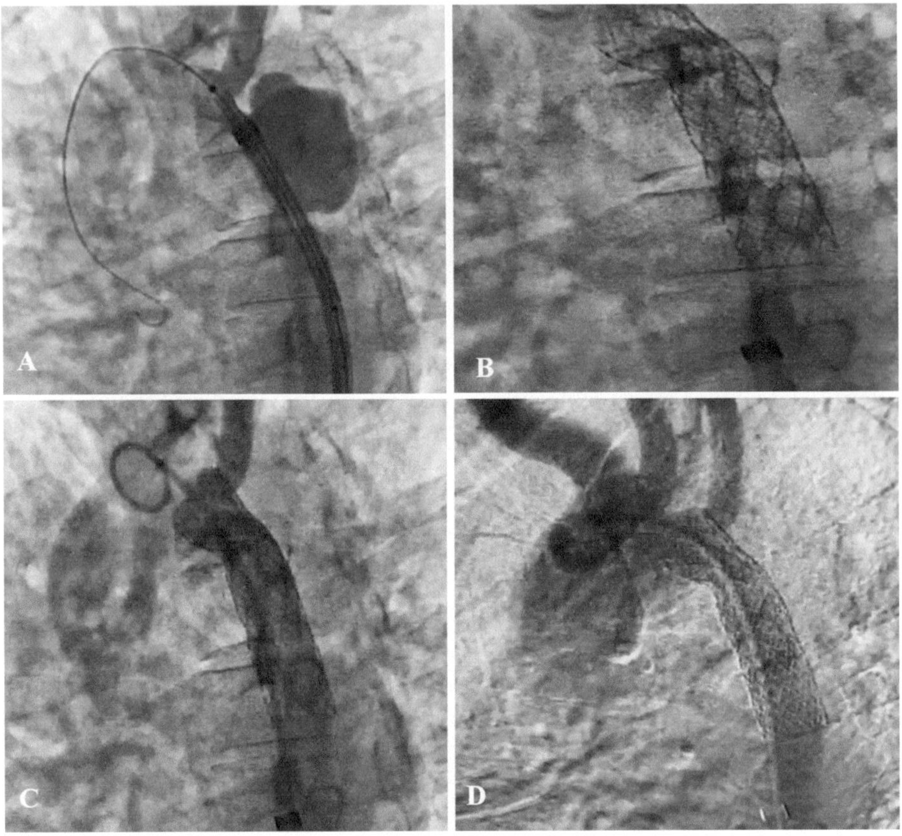

Fig. 12.10 (**a**) Aortic arch injection in a deep left anterior oblique projection confirmed the presence of a large saccular pseudoaneurysm just distal to the left subclavian artery at the site of the coarctation. (**b**) Following aortography, the correct positioning of the stent was verified. To achieve optimal stent placement, we covered the proximal end of the stent with the delivery sheath and gradually expanded the distal portion of the stent to its full size. Subsequently, by retracting the sheath from the stent catheter, the remainder of the stent was deployed across the coarctation segment. (**c**, **d**) The final aortography illustrated the proper positioning of the stent, with no residual pseudoaneurysm or endoleak observed

Discussion

The standard treatment for native coarctation in infants and young children is surgical repair, whereas percutaneous coarctoplasty is the first-line therapy for older children, adolescents, and adults [15]. Several surgical techniques are available. Nonetheless, end-to-end anastomosis and subclavian flap repair carry a high risk of recoarctation (10%), while patch aortoplasty is associated with a significant risk of aneurysm formation (10%) [16]. Postoperative survival rates decline to 70% by 40 years [17]. Cardiovascular complications account for more than 70% of late deaths. Approximately one-third of these deaths are attributed to coronary artery disease and systemic hypertension. Notably, only 32% of patients remain normotensive 30 years after surgical repair [18].

Endovascular Repair

Balloon Angioplasty

Balloon angioplasty has been a widely accepted technique for relieving coarctation for over three decades. The procedure involves creating controlled tears in the intima and part of the media, leading to an increase in vessel diameter. Balloon angioplasty is particularly safe and effective in infants aged 1–6 months with discrete narrowing and no evidence of arch hypoplasia [19]. It is also considered for critically ill patients of any age presenting with heart failure due to severe ventricular dysfunction, mitral regurgitation, or low cardiac output. Still, the broader application of this technique across all age groups has declined in popularity due to the high incidence of future aneurysm formation (up to 9%) [16] and the widespread availability of stents in most centers. This may be attributed to the fragility of ductal tissue. An aneurysm is defined as an area of dilatation measuring at least 150% of the aortic diameter at the level of the diaphragm or a discrete saccular dilation at the coarctation site that is at least 110% of the diameter of the adjacent lumen, provided such dilation was not present prior to the intervention [20]. Recoarctation is a frequent complication following balloon angioplasty, particularly in neonates, most of whom require repeat angioplasty within 5–12 weeks after the initial procedure. Consequently, balloon angioplasty is not recommended for infants under 4 months of age, especially when the lesion is associated with arch hypoplasia.

Compared with surgical repair, balloon angioplasty is equally effective in reducing the pressure gradient immediately after the procedure. However, it carries a higher risk of recoarctation and aneurysm formation [21].

Stenting

While balloon angioplasty is effective in alleviating vascular obstructions, it has certain limitations. Elastic recoil of the vessel wall and intimal dissection may prevent adequate relief of the narrowing. Stent placement following balloon angioplasty or surgery reduces complications, improves luminal diameter, achieves a minimal residual gradient, and sustains hemodynamic benefits [22]. However, stent placement is generally not recommended for patients weighing <25 kg due to the small size of the aorta and the risk of femoral artery injury from the large sheath required for stent delivery. Children who undergo aortic stent placement are more likely to require planned reinterventions, as the stent often needs to be dilated to accommodate their growth. In our practice, we consider stent placement only for

Table 12.1 Characteristics of stents currently available in the United States

Characteristics	Palmaz P188 /308	Palmaz 30/4010	Genesis 19/29/39	ITLD 16/26	Mega 16/26/36	Mega 16/26/36
Nominal expansion sizes	18-7mm	26-12mm	20-5mm	20-7mm	20-7mm	25-9mm
Final wall thickness	0.0055"	0.0103"	0.0095"	0.0076"	0.0093	0.0105
Minimal crimp ID	0.093"	N /T	0.066"	N /A	N /A	N /A
Del sheath (8-5mm)	8Fr	10Fr	6Fr	8Fr	8Fr	9Fr
Del sheath (12-9mm)	9Fr	10Fr	7Fr	9Fr	9Fr	10Fr
Del sheath (16-14mm)	10Fr	11Fr	8Fr	10Fr	10Fr	11Fr
Cell	Closed	Closed	Closed	Open	Open	Open
Pre -delivery flexibility	None	None	Yes	Yes	Yes	Yes

patients who are large enough to accommodate a stent that can be expanded to adult size. Several endovascular stents are commercially available, but very few can be expanded to the average diameter of a large adult aorta (21.1 ± 3.2 mm for women and 26.1 ± 4.3 mm for men (Table 1). The choice of stent depends on the anatomy of the coarctation, the patient's size, the operator's preference, and availability. Balloon-expandable bare metal stents are the most commonly used and are made from stainless steel (Palmaz Genesis, Johnson & Johnson; Mega LD and Maxi LD series, ev3), platinum-iridium alloy (Cheatham-Platinum stent, NED), or chromium-cobalt alloy (AndraStent XL and XXL, Andramed) (Tables 12.1 and 12.2).

Chromium-cobalt alloy is stronger than stainless steel, allowing for thinner struts that provide a lower crimped profile without sacrificing radial strength. Closed-cell stents, such as the Palmaz Genesis, are strong and rigid but exhibit significant foreshortening. In contrast, open-cell stents (e.g., Mega LD and Maxi LD) are less rigid,

Table 12.2 Compatibility chart for BIB, bare-metal CP stents, and covered CP stents

BIB Balloon Diameter (mm)	Sheath Size	Bare CP Stent Diameter (mm)	Sheath Size	Covered CP Stent Diameter (mm)	Covered CP Stent Diameter (mm)
12-14	8F	12-14	10 F	12-16	12F
15-16	9F	15-16	11 F	18-24	14F
18-20	IOF	18-24	12F	26	16F
22-24	LIF	26-30	16F	28-30	18F
26-30	16F				

foreshorten minimally, conform better to the anatomy, and allow access to side branches. A hybrid open-closed cell stent (AndraStent) combines the benefits of both designs. Balloon-expandable expanded polytetrafluoroethylene (ePTFE)-covered stents are available in both closed-cell (CP stent) and open-cell designs (Advanta V12 LD, Atrium Medical).

It has been suggested that covered stents may reduce the risk of aneurysm formation. However, a randomized trial involving 120 patients with severe native coarctation found no difference in the rates of recoarctation or pseudoaneurysm formation after 31 months of follow-up between patients who received bare metal stents and those who received covered stents [23]. Despite this, covered stents offer the advantage of shielding the endoluminal aspect of the aorta from stretch-induced wall trauma, which is particularly beneficial in the rare event of aortic rupture.

Data from a multicenter case series of over 500 patients, treated between 1989 and 2005 and aged over 4 years, demonstrated the efficacy and relative safety of stent placement for both native and recurrent coarctation [24]. The success rate of stenting in these cases was 98%, defined as a reduction in the gradient to $\leq$20 mm Hg or a post-stent coarctation-to-descending aorta diameter ratio $\geq$ 0.8. Acute complications occurred in 14% of patients, including two procedure-related deaths.

According to the Congenital Cardiovascular Interventional Study Consortium (CCISC) experience [25], the most commonly used stents for treating CoA in the United States are off-label applications of the Genesis XD (Cordis Corp., Warren, NJ), the EV3 Maxi (EV3, Plymouth, MN), the Palmaz "10" series stent (Cordis, Warren, NJ), and the CP stent (NuMED Corp., Hopkinton, NJ), as part of the ongoing Coarctation of the Aorta Stent Trial (COAST). The ideal stent for COA treatment should be dilatable from 12 mm to 22 mm, exhibit minimal foreshortening, provide adequate radial strength at maximal dilation, and resist fracture. The Palmaz XL stents, known for their high radial strength and fracture resistance (the CCISC has reported no fractures in the XL series), have been primarily limited to adult use due to their high profile. The EV3 Mega and Maxi LD stents, with their open-cell design, minimal foreshortening, and significant flexibility, are the preferred choice for treating transverse aortic arch obstruction, particularly when overlapping with head and neck vessels is anticipated [26]. The final stent diameter is determined based on the proximal arch diameter (transverse or distal arch) and should not exceed the diameter of the aorta at the level of the diaphragm. Additionally, the ratio of the stent diameter to the narrowest coarctation segment before intervention should be <3.5 [20].

Stent Types

Both balloon-expandable and self-expandable stents have been utilized for the endovascular treatment of CoA. Sadeghipour et al. [27] conducted a randomized trial comparing balloon-expandable CP stents with self-expandable nitinol stents in coarctoplasty among 92 eligible patients. Their findings confirmed that both balloon-expandable stents and self-expandable stents are safe and effective for treating native coarctation. Similarly, in a cohort of patients treated endovascularly, Firouzi et al. [28] demonstrated that both stent types are safe and effective for CoA treatment. Notably, after propensity-score matching, Firouzi and colleagues reported that self-expandable stents had slightly better outcomes.

In a randomized controlled trial by Sohrabi et al. [23], a comparison between covered and non-covered stents in non-interrupted aortas revealed no significant superiority for covered stents.

To our knowledge, no studies have compared the efficacy of covered versus non-covered stents in patients with IAA. Given the potential risks during wiring and initial ballooning, most experts recommend the use of covered stents for treating IAA. Our group has employed covered stents in these patients under the following conditions:

1. critical or sub-atretric obstructions, defined as a minimum diameter of 2–3 mm at the coarctation site on angiography; and
2. coarctation associated with atresia of the aortic lumen (interrupted) [29].

Balloon angioplasty with bailout stenting is the preferred approach for treating CoA in children, while primary stenting remains the preferred modality for adults. In primary stenting, routine predilation is not recommended except in cases of sub-atretric or interrupted aorta. Care should be taken to avoid using balloons longer than 3–4 cm. The BIB catheter is the preferred delivery system when the initial stent dilation is ≥ 18 mm, as it allows for controlled stent expansion, adjustment of stent position after inner balloon inflation, and reduced stent foreshortening (Table 12.3). For stents initially deployed on balloon catheters ≤ 16 mm, the Z-MED II balloon catheters (NuMED Inc., Hopkinton, NY) are the most commonly used.

Recently, the BIB (NuMED, NY, USA) catheter has gained popularity. This system consists of an inner and an outer balloon, with the outer balloon being 1 cm longer and twice the diameter of the inner balloon. The BIB catheter is available in diameters ranging from 1230 mm and lengths from 2.5 to 5.5 cm, and it is compatible with 8–11 Fr introducer sheaths (Fig. 12.11).

For CP stents, the inner balloon is inflated first, and the stent position is reassessed. If necessary, the stent can be repositioned before inflating the outer balloon.

Table 12.3 The foreshortening chart for CP stents

INFLATED BALLOON DIAMETER	CP8Z16 (LENGTH AFTER EXPANSION) SHORTENING)	CP8Z22 (LENGTH AFTER EXPANSION) (%SHORTENING)	CP8Z28 (LENGTH AFTER EXPANSION) (%SHORTENING)	CP8Z34 (LENGTH AFTER EXPANSION) (%SHORTENING)	CP8Z39 (LENGTH AFTER EXPANSION) (%SHORTENING)	CP8Z45 (LENGTH AFTER EXPANSION) (%SHORTENING)
12 mm	1.61cm(2.8%)	2.18cm(%0.8)	2.62cm(%4.4)	3.23cm(%3.1)	3.72cm(%1.9)	4.17cm(%3.8)
14 mm	1.54cm(6.5%)	2.08cm(%5.4)	2.56cm(%6.8)	3.15cm(%5.4)	3.66cm(%3.6)	3.97cm(%8.4)
15 mm	1.51cm(8.5%)	2.02cm(%7.9)	2.51cm(%8.6)	3.10cm(%7.0)	3.54cm(%6.6)	3.94cm(%9.2)
16 mm	1.48cm(10.6%)	1.98cm(%10.1)	2.45cm(%10.7)	3.00cm(%9.8)	3.48cm(%8.2)	3.84cm(%11.4)
18 mm	1.43cm(13.7%)	1.89cm(%14.0)	2.38cm(%13.3)	2.88cm(%13.5)	3.20cm(%15.6)	3.71cm(%14.5)
20 mm	1.32cm(20.0%)	1.80cm(%17.9)	2.30cm(%16.3)	2.63cm(%20.9)	2.96cm(%21.9)	3.27cm(%24.7)
22 mm	1.23cm(25.4%)	1.67cm(%23.9)	2.09cm(%24.0)	2.46cm(%26.0)	2.85cm(%25.0)	3.15cm(%27.3)
24 mm	1.05cm(36.4%)	1.46cm(%33.8)	1.91cm(%30.3)	2.07cm(%37.9)	2.27cm(%40.1)	2.83cm(%34.9)

Fig. 12.11 The image showcases a BIB (NuMED, NY, USA) balloon

It is recommended to reinflate the outer balloon to ensure optimal stent apposition. Further dilation is typically unnecessary, and it may be preferable to leave a slight waist if the gradient has resolved. To ensure optimal stent delivery and prevent malposition or migration, RV pacing may be employed. While this technique is rarely needed for standard moderate-to-severe coarctations, it is essential for stent treatment of transverse aortic arch narrowing (e.g., gothic arch types) and patients with mild CoA.

If the distal stent extends into a dilated segment, it should not be post-dilated, as post-dilation on either side of the waist is not recommended to avoid aortic wall injury. To remove the balloon, gently advance the sheath over the balloon and then withdraw the sheath. Excessive manipulation of the balloon, wire, or catheter after stent deployment should be avoided, as such maneuvers risk dislodging the stent.

In most cases, any bare-metal stent, covered stent, or self-expanding stent may be suitable, depending on the operator's experience and comfort level. Aggressive pre- or postdilation should be avoided, as it can increase the risk of aortic trauma. If necessary, the stent can be redilated after 3 months. At each stage of the procedure, the gradient across the coarctation segment should be carefully documented. A successful procedure is defined by achieving a peak-to-peak systolic gradient <10 mm Hg and improving the vessel caliber to >80% of the normal adjacent aortic arch.

Procedural Technique

A step-by-step approach is outlined for each case (Sections "Cases 1, 2, 3, 4, and 5").

Transcatheter Approach to the Isolated IAA [29]

A Step-by-Step Approach

Here, we present our institutional step-by-step approach to the endovascular treatment of IAA. Notably, the percutaneous reconstruction of IAAs is typically agreed upon by the heart team's decision-making members, which include an adult congenital cardiologist, an interventional cardiologist, a radiologist, and an echocardiographer.

1. **Comprehensive Assessment**: Evaluate transthoracic or transesophageal echocardiographic data and computed tomography angiography. Key details to assess include:
    - The type of IAA;
    - Any residual lumens;
    - Blind endings within the transverse or distal arch;
    - The length of the interruption;
    - Collateral vessels;
    - The size of the aortic arch;
    - The optimal angiographic projection at the level of the diaphragm for aligning the proximal and distal ends of the interrupted segment (the most challenging aspect of recanalization);
    - The vascular origin of the subclavian arteries;
    - The distance between the left subclavian artery and the interrupted segment; and
    - Any additional cardiac anomalies.
    The radiological working view should also be determined.
2. **Surgical Backup**: Notify the on-call surgical team about the procedure due to potential complications, such as aortic rupture.

3. **Vascular Access**: Obtain right radial artery and right femoral artery access for simultaneous antegrade and retrograde aortography. Simultaneous injection is mandatory at the start of the procedure. Perform the injection in the lateral view (deep left anterior oblique projection) or, preferably, in the working view suggested by CT angiography. This ensures precise identification of the proximal and distal ends and delineates the projection that best aligns the interrupted aorta.
4. **Ultrasound-Guided Puncture**: Due to absent or diminished femoral pulses, perform arterial puncture under ultrasound guidance to ensure cannulation of the common femoral artery. A central point of entry is critical, given the size of the sheath required.
5. **Crossing the Interrupted Segment**: If an IAA is confirmed, avoid crossing the interrupted distal end with a 0.035-inch guidewire, as this may cause dissection. Depending on the location of the tapered end, choose either a retrograde or antegrade approach to cross the interrupted segment using dedicated instruments. At our center, the approach for crossing the interrupted segment mirrors that used for chronic total occlusion revascularization, with a high success rate.
6. **Choosing the Approach**: Select the tapered end for the initial approach. If both ends are tapered, the antegrade approach is preferred.

The Antegrade Approach

When an acceptable landing zone is present—defined as a distance between the left subclavian artery and the interrupted segment greater than the distal curve of the Judkins right coronary catheter or more than 10 mm—the left radial artery access is preferred. If the distance between the left subclavian artery and the interrupted segment is insufficient, the right radial artery approach is recommended. We favor the left radial approach because navigating the catheter through the right subclavian artery carries a risk of injury or embolic events to the carotid arteries.

Multipurpose catheters or Judkins right coronary catheters (6 Fr) are preferred over guiding catheters due to their superior kink resistance and compatibility with balloons up to 2.5 mm. Furthermore, these catheters can be firmly advanced into the proximal blind end using 0.014-inch chronic total occlusion coronary guidewires. These guidewires offer better torque control, steerability, tapered tips, and high tip-load, enabling them to easily enter micro-channels in the working projection suggested by prior imaging, typically the right anterior oblique or right anterior oblique caudal view.

Attempts are made to perforate the blind end under close hemodynamic monitoring, advancing the wire into the descending aorta. The stenotic segment is sequentially dilated using balloons up to 9–10 mm to facilitate catheter crossing through the interrupted segment. The catheter is then advanced into the descending aorta using the balloon-kissing technique. During all ballooning steps, it is critical to have a covered stent readily available in case of aortic rupture. The disappearance of the waist on the balloon confirms a compliant lesion, which is suitable for stenting. If a significant waist persists, a slightly smaller balloon is selected, and complete stent expansion is deferred until a subsequent catheterization, typically around 6 months later.

Once the catheter has crossed the coarctation site, the coronary guidewire is exchanged for a 260-cm 0.035-inch J-tip guidewire, which is then snared from the femoral artery. Snaring is followed by routine left heart catheterization.

The Retrograde Approach

If the tapered blind end is located in the distal segment, a multipurpose catheter (6 Fr) is advanced into the distal blind end using a chronic total occlusion coronary guidewire in the appropriate working projection determined by prior imaging. Attempts are made to perforate the blind end, with a simultaneous antegrade aortogram performed in the adjacent segment to confirm proper alignment of the tract. The wire is then advanced into the ascending aorta. Following this, similar to the antegrade approach, the stenotic segment is sequentially dilated with progressively larger balloons to enable catheter crossing through the interrupted segment. The catheter is advanced into the aortic arch. Finally, the coronary guidewire is exchanged for a 0.035-inch-long guidewire to complete the procedure.

The Transcatheter Coarctoplasty Outcome

Recent multicenter reports from the Congenital Cardiovascular Interventional Study Consortium (CCISC) registry [25] have demonstrated excellent acute, intermediate, and long-term success rates. In a study of 565 procedures with a median patient age of 15 years (mean age: 18), Forbes et al. reported procedural success in 97.9% of cases, defined as a gradient <20 mm Hg and a coarctation-to-descending aorta (COA/DAO) ratio > 0.8.25 Significant improvements were observed in pre- versus post-procedure COA dimensions, systolic gradient, and COA/DAO dimensions.

A recent report by Holzer et al. [30] involving a cohort of 302 patients showed acute procedural success in 96%, intermediate success in 86% (with 41% of patients completing follow-up), and long-term success in 77% of patients. Procedural success was defined as an upper-to-lower limb (UL/LL) systolic gradient <20 mm Hg, absence of significant recurrent obstruction, and freedom from unplanned reintervention. At long-term follow-up, persistent hypertension (systolic BP >95th percentile) was noted in 23% of patients, and 32% were on antihypertensive medications.

Regarding interventional therapy and outcomes for complex aortic arch obstruction, Holzer and colleagues reported the largest series to date, involving 40 patients with aortic arch stent placement. Acute procedural success was achieved in 90% of cases, and neck vessels were crossed in 75% without adverse events. [26]

Comparison Between Stent/Balloon Angioplasty and Surgery

Carr conducted a comparison of balloon angioplasty and stent placement (n = 633) versus surgical repair (n = 213) by reviewing publications from 1995 to 2005. Stent therapy was associated with lower morbidity than surgery (9% vs. 11%, with ranges of 1%–20% vs. 1%–25%, respectively). However, restenosis rates were higher in the stent group at 11% (range: 0%–25%) compared with 2% (range: 0%–9%) in the surgical group. Post-procedure freedom from hypertension was similar between the two groups (61% vs. 65%). Notably, long-term data on hypertension and aortic wall imaging were significantly lacking in the surgical group [31].

Hager et al. [32] published the longest surgical follow-up data (27 years) in a cohort of 404 patients. They observed that 57% of patients had persistent hypertension, with only 13% attributed to restenosis. Risk factors for long-term hypertension included surgical repair using prosthetic material, male sex, residual gradient, and older age at follow-up. Despite favorable anatomical outcomes, this report underscores the prevalence of long-term hypertension, influenced by various factors beyond arch reobstruction.

The Congenital Cardiovascular Interventional Study Consortium (CCISC) registry was established to specifically address some of these factors. Recent data published by Forbes et al. [33], comparing all three treatment modalities in a cohort of 350 patients (stent [n = 217], angioplasty [n = 67], and surgery [n = 72]), demonstrated significant acute improvements in systolic BP and upper-to-lower extremity gradient. Stent patients experienced fewer acute complications than surgical and balloon angioplasty patients (2.3%, 8.1%, and 9.8%, respectively; $P < 0.001$).

At intermediate-term follow-up, there were no significant differences in resting hypertension. However, the COA-to-descending aorta ratio was significantly higher in the surgical group (0.98) than in the stent (0.80) and balloon angioplasty (0.79) groups. Aortic wall injury was observed in 12.6% of surgical patients, 7.1% of stent patients, and 43.6% of balloon angioplasty patients.

Improved long-term follow-ups across all three groups are necessary before definitive conclusions can be drawn regarding the comparative outcomes of these treatment modalities.

Complications [34]

Major complications occur in approximately 15% of cases during the procedure, including:

- Intimal tearing;
- Dissection;
- Perforation (Section "Case 3");

- Stent migration;
- Cerebrovascular accident (<3%);
- Tamponade (Section "Case 2"); and
- Vascular complications (<1%), which are the most critical.

Late complications, such as:

- Aneurysm and pseudoaneurysm formation (Section "Case 5");
- Recoarctation;
- Restenosis; and
- Stent fracture, are rare.

Conclusions

Although the surgical approach remains the gold standard for treating coarctation in neonates and adolescents, recent advancements in equipment and expertise have made transcatheter coarctoplasty, particularly primary stenting, the preferred treatment for young adults and adults. However, more long-term follow-up is needed to compare the surgical approach with transcatheter techniques in terms of risks such as recoarctation, aneurysm formation, and restenosis.

References

1. Topol EJ, Teirstein PS, editors. Textbook of interventional cardiology. 8th ed. Amsterdam: Elsevier Inc; 2019. p. 959–62.
2. Rao PS. Coarctation of the aorta. Curr Cardiol Rep. 2005;7(6):425–32. https://doi.org/10.1007/s11886-005-0061-9. PMID: 16256011
3. Campbell M. Natural history of coarctation of the aorta. Br Heart J. 1970;32(5):633–40. https://doi.org/10.1136/hrt.32.5.633. PMID: 5470045.
4. Crafoord C, Nylin G. Congenital coarctation of the aorta and its surgical treatment. J Thorac Surg. 1945;14(5):347–61. https://doi.org/10.1016/S0096-5588(20)31801-8.
5. Lock JE, Bass JL, Amplatz K, Fuhrman BP, Castaneda-Zuniga W. Balloon dilation angioplasty of aortic coarctations in infants and children. Circulation. 1983;68(1):109–16. https://doi.org/10.1161/01.CIR.68.1.109. PMID: 6222843
6. O'Laughlin MP, Slack MC, Grifka RG, Perry SB, Lock JE, Mullins CE. Implantation and intermediate-term follow-up of stents in congenital heart disease. Circulation. 1993;88(2):605–14. https://doi.org/10.1161/01.CIR.88.2.605. PMID: 8339424
7. Gunn J, Cleveland T, Gaines P. Covered stent to treat co-existent coarctation and aneurysm of the aorta in a young man. Heart. 1999;82(3):351. https://doi.org/10.1136/hrt.82.3.351. PMID: 10455086
8. Backer CL, Mavroudis C. Congenital Heart Surgery Nomenclature and Database Project: patent ductus arteriosus, coarctation of the aorta, interrupted aortic arch. Ann Thorac Surg. 2000;69(4 Suppl):S298–307. https://doi.org/10.1016/S0003-4975(99)01282-5. PMID: 10798433

9. Hoffman JI, Kaplan S. The incidence of congenital heart disease. J Am Coll Cardiol. 2002;39(12):1890–900. https://doi.org/10.1016/S0735-1097(02)01886-7. PMID: 12084585
10. Varghese R, Saheed SB, Omoregbee B, Ninan B, Pavithran S, Kothandam S. Surgical repair of interrupted aortic arch and interrupted pulmonary artery. Ann Thorac Surg. 2015;100(6):e139–40. https://doi.org/10.1016/j.athoracsur.2015.06.098. PMID: 26652555
11. Ramirez Alcantara J, Mendez MD. Interrupted aortic arch. In: StatPearls. Treasure Island: StatPearls Publishing; 2020. PMID: 32644592.
12. Sai Krishna C, Bhan A, Sharma S, Kiran U, Venugopal P. Interruption of aortic arch in adults: surgical experience with extra-anatomic bypass. Tex Heart Inst J. 2005;32(2):147–50. PMID: 16107103
13. Celoria GC, Patton RB. Congenital absence of the aortic arch. Am Heart J. 1959;58(3):407–13. https://doi.org/10.1016/0002-8703(59)90147-8. PMID: 14437010
14. Baumgartner H, De Backer J, Babu-Narayan SV, Budts W, Chessa M, Diller GP, et al. 2020 ESC guidelines for the management of adult congenital heart disease: the Task Force for the management of adult congenital heart disease of the European Society of Cardiology (ESC). Eur Heart J. 2021;42(6):563–645. https://doi.org/10.1093/eurheartj/ehaa554. PMID: 32860028
15. Sinha SC. Transcatheter intervention for coarctation of aorta: current status. Indian J Clin Cardiol. 2021;2(1):44–50. https://doi.org/10.1177/2632463620978008.
16. Walhout RJ, Suttorp MJ, Mackaij GJ, Ernst JM, Plokker HW. Long-term outcome after balloon angioplasty of coarctation of the aorta in adolescents and adults: is aneurysm formation an issue? Catheter Cardiovasc Interv. 2009;73(4):549–56. https://doi.org/10.1002/ccd.21899. PMID: 19235242
17. Fawzy ME, Awad M, Hassan W, Al Kadhi Y, Shoukri M, Fadley F. Long-term outcome (up to 15 years) of balloon angioplasty of discrete native coarctation of the aorta in adolescents and adults. J Am Coll Cardiol. 2004;43(6):1062–7. https://doi.org/10.1016/j.jacc.2003.11.034. PMID: 15028367
18. Benjamin G, Smith SA, Qureshi. Coarctation of the aorta. In: Percutaneous interventional cardiovascular medicine: the PCR-EAPCI textbook. Toulouse: Europa Edition; 2019.
19. Feltes TF, Bacha E, Beekman RH 3rd, Cheatham JP, Feinstein JA, Gomes AS, et al. Indications for cardiac catheterization and intervention in pediatric cardiac disease: a scientific statement from the American Heart Association. Circulation. 2011;123(22):2607–52. https://doi.org/10.1161/CIR.0b013e31821b1f10. PMID: 21553329
20. Forbes TJ, Moore P, Pedra CA, Zahn EM, Nykanen D, Hijazi ZM, et al. Intermediate follow-up following intravascular stenting for treatment of coarctation of the aorta. Catheter Cardiovasc Interv. 2007;70(4):569–77. https://doi.org/10.1002/ccd.21191. PMID: 17896403
21. Cowley CG, Orsmond GS, Feola P, McQuillan L, Shaddy RE. Long-term, randomized comparison of balloon angioplasty and surgery for native coarctation of the aorta in childhood. Circulation. 2005;111(25):3453–6. https://doi.org/10.1161/CIRCULATIONAHA.104.510198. PMID: 15967831
22. Chessa M, Carrozza M, Butera G, Piazza L, Negura DG, Bussadori C, et al. Results and mid-long-term follow-up of stent implantation for native and recurrent coarctation of the aorta. Eur Heart J. 2005;26(24):2728–32. https://doi.org/10.1093/eurheartj/ehi443. PMID: 16141259
23. Sohrabi B, Jamshidi P, Yaghoubi A, Habibzadeh A, Hashemi-Aghdam Y, Moin A, et al. Comparison between covered and bare Cheatham-Platinum stents for endovascular treatment of patients with native post-ductal aortic coarctation: immediate and intermediate-term results. JACC Cardiovasc Interv. 2014;7(4):416–23. https://doi.org/10.1016/j.jcin.2013.11.018. PMID: 24630879
24. Forbes TJ, Garekar S, Amin Z, Zahn EM, Nykanen D, Moore P, et al. Procedural results and acute complications in stenting native and recurrent coarctation of the aorta in patients over 4 years of age: a multi-institutional study. Catheter Cardiovasc Interv. 2007;70(2):276–85. https://doi.org/10.1002/ccd.21164. PMID: 17630664

25. Egan M, Holzer RJ. Comparing balloon angioplasty, stenting and surgery in the treatment of aortic coarctation. Expert Rev Cardiovasc Ther. 2009;7(11):1401–12. https://doi.org/10.1586/erc.09.121. PMID: 19900020
26. Holzer RJ, Chisolm JL, Hill SL, Cheatham JP. Stenting complex aortic arch obstructions. Catheter Cardiovasc Interv. 2008;71(3):375–82. https://doi.org/10.1002/ccd.21379. PMID: 18307231
27. Sadeghipour P, Mohebbi B, Firouzi A, Khajali Z, Alemzadeh-Ansari MJ, Pouraliakbar H, et al. Balloon-expandable Cheatham-Platinum stents versus self-expandable Nitinol stents in coarctation of aorta. JACC Cardiovasc Interv. 2022;15(3):308–17. https://doi.org/10.1016/j.jcin.2021.11.017. PMID: 35144788
28. Firoozi A, Mohebbi B, Noohi F, Bassiri H, Mohebbi A, Abdi S, et al. Self-expanding versus balloon-expandable stents in patients with isthmic coarctation of the aorta. Am J Cardiol. 2018;122(6):1062–7. https://doi.org/10.1016/j.amjcard.2018.06.013. PMID: 30072126
29. Firouzi A, Hosseini Z, Shafiei A, Mohebbi B, Noohi F, Alemzadeh-Ansari MJ, et al. Paradigm shift in management of interrupted aortic arch in adulthood. Curr Probl Cardiol. 2021;46(3):100717. https://doi.org/10.1016/j.cpcardiol.2020.100717. PMID: 33162052
30. Holzer R, Qureshi S, Ghasemi A, Vincent J, Sievert H, Gruenstein D, et al. Stenting of aortic coarctation: acute, intermediate, and long-term results of a prospective multi-institutional registry—Congenital Cardiovascular Interventional Study Consortium (CCISC). Catheter Cardiovasc Interv. 2010;76(4):553–63. https://doi.org/10.1002/ccd.22587. PMID: 20853353
31. Carr JA. The results of catheter-based therapy compared with surgical repair of adult aortic coarctation. J Am Coll Cardiol. 2006;47(6):1101–7. https://doi.org/10.1016/j.jacc.2005.10.033. PMID: 16545638
32. Hager A, Kanz S, Kaemmerer H, Schreiber C, Hess J. Coarctation Long-term Assessment (COALA): significance of arterial hypertension in a cohort of 404 patients up to 27 years after surgical repair of isolated coarctation of the aorta, even in the absence of restenosis and prosthetic material. J Thorac Cardiovasc Surg. 2007;134(3):738–45. https://doi.org/10.1016/j.jtcvs.2007.04.027. PMID: 17723827
33. Forbes TJ, Kim DW, Du W, Turner DR, Holzer R, Amin Z, et al. Comparison of surgical, stent, and balloon angioplasty treatment of native coarctation of the aorta: an observational study by the CCISC (Congenital Cardiovascular Interventional Study Consortium). J Am Coll Cardiol. 2011;58(25):2664–74. https://doi.org/10.1016/j.jacc.2011.08.053. PMID: 22152957
34. Meadows JJ, Minahan M, McElhinney DB, McEnaney K, Ringel R, Investigators COASTII. Intermediate outcomes in the prospective, multicenter Coarctation of the Aorta Stent Trial (COAST II). Circulation. 2023;147(16):1223–33. https://doi.org/10.1161/CIRCULATIONAHA.122.060260. PMID: 36876491

Chapter 13
Percutaneous Pulmonary Valve Commissurotomy

Ata Firouzi, Zahra Hosseini, and Mohammadreza Baay

Abstract Congenital isolated PS represents 5%–10% of all congenital heart defects. In most cases, the stenosis results from the fusion of commissures, which causes the valve leaflets to "dome" and leads to a narrowed annulus. Most adult patients with chronic stenosis exhibit infundibular hypertrophy and post-stenotic dilation. Currently, balloon valvuloplasty is the preferred treatment for individuals with severe stenosis or those who are symptomatic. Long-term studies indicate that outcomes between PTPC and surgical intervention are comparable, with minimal complications reported.

Keywords Valvular pulmonary stenosis · Thick and dome-shaped leaflets Infundibular hypertrophy · Post-stenotic dilatation · Long-term outcomes

Abbreviations

LV	Left ventricle
NYHA	New York Heart Association
PS	Pulmonary valve stenosis
PTPC	Percutaneous pulmonary valve commissurotomy
RA	Right atrium
RV	Right ventricle
TTE	Transthoracic echocardiography

A. Firouzi · Z. Hosseini (✉) · M. Baay
Cardiovascular Intervention Research Center, Rajaie Cardiovascular Institute, Iran University of Medical Sciences, Tehran, Iran

A. Firouzi et al. (eds.), *Mastering Skills in Complex Cardiovascular Interventions*, https://doi.org/10.1007/978-3-031-97321-5_13

Table 13.1 Isolated PS severity

Mild	Peak gradient <36 mm Hg (peak velocity <3 m/s)
Moderate	Peak gradient 36–64 mm Hg (peak velocity 3–4 m/s)
Severe	Peak gradient 64 mm Hg (peak velocity >4 m/s); mean gradient >35 mm Hg

Introduction

Congenital isolated PS accounts for 5%–10% of all congenital heart disease [1]. In most cases, the stenosis results from the fusion of commissures, causing "doming" of the valve leaflets with a narrow central opening but a preserved mobile valve base. When RV function and transvalvular flow are normal, PS is classified as mild if the peak gradient across the obstruction is <36 mm Hg, moderate if the gradient is 36–64 mm Hg, and severe if the gradient exceeds 64 mm Hg (Table 13.1). The severity of symptoms in adult patients depends on the degree of valvular stenosis, the presence of subvalvular or peripheral stenosis, the extent of RV dysfunction, tricuspid regurgitation, and any associated defects such as atrial septal defect or ventricular septal defect. Patients with isolated mild PS are typically asymptomatic, and the condition is generally non-progressive. Nonetheless, moderate PS may progress at the valvular or subvalvular level due to compensatory myocardial hypertrophy. Those with severe PS often present with exertional dyspnea, chest pain, palpitations, cyanosis, syncope, and, in advanced cases, signs and symptoms of RV failure. Intervention is usually necessary for patients with moderate to severe PS to alleviate symptoms [2].

Since the first successful PTPC performed by Kan et al. [3] in 1982, this procedure has replaced surgical methods and is now the preferred treatment for patients with PS, regardless of age or valve morphology.

PTPC consistently provides a safe and significant improvement in valvular obstruction, with less trauma and equivalent sustained gradient relief compared to surgical intervention. Although numerous studies have assessed the success rate of PTPC, long-term data remain limited. A successful procedure is defined as achieving a transvalvular gradient of <30 mm Hg with no more than moderate pulmonary regurgitation.

Diagnostic Workup

The diagnostic findings on physical examination include a systolic ejection murmur in the second left intercostal space, the severity of which correlates with its duration, a late systolic peaking murmur, and a diminished P2. In severe cases, an ECG may show evidence of RV hypertrophy and RA enlargement. TTE is an essential diagnostic tool. In severe cases, TTE reveals RV and RA enlargement with a D-shaped septum, moderate-to-severe tricuspid regurgitation that can help estimate RV systolic pressure, a fused and domed pulmonary valve with a narrowed orifice (annulus diameter), and post-stenotic pulmonary artery dilatation. Other anomalies, such as atrial septal defect, patent foramen ovale, ventricular septal defect, patent ductus arteriosus, subvalvular or peripheral pulmonary stenosis, aortic stenosis, and evaluation of all other valves, should also be considered. The key finding for selecting the appropriate balloon size is the hinge-to-hinge annulus diameter.

Challenging Cases

Case 1

The patient was a 28-week pregnant woman (G2P1) referred for evaluation due to exertional dyspnea (NYHA functional class III), palpitations, and one episode of syncope. On physical examination, her hemodynamic status was stable (blood pressure: 90/60 mm Hg, heart rate: 100 bpm, and SpO_2: 93%). Jugular venous pressure was elevated, and a harsh systolic murmur with a thrill was audible in the left second intercostal space. ECG showed sinus rhythm, RV hypertrophy criteria, and RA enlargement. TTE revealed normal LV size and systolic function, severe RV enlargement with moderate dysfunction and a D-shaped septum, a normal aortic valve with mild aortic insufficiency, and a normal mitral valve with mild-to-moderate mitral regurgitation. Moderate-to-severe tricuspid regurgitation was noted, with a tricuspid regurgitation gradient of 120 mm Hg. The pulmonary valve leaflets exhibited a domed shape with restricted motion and severe stenosis (annulus diameter: 18 mm and peak pressure gradient: 100 mm Hg), along with post-stenotic pulmonary artery dilatation. No evidence of subvalvular or peripheral pulmonary stenosis was observed. The only additional anomaly detected was a stretched patent foramen ovale with a right-to-left shunt.

Given the patient's symptoms and echocardiography findings, the heart team, in consultation with a gynecologist and after discussing the risks of radiation exposure to both the mother and fetus with the patient and her family, decided to proceed with transcatheter pulmonary valve commissurotomy. Additional concerns during the procedure included the potential procedural stress on the fetus, the risk of hemodynamic instability during the intervention, and radiation exposure. Ionizing radiation poses significant risks to the fetus or neonate, including growth retardation, mental retardation, structural abnormalities, microcephaly, and seizures. These complications are not observed at doses below 50 mGy but are expected at doses ≥100 mGy [4]. Termination of pregnancy is strongly recommended if radiation exposure exceeds 10 rads [5].

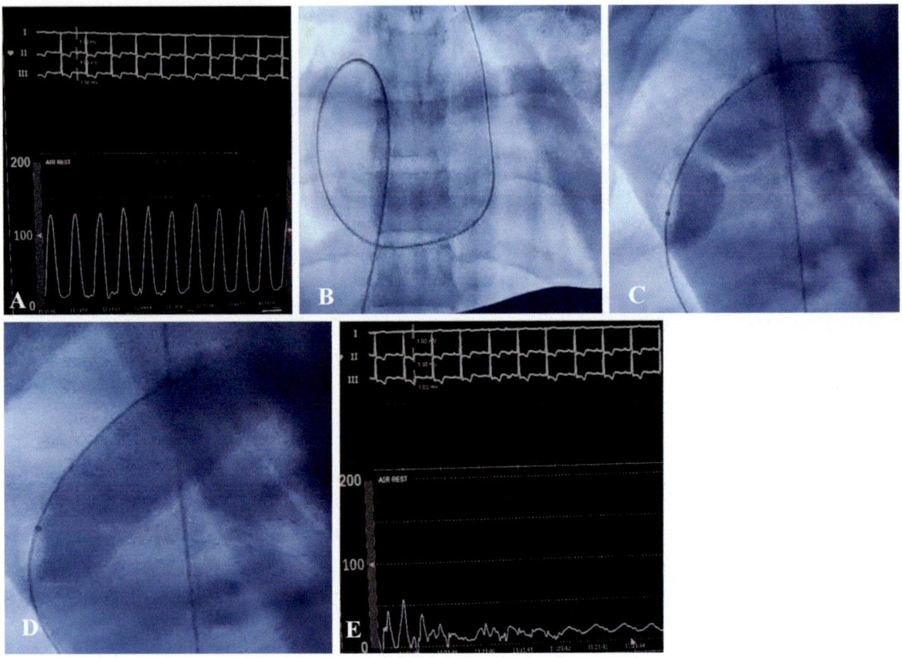

Fig. 13.1 (**a**–**e**) TTE, following the procedure, revealed a transvalvular gradient of 25 mm Hg, no subvalvular gradient, and mild-to-moderate pulmonary regurgitation

Right Heart Catheterization (Fig. 13.1a)

The procedure yielded the following:

- Blood pressure: 90/60 mm Hg;
- RV pressure: 120/20 mm Hg;
- Pulmonary artery pressure: 25/10 mm Hg; and
- Mean RA pressure: 20 mm Hg.

Procedural Technique

After inserting 6 Fr sheaths into the right common femoral artery and vein, removable lead sheeting (from a catheterization laboratory table shield) was placed on the procedure table beneath the patient. This shielding extended from above the umbilicus to below the mid-thigh. The removable lead was chosen to allow for emergency imaging of the abdomen and pelvis if needed. In addition, standard gonadal shielding was positioned to further protect the uterus and fetus. The procedure commenced with a cardiac anesthesiologist, cardiac surgeon, gynecologist, and echocardiographer in attendance.

Using minimal fluoroscopy and without RV injection, a Numed II 24 × 40 mm balloon was selected based on echocardiography data (pulmonary valve hinge-to-hinge annulus diameter: 18 mm). After crossing the pulmonary valve with a 0.035-inch, 260 cm long guidewire, a pigtail catheter was advanced over the wire (Fig. 13.1b). The guidewire was then exchanged for a 0.035-inch Amplatzer Super Stiff 260 cm long guidewire positioned in the left pulmonary artery. After aligning the valve and balloon in the lateral view, the balloon was inflated with diluted contrast using a PTMC syringe until the waist completely disappeared (Fig. 13.1c, d).

The most critical aspect of this step is ensuring the balloon remains stable in position (without slippage). Each inflation-deflation cycle should not exceed 10 seconds, with continuous hemodynamic monitoring. In this case, to minimize radiation exposure, only one inflation was performed. After the procedure, the final transvalvular gradient was measured using a multipurpose catheter, revealing a gradient of 20 mm Hg (Fig. 13.1e) and no subvalvular gradient [6].

Case 2

The patient was a 45-year-old man referred for evaluation due to exertional dyspnea (NYHA functional class II) and pleuritic chest pain. On physical examination, his hemodynamics were stable. A harsh mid-systolic murmur was audible in the left second intercostal space and along the left sternal border, accompanied by an RV heave. ECG revealed RV hypertrophy criteria with a strain pattern in the right precordial leads. Echocardiography findings included normal LV size and function, moderate-to-severe RV enlargement with mild-to-moderate dysfunction, and normal aortic and mitral valves. The pulmonary valve leaflets appeared dome-shaped and thickened, with restricted motion (annulus diameter: 19 mm). The transvalvular gradient was 110 mm Hg, with no evidence of subvalvular or peripheral pulmonary artery stenosis. Moderate tricuspid regurgitation was noted, with a tricuspid regurgitation gradient of 130 mm Hg. No atrial septal defect, ventricular septal defect, patent ductus arteriosus, or other anomalies were detected.

Based on the patient's symptoms and echocardiography findings, he was deemed a candidate for percutaneous balloon valvuloplasty.

Right Heart Catheterization

The procedure revealed the following information:

- RV systolic pressure: 130/15 mm Hg;
- Pulmonary arterial pressure: 25/10 mm Hg;
- Transvalvular gradient: 105 mm Hg;
- Mean RA pressure: 18 mm Hg.
- Cardiac index: 2.3 L/min; and
- SO_2: 95%.

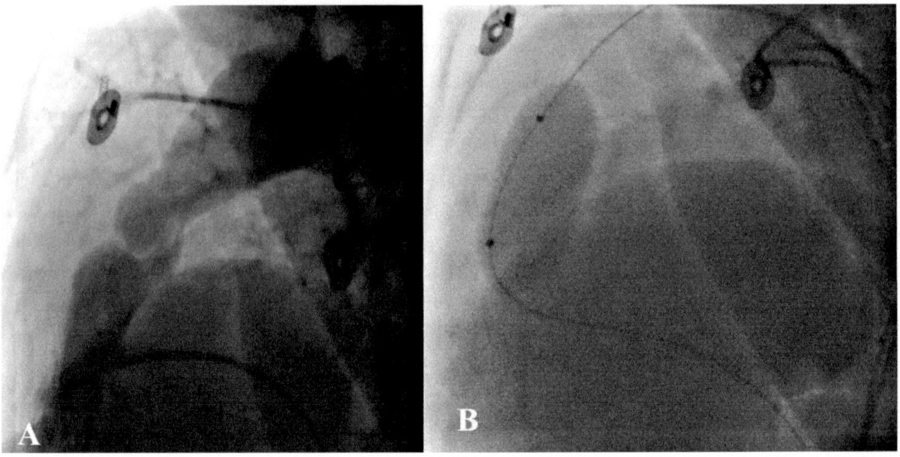

Fig. 13.2 (**a**) RV injection in the lateral projection demonstrated thick, dome-shaped pulmonary valve leaflets along with post-stenotic pulmonary artery dilation. (**b**) Following valvuloplasty with an Atlas 24 × 40 balloon, the waist completely disappeared

Procedural Technique [7]

After inserting 6 Fr femoral arterial and venous sheaths and performing right heart catheterization under local anesthesia, RV injection in anteroposterior and lateral projections revealed severe RV hypertrophy and enlargement, along with thick, dome-shaped pulmonary valve leaflets and post-stenotic pulmonary artery dilatation (Fig. 13.2a).

Fluoroscopy measured the hinge-to-hinge annulus diameter at 20 mm (compared with 19 mm on TTE). An Atlas 24 × 40 mm balloon was selected. The pulmonary valve was crossed with a 0.035-inch, 260 cm long guidewire, and a pigtail catheter was advanced over it. The wire was then exchanged for a 0.035-inch Amplatzer Super Stiff 260 cm long guidewire positioned in the left pulmonary artery. After aligning the valve and balloon in the lateral view, the balloon was inflated with diluted contrast using a PTMC syringe until the waist disappeared (Fig. 13.2b).

Following the procedure, the final transvalvular gradient was measured using a multipurpose catheter. Initially, the gradient was >30 mm Hg, so the procedure was repeated with higher-pressure inflation (this can be repeated up to eight times) or with larger balloons. The final transvalvular gradient achieved was 25 mm Hg.

Discussion

Isolated pulmonary stenosis is a common congenital heart disease, with most cases diagnosed during childhood. Until 1982, surgical valvotomy was the standard treatment, when Kan et al. [8] introduced the technique of percutaneous valvuloplasty.

PTPC has been established as an effective and safe procedure with a high success rate for treating severe pulmonary stenosis in adult patients, demonstrating sustainable immediate, short-, mid-, and long-term outcomes [9, 10]. Several studies have confirmed the efficacy and feasibility of this procedure in adults, making PTPC a viable alternative to surgery with favorable long-term results. Pulmonary stenosis may also be associated with infundibular hypertrophy and post-stenotic dilation, particularly in older patients with long-standing pulmonary stenosis or higher degrees of valvular stenosis [11]. Post-stenotic dilation and infundibular hypertrophy are more commonly observed in patients with higher residual transvalvular gradients following valvuloplasty [6]. Moreover, patients with dysplastic valvular pulmonary stenosis tend to have a less favorable response to treatment. Some studies suggest that infundibular hypertrophy may regress following surgical valvotomy [12]. Long- and mid-term studies have demonstrated that this complication of severe, long-standing stenosis in adults may improve after a successful PTPC [13, 14]. In one study, the use of beta-blockers in patients with significant infundibular stenosis after the procedure reduced the gradient within 3–6 months [15]. This finding suggests that beta-blockers can further reduce the gradient in adult patients with severe pulmonary stenosis and infundibular hypertrophy, leading to improved NYHA functional class. A long-term study by Fawzy et al. [13] demonstrated that severe tricuspid regurgitation regressed following a successful PTPC. In addition, Blom et al. [16] reported rapid improvement in RV systolic function but delayed improvement in RV diastolic function due to the slow regression of RV mass.

The threshold for defining a successful PTPC in adult patients remains controversial. However, in our study, we demonstrated that the optimal cut-off is a post-valvuloplasty transvalvular gradient <25 mm Hg [6].

Complications

Several studies and case reports have documented rare complications associated with PTPC, such as pulmonary edema in elderly patients [13]. Other infrequent complications include seizures, cardiac arrest, cerebrovascular accidents, balloon rupture, RV outflow tract perforation, pulmonary artery dissection, endocarditis, pericardial effusion (which typically follows a benign course), and rupture of the tricuspid valve papillary muscle.

Conclusion

Experience demonstrates that percutaneous balloon valvuloplasty is a safe and effective treatment for congenital pulmonary stenosis, applicable to patients of any age with favorable short- and long-term outcomes. Freedom from re-intervention is 95%, 88%, and 84% at 5, 10, and 20 years, respectively [3]. The most critical factor for success is appropriate balloon selection based on the annulus diameter. We believe that our recent, more aggressive approach has led to improved outcomes. Although definitive indications have not yet been established, this procedure is currently our treatment of choice for congenital pulmonary stenosis.

References

1. Hong D, Xiao-Yun S, Qian M, Jian-Yong Z, Jun-Zhong T, Jian-Hua L, et al. Immediate therapeutic outcomes and medium-term follow-up of percutaneous balloon pulmonary valvuloplasty in infants with pulmonary stenosis: a single-center retrospective study. Chin Med J. 2017;130(23):2785–90. https://doi.org/10.4103/0366-6999.219144. PMID: 29176192
2. Lanjewar C, Phadke M, Singh R, Sabnis G, Jadhav D, Kerkar P. Percutaneous balloon valvuloplasty with Inoue balloon catheter technique for pulmonary stenosis in adolescents and adults. Indian Heart J. 2017;69(2):176–81. https://doi.org/10.1016/j.ihj.2016.08.008. PMID: 28460766
3. Merino-Ingelmo R, Santos-de Soto J, Coserria-Sánchez JF, Descalzo-Señoran A, Fernández-González R, Valverde-Pérez I. Long-term results of percutaneous balloon valvuloplasty in pulmonary stenosis in the pediatric population. Rev Esp Cardiol (Engl Ed). 2014;67(5):374–9. https://doi.org/10.1016/j.rec.2013.09.021. PMID: 24774706
4. Shaw P, Duncan A, Vouyouka A, Ozsvath K. Radiation exposure and pregnancy. J Vasc Surg. 2011;53(1 Suppl):28S–34S. https://doi.org/10.1016/j.jvs.2010.05.140. PMID: 20869192
5. Hameed AB, Mehra A, Rahimtoola SH. The role of catheter balloon commissurotomy for severe mitral stenosis in pregnancy. Obstet Gynecol. 2009;114(6):1336–40. https://doi.org/10.1097/AOG.0b013e3181bf6786. PMID: 19935038
6. Khajali Z, Firouzi A, Hosseini Z, Alemzadeh-Ansari MJ, Noohi F, Saedi S, et al. The result of percutaneous transcatheter pulmonary valvuloplasty in adult patients. J Cardiol Cardiovasc Ther. 2021;17(1):555951. https://doi.org/10.19080/JOCCT.2021.17.555951.
7. Firouzi A, Hosseini Z. Percutaneous pulmonary valve commissurotomy (PTPC). In: Maleki M, Alizadehasl A, editors. Case-based clinical cardiology. London: Springer; 2021. p. 615–21. https://doi.org/10.1007/978-1-4471-7496-7_67.
8. Kan JS, White RI Jr, Mitchell SE, Gardner TJ. Percutaneous balloon valvuloplasty: a new method for treating congenital pulmonary-valve stenosis. N Engl J Med. 1982;307(9):540–2. https://doi.org/10.1056/NEJM198208263070906. PMID: 7099226
9. Kaul UA, Singh B, Tyagi S, Bhargava M, Khalilullah M, Choudhary SK, et al. Long-term results after balloon pulmonary valvuloplasty in adults. Am Heart J. 1993;126(5):1152–5. https://doi.org/10.1016/0002-8703(93)90674-9. PMID: 8237759
10. Behjati-Ardakani M, Moshtaghion SH, Rajaei S. Long-term results of balloon pulmonary valvuloplasty in children with congenital pulmonary stenosis. Iran J Pediatr. 2013;23(1):32–6. PMID: 23550072
11. Idrizi S, Kabashi A, Ahmetgjekaj I, Haliti E, Ismaili-Jaha V, Muçaj S, et al. Interventional treatment of pulmonary stenosis: a single-center experience. Open Access Maced J Med Sci. 2015;3(3):408–12. https://doi.org/10.3889/oamjms.2015.082. PMID: 27275257
12. Arendrup HC, Kruse-Andersen S, Alstrup P. Regression of infundibular hypertrophy after pulmonary valvulotomy without myocardial resection. Scand J Thorac Cardiovasc Surg. 1983;17(3):243–8. https://doi.org/10.3109/14017438309101046. PMID: 6658385
13. Fawzy ME, Mercer JL, Dunn B, al-Amri M, Andaya W, Galal O, et al. Long-term results (up to 17 years) of pulmonary balloon valvuloplasty in adults and its effects on concomitant severe infundibular stenosis and tricuspid regurgitation. Am Heart J. 2007;153(3):433–8. https://doi.org/10.1016/j.ahj.2006.11.021. PMID: 17307424

14. Fawzy M, Bakhsh M, Galal O, Dunn B, Shaikh A, Khan B, et al. Regression of infundibular pulmonary stenosis after successful balloon pulmonary valvuloplasty in adults. Catheter Cardiovasc Diagn. 1990;21(2):77–81. https://doi.org/10.1002/ccd.1810210205. PMID: 2225046
15. Thapar MK, Rao PS. Significance of infundibular obstruction following balloon valvuloplasty for valvar pulmonic stenosis. Am Heart J. 1989;118(1):99–103. https://doi.org/10.1016/0002-8703(89)90077-0. PMID: 2741798
16. Romeih S, Kroft LJ, van der Plas MN, de Roos A, Hazekamp MG, Mulder BJ, et al. Delayed improvement of right ventricular diastolic function and regression of right ventricular mass after percutaneous pulmonary valve implantation in patients with congenital heart disease. Am Heart J. 2009;158(1):40–6. https://doi.org/10.1016/j.ahj.2009.04.023. PMID: 19540390

Chapter 14
Transcatheter Pulmonary Arteriovenous Malformation Closure

Ata Firouzi, Zahra Khajali, and Zahra Hosseini

Abstract PAVMs or fistulae are abnormal vascular structures characterized by low resistance and high flow, which bypass the pulmonary capillary system. In this condition, the arterial system is directly connected to the venous system, resulting in the creation of an intrapulmonary right-to-left shunt. This allows systemic venous blood to circumvent gas exchange and processing by the pulmonary capillary bed, leading to hypoxemia.

HHT is the primary underlying cause. Given the high risk of systemic thromboembolism and hypoxemia in patients with significant shunts, all treatment options—whether transcatheter or surgical—should be considered. Currently, transcatheter feeding artery occlusion (for arteries with a diameter > 3 mm) is regarded as the gold standard for treatment. Due to the considerable risk of recanalization, the occlusive device should be placed as distally as possible. In many instances, multiple fistulae are identified, necessitating a multi-staged approach. Meticulous follow-up is essential for these patients, particularly during pregnancy.

Keywords Pulmonary arteriovenous malformation · Right-to-left shunt
Hypoxemia · Systemic embolism · Hereditary hemorrhagic telangiectasia
Transcatheter closure · Variety occlusive device · Recanalization · Follow-up

Abbreviations

ADO	Amplatzer duct occluder
CT	Computed tomography
HHT	Hereditary hemorrhagic telangiectasia
PAVM	Pulmonary arteriovenous malformation
SO_2	Oxygen saturation

A. Firouzi · Z. Khajali · Z. Hosseini (✉)
Cardiovascular Intervention Research Center, Rajaie Cardiovascular Institute, Iran University of Medical Sciences, Tehran, Iran

A. Firouzi et al. (eds.), *Mastering Skills in Complex Cardiovascular Interventions*, https://doi.org/10.1007/978-3-031-97321-5_14

Introduction

PAVMs represent a rare cardiovascular anomaly, occurring in approximately 1 in 2600 individuals. These abnormal vascular structures, also known as fistulae, are characterized by low resistance and high flow, bypassing the pulmonary capillary system. In this condition, the arterial and venous systems are directly connected, creating an intrapulmonary right-to-left shunt. As a result, systemic venous blood circumvents gas exchange and pulmonary capillary processing, leading to hypoxemia. Most PAVM cases are congenital, often associated with HHT or polysplenia syndrome, although isolated instances have also been documented. Additional causes include acquired conditions such as schistosomiasis, tuberculosis, trauma, juvenile hepatic cirrhosis, and the Glenn operation—whether in its classic form or its modern adaptation, the bidirectional cavopulmonary anastomosis (60%–100%) [1]. The absence of symptoms does not rule out a PAVM diagnosis, as case series have shown that 13%–55% of adult and pediatric patients with PAVMs are clinically asymptomatic. The most common clinical feature is cyanosis, which is frequently observed in cases involving large PAVMs, often starting from the newborn period. It is important to note that while hypoxemia caused by right-to-left shunting can be severe, the most notable clinical finding in this patient group is the presence of asymptomatic hypoxemia [2]. A gradual increase in cyanosis is often observed due to the opening of new fistulae, the enlargement of existing ones, or the development of polycythemia. PAVM progression is commonly noted during pregnancy. Other clinical manifestations include hemoptysis (resulting from PAVM hemorrhage), hemothorax (particularly in pregnant patients and those with pulmonary hypertension), dyspnea (occurring in 15% of cases), platypnea (more pronounced in patients with HHT and pregnant women), migraine, and pleuritic chest pain (seen in 10% of cases) [3]. Neurological sequelae and other systemic thromboembolic events (paradoxical emboli) are common, occurring in up to 33% of patients with HHT. The most significant risk factors for stroke include low serum iron levels—often due to inadequate iron intake to compensate for HHT-related hemorrhagic losses—and low pulmonary artery pressure, which is thought to promote the paradoxical embolic process. In addition, serum fibrinogen, the primary circulating plasma protein responsible for platelet adhesion, plays a role [4]. Both cerebral infarctions and cerebral abscesses may develop. In advanced cases, high cardiac output states and pulmonary hypertension can rarely occur. Patients with long-standing cyanosis may exhibit clubbing. Hemorrhages can also arise from other areas, such as the gastrointestinal, nasopharyngeal, and urogenital tracts. Bacterial endarteritis has also been reported.

Challenging Cases

Case 1

The patient was a 40-year-old woman who presented to our clinic with central cyanosis, dyspnea, and two episodes of hemoptysis. Her SO_2 level at room air was 89%, and her hematocrit level was 57%. A chest X-ray raised suspicion of PAVM, while ECG results were normal. During catheterization, a solitary PAVM in the right lung was identified.

Under local anesthesia, 6 Fr sheaths were inserted into the right femoral artery and vein, followed by a complete intravenous heparin infusion. Right heart catheterization was performed, revealing a mean pulmonary artery pressure of 18 mm Hg. Angiography of the pulmonary artery trunk, right pulmonary artery, and right upper lobar artery was conducted in anteroposterior and left anterior oblique projections to confirm the presence of PAVM. A selective right pulmonary artery injection using a pigtail catheter demonstrated rapid contrast opacification of the pulmonary vein and left atrium (Fig. 14.1).

The location and diameter of the feeding artery, along with the number of fistula tracts, were identified. A single feeding artery, measuring 4 mm in diameter, was found in the right upper lobe, accompanied by a tortuous tract. Given the risk of collateralization, PAVM recanalization is common if embolization is performed too proximally. To minimize lesion recurrence, distal embolization—either at the far end of the feeding artery or within the PAVM itself—is recommended.

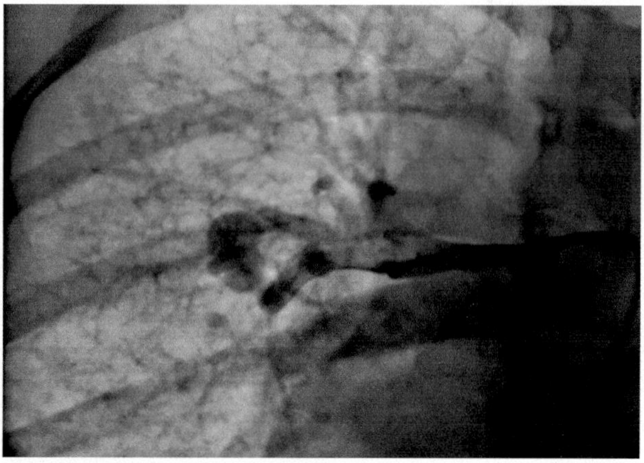

Fig. 14.1 Right pulmonary artery injection shows a PAVM in the middle lobe of the right lung

Using venous access, a straight non-hydrophilic guidewire (0.035 inches, 260 cm) was advanced through a multipurpose catheter (6 Fr) into the lobar branch of the right upper pulmonary artery. The guidewire was gently maneuvered as distally as possible through the dedicated artery. Subsequently, the catheter was advanced over the wire to the distal portion of the artery. Selective angiography clearly visualized the feeding artery and the saccular fistula.

The catheter was withdrawn while the guidewire remained in place. A hydrophilic delivery sheath (Cook 7 Fr) was then gently advanced to the target location. A detachable coil (Cook 6 × 20 mm) was introduced and guided to the distal tip of the sheath. (Typically, a coil oversized by 25%–30% relative to the feeding artery diameter is selected.) The coil's correct positioning was confirmed through angiography. Once verified, the coil was fully deployed and detached (Fig. 14.2).

Final angiography revealed that the PAVM was not fully occluded. An ADO type I device (6 × 8 mm) was selected to achieve complete fistula closure. The next step involved advancing the loaded delivery catheter (after thorough de-airing) through the sheath until its tip reached a position just proximal to the deployed coil. Angiography confirmed the correct placement of the device, after which the sheath was slowly retracted, allowing the device to fully deploy (Fig. 14.3). Moreover, no residual shunt or oozing was detected (Fig. 14.4), and the procedure was concluded. The patient's SO_2 immediately improved from 89% to 94%. It is crucial to emphasize that during the procedure, blood should be aspirated before each injection—even when using a microcatheter—to ensure a solid-liquid column within the catheter. Wires should be removed either under saline in a basin or with a continuous saline drip into the catheter hub. Any air or clot entering a PAVM can directly pass into the left-sided circulation and subsequently reach the brain, potentially causing a transient ischemic attack or more severe complications.

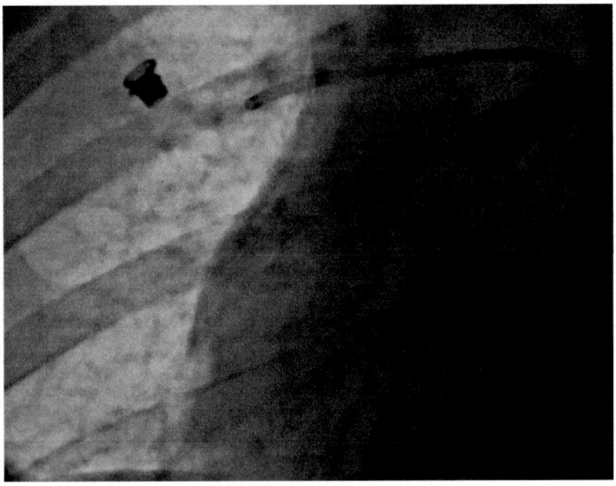

Fig. 14.2 The image shows coil implantation in the malformation

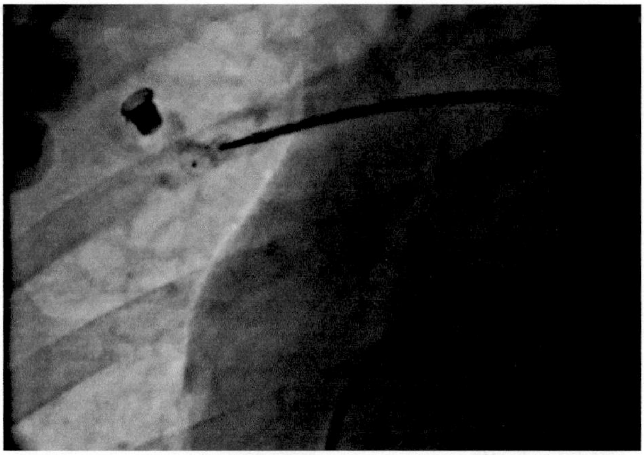

Fig. 14.3 The image shows the implantation of an ADO type I device

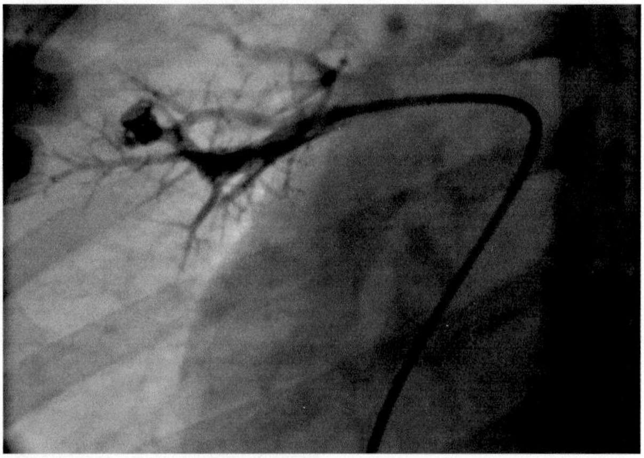

Fig. 14.4 The final angiography showed no residual shunt or oozing

Case 2

A 19-year-old female patient presented to our clinic with profound cyanosis. Her SO_2 level at room air was 75%. She had a history of a pelvic abscess, which had been surgically removed. No prior cardiac evaluation had been conducted, and her family history was unremarkable.

On physical examination, cyanosis was observed in the nail beds, lips, and mucous membranes, along with digital clubbing. She also exhibited knee swelling due to hyperostosis. Her hematocrit level was 65%, and ECG results were normal.

A chest X-ray revealed multiple round, vascular-type densities in the right lung, particularly in the middle and lower lobes (Fig. 14.5). Contrast echocardiography showed filling of the left heart chambers after three injection cycles (Fig. 14.6).

A CT scan revealed diffuse, segmental, and smooth dilatation of the pulmonary artery branches and veins, accompanied by tortuosity. Multiple feeding arteries and veins were also visualized (Fig. 14.7).

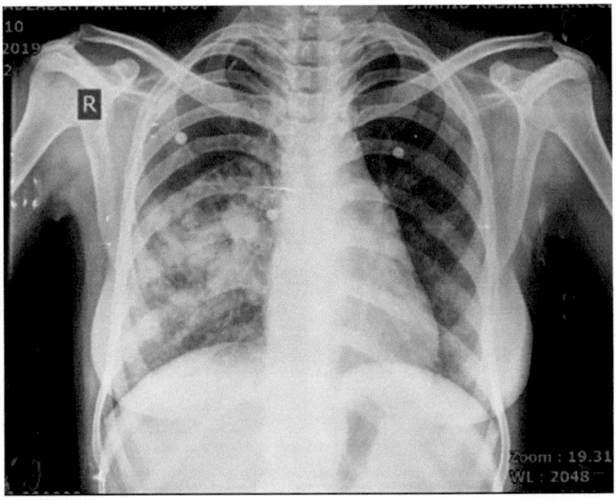

Fig. 14.5 The chest X-ray illustrated multiple vascular-type round densities in the right lung, especially in the middle and lower lobes

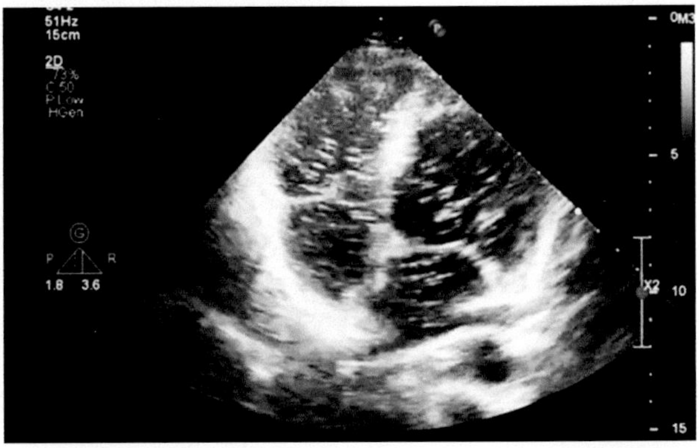

Fig. 14.6 All the chambers were filled during the contrast injection

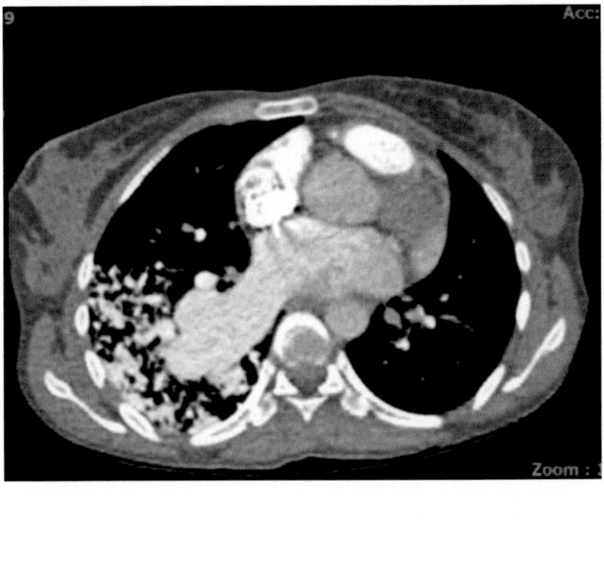

Fig. 14.7 The image demonstrates multiple round, homogeneous masses

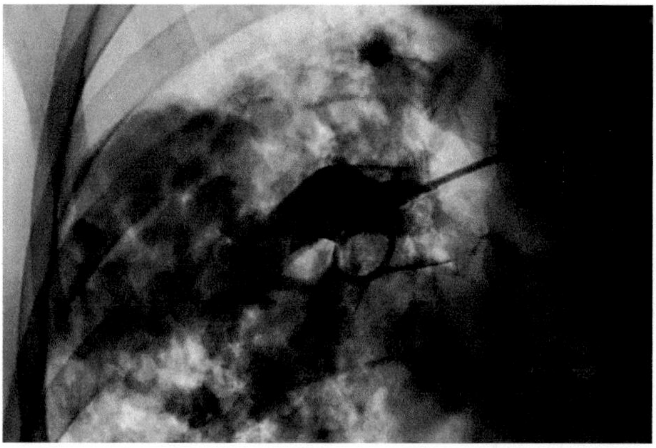

Fig. 14.8 Right pulmonary artery injection showed multiple fistulae and feeding arteries connected to the right lower pulmonary vein

During catheterization, an injection into the right pulmonary artery revealed multiple PAVMs between the lower branch of the right pulmonary artery and the right lower pulmonary vein, with multiple feeding arteries.

First, a hydrophilic guidewire (0.035 inch, 260 cm) was gently advanced through a multipurpose catheter to the distal portion of the largest feeding artery. The catheter was then passed over the guidewire, and the wire was exchanged for a Super Stiff guidewire. An introducer sheath was advanced over the stiff guidewire into the right pulmonary artery, after which the guidewire was removed. Next, an Occlutech muscular ventricular septal defect occluder (8 mm) was mounted on a delivery cable, advanced through the introducer sheath, and deployed in one of the feeding arteries. This process was repeated using two additional devices (two 6 × 8 mm ADO type II devices) in two other feeding arteries. The final right pulmonary artery injection showed a significant reduction in malformation communication and feeding artery flow. Moreover, systemic SO_2 increased from 75% to 88%. The procedure was concluded, and the closure of the remaining lesions was deferred to a subsequent session (Fig. 14.8).

Diagnostic Workup

On physical examination, central cyanosis (which may be less pronounced in patients with HHT due to repeated blood loss and anemia), peripheral telangiectasia, desaturation, and clubbing may be observed. Further, systolic or continuous murmurs may be heard over the area of the fistula.

A chest X-ray may reveal one or more rounded or multilobular opacities if the fistulae are sufficiently large. Cardiac size is typically normal, though cardiomegaly may rarely occur (in cases of very large fistulae causing a hyperdynamic state).

The most sensitive noninvasive screening test for PAVMs, whether in the context of complex congenital heart diseases or in patients with HHT, is contrast echocardiography. A positive test is defined by the appearance of bubbles in the pulmonary vein and left atrium at least three cardiac cycles after their appearance in the right atrium, ruling out intracardiac communications as the cause of right-to-left shunting [5].

Contrast echocardiographic studies often yield positive results even in normal lungs without detectable impairment of gas exchange. Approximately 8% of the general population exhibits this phenomenon at rest, with the percentage increasing to 90% during exercise or other high cardiac output states. Careful grading of the number of bubbles visible in the left ventricle can help differentiate PAVMs from normal variations. Grade III shunts (>100 bubbles per frame) have a positive predictive value of 92.5% for CT-evident PAVMs, compared with a positive predictive value of 13.4% for the lowest positive contrast shunt grade I (<30 bubbles per frame) [6].

Radionuclide testing with the aid of Tc99m-labeled albumin microspheres is another valuable diagnostic tool, providing a quantitative measurement of right-to-left shunts by determining the fraction of the injected microsphere dose that reaches the kidneys.

Chest CT can accurately identify PAVMs. Recently, contrast-enhanced magnetic resonance angiography was compared to CT, demonstrating a sensitivity of 75% and a specificity of 100%. Furthermore, all significant PAVMs with a feeding vessel diameter exceeding 3 mm were detected [7]. These findings suggest that contrast-enhanced magnetic resonance angiography is a highly useful, noninvasive, and non-ionizing method for diagnosing and precisely localizing PAVMs anatomically (Fig. 14.9).

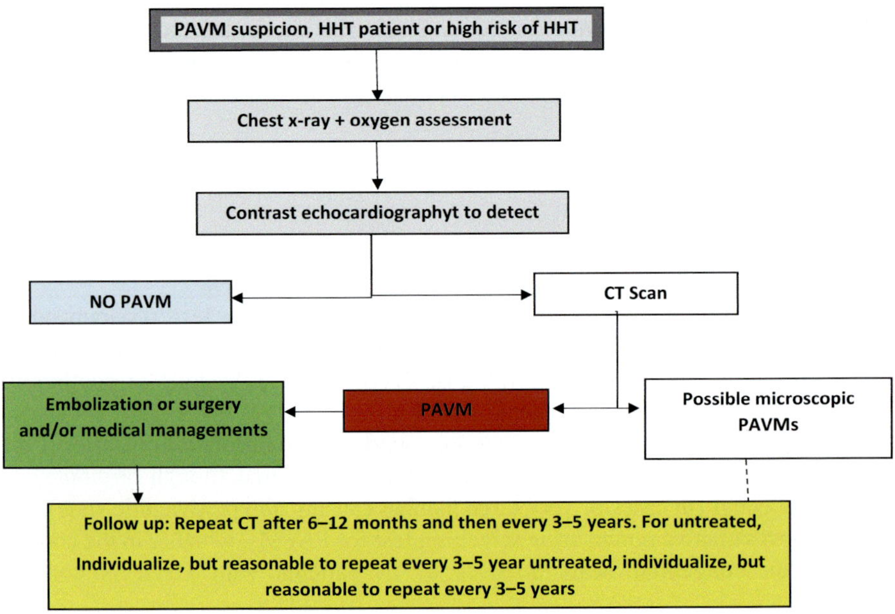

Fig. 14.9 The diagram presents a diagnostic approach to PAVM detection

Cardiac catheterization and pulmonary angiocardiography are the most definitive diagnostic tests. They provide precise information about the location and anatomy of PAVMs and can guide therapeutic embolization. Pulmonary artery pressure is typically normal or low due to reduced pulmonary vascular resistance, particularly when the fistulae are large or diffuse. During injection into the appropriate pulmonary artery, the feeding vessel(s) is visualized, along with rapid opacification of the pulmonary veins and left atrium. Smaller fistulae may only be opacified if a selective injection is performed in the specific branch of the pulmonary artery supplying the PAVM.

Management

If left untreated, PAVMs have complication rates reported to reach 50% or higher, particularly during pregnancy. Diffuse forms are associated with even greater complications, with neurological morbidity reaching 70% in untreated cases. Thus, current guidelines recommend regular screening for families with HHT [8].

Three primary management approaches are available for different types of PAVMs. Transcatheter embolization therapy employs various devices, such as coils, detachable balloons, and specialized devices designed for occluding atrial septal defects, patent ductus arteriosus, and vascular plugs. Surgical options depend on the location and extent of PAVMs and include vascular ligation, local excision, lobectomy, and pneumonectomy, performed via video-assisted thoracoscopic surgery or open thoracotomy. The morbidity and mortality rates for surgical intervention are comparable to those of other thoracic surgeries. Lung transplantation is reserved for patients with refractory, often bilateral and diffuse disease, or those at high risk of fatal complications [9]. In rare cases, medical therapy, including pharmacological management, may be considered.

Transcatheter Embolization Therapy

Currently, embolization therapy is the preferred treatment for PAVMs, provided there are no contraindications such as severe pulmonary hypertension, pulmonary infection, renal failure, or early pregnancy. Treatment is recommended for patients with PAVMs featuring feeding arteries >3 mm in diameter, even if they are asymptomatic [10].

Amplatzer vascular plugs have quickly become the preferred choice for embolization over coils. Their advantages include the ability to occlude large-diameter feeding arteries with a single plug, easier placement at the neck of the sac, and occlusion over a shorter vessel length. This reduces the risk of compromising blood supply to normal lung tissue.

Complications

Imaging findings indicative of successful treatment include a reduction in the diameter of the draining vein, at least a 70% decrease in sac size, and the absence of contrast enhancement [11].

The most common postprocedural complication, occurring in approximately 10% of patients, is self-limiting pleuritic chest pain caused by thrombosis in the feeding artery or sac and/or pulmonary infarction (2%). Pleurisy rates are often higher in patients with feeding vessels larger than 8 mm. Pleural effusions occur in 30% of cases. Additionally, systemic arterial embolization of clots, air, or embolic devices occurs in fewer than 2.3% of cases and may present as transient ischemic attack, angina, or bradycardia [12].

Recanalization is estimated to occur in 10%–25% of cases, with higher rates reported in pediatric patients. The risk of reperfusion due to the recanalization of the embolized lesion depends on several factors, including angioarchitecture, the distance from the coil to the sac (>1 cm), the number of coils used, and the diameter of the feeding artery [13]. Furthermore, the potential for massive hemoptysis resulting from systemic arterial collaterals that may develop in the months following embolization remains a significant concern. This situation necessitates the use of dual systemic (bronchial) and pulmonary angiographic approaches for effective embolization.

Conclusions

PAVMs are a complex condition with diverse etiologies. HHT is frequently associated with this disorder. In most cases, treatment involves transcatheter embolization using various devices. Although recanalization and complications can occur, outcomes are generally favorable. Surgical intervention is reserved for very large or extensively diffuse fistulae. Medical therapy plays a limited, sometimes adjunctive, role in managing diffuse PAVMs [14].

References

1. Kim SJ, Bae EJ, Cho DJ, Park IS, Kim YM, Kim WH, et al. Development of pulmonary arteriovenous fistulas after bidirectional cavopulmonary shunt. Ann Thorac Surg. 2000;70(6):1918–22. https://doi.org/10.1016/S0003-4975(00)01806-8. PMID: 11156095
2. Allen SW, Whitfield JM, Clarke DR, Sujansky E, Wiggins JW. Pulmonary arteriovenous malformations in the newborn: a familial case. Pediatr Cardiol. 1993;14(1):58–61. https://doi.org/10.1007/BF00794852. PMID: 8456036
3. Santhirapala V, Williams LC, Tighe HC, Jackson JE, Shovlin CL. Arterial oxygen content is precisely maintained by graded erythrocytotic responses in settings of high/normal serum iron levels and predicts exercise capacity: an observational study of hypoxaemic patients with pul-

monary arteriovenous malformations. PLoS One. 2014;9(3):e90777. https://doi.org/10.1371/journal.pone.0090777. PMID: 24601907

4. Shovlin CL, Jackson JE, Bamford KB, Jenkins IH, Benjamin AR, Ramadan H, et al. Primary determinants of ischaemic stroke/brain abscess risks are independent of severity of pulmonary arteriovenous malformations in hereditary haemorrhagic telangiectasia. Thorax. 2008;63(3):259–66. https://doi.org/10.1136/thx.2007.087452. PMID: 18024538
5. Kjeldsen AD, Oxhøj H, Andersen PE, Elle B, Jacobsen JP, Vase P. Pulmonary arteriovenous malformations: screening procedures and pulmonary angiography in patients with hereditary hemorrhagic telangiectasia. Chest. 1999;116(2):432–9. https://doi.org/10.1378/chest.116.2.432. PMID: 10453873
6. Gazzaniga P, Buscarini E, Leandro G, Reduzzi L, Grosso M, Pongiglione G, et al. Contrast echocardiography for pulmonary arteriovenous malformations screening: does any bubble matter? Eur J Echocardiogr. 2009;10(4):513–8. https://doi.org/10.1093/ejechocard/jen324. PMID: 19252180
7. Khalil A, Farres MT, Mangiapan G, Tassart M, Bigot JM, Carette MF. Pulmonary arteriovenous malformations. Chest. 2000;117(5):1399–403. https://doi.org/10.1378/chest.117.5.1399. PMID: 10807828
8. Faughnan ME, Lui YW, Wirth JA, Pugash RA, Redelmeier DA, Hyland RH, et al. Diffuse pulmonary arteriovenous malformations: characteristics and prognosis. Chest. 2000;117(1):31–8. https://doi.org/10.1378/chest.117.1.31. PMID: 10631195
9. Swanson KL, Prakash UBS, Stanson AW. Pulmonary arteriovenous fistulas: Mayo Clinic experience, 1982–1997. Mayo Clin Proc. 1999;74(7):671–80. https://doi.org/10.4065/74.7.671. PMID: 10405695
10. Dutton JA, Jackson JE, Hughes JM, Whyte MK, Peters AM, Ussov W, et al. Pulmonary arteriovenous malformations: results of treatment with coil embolization in 53 patients. AJR Am J Roentgenol. 1995;165(5):1119–25. https://doi.org/10.2214/ajr.165.5.7572488. PMID: 7572488
11. White RI Jr, Pollak JS, Wirth JA. Pulmonary arteriovenous malformations: diagnosis and transcatheter embolotherapy. J Vasc Interv Radiol. 1996;7(6):787–804. https://doi.org/10.1016/S1051-0443(96)70851-5. PMID: 8957744
12. Lee DW, White RI Jr, Egglin TK, Pollak JS, Fayad PB, Wirth JA, et al. Embolotherapy of large pulmonary arteriovenous malformations: long-term results. Ann Thorac Surg. 1997;64(4):930–40. https://doi.org/10.1016/S0003-4975(97)00804-7. PMID: 9354507
13. Woodward CS, Pyeritz RE, Chittams JL, Trerotola SO. Treated pulmonary arteriovenous malformations: patterns of persistence and associated retreatment success. Radiology. 2013;269(3):919–26. https://doi.org/10.1148/radiol.13122153. PMID: 23985274
14. Puskas JD, Allen MS, Moncure AC, Wain JC, Hilgenberg AD, Wright C, et al. Pulmonary arteriovenous malformations: therapeutic options. Ann Thorac Surg. 1993;56(2):253–8. https://doi.org/10.1016/0003-4975(93)91157-7. PMID: 8347002

Part III
Transcatheter Valvular Intervention

Chapter 15
Percutaneous Transluminal Mitral Commissurotomy

Ata Firouzi, Zahra Hosseini, and Mohammadreza Baay

Abstract Globally, rheumatic fever is the leading cause of MV stenosis. The onset of clinical symptoms indicating MV obstruction following an initial episode of rheumatic fever can vary significantly, emerging anywhere from a few years to over 20 years later. Common symptoms experienced by patients in the early stages of the disease include exertional dyspnea, orthopnea, pulmonary edema, hemoptysis, palpitations (atrial fibrillation), and systemic embolic events, with cerebral emboli occasionally presenting as the initial symptom. Prolonged severe MS and elevated LA pressure can result in pulmonary hypertension, which in turn may lead to RV failure and severe TR. At this stage, patients often present with fatigue, peripheral edema, and signs of low cardiac output. Currently, percutaneous valvuloplasty utilizing a specialized balloon design (the Inoue balloon) is the preferred treatment approach for suitable patients, offering long-term outcomes comparable to OMVC. While procedure-related complications are generally minimal, rare but potentially devastating events like cerebral thromboembolism can occur. Multiple studies have demonstrated the safety and life-saving benefits of this procedure during pregnancy.

Keywords Rheumatic fever · Mitral stenosis · Pulmonary edema · Low cardiac output · Inoue balloon · Percutaneous valvuloplasty · OMVC · Thromboembolic events · Pregnancy

A. Firouzi · Z. Hosseini (✉) · M. Baay
Cardiovascular Intervention Research Center, Rajaie Cardiovascular Institute, Iran University of Medical Sciences, Tehran, Iran

A. Firouzi et al. (eds.), *Mastering Skills in Complex Cardiovascular Interventions*, https://doi.org/10.1007/978-3-031-97321-5_15

Abbreviations

AV	Atrioventricular
LA	Left atrium
LAA	Left atrial appendage
LV	Left ventricle
MR	Mitral regurgitation
MS	Mitral stenosis
MV	Mitral valve
MVA	Mitral valve area
OMVC	Open mitral valve commissurotomy
PAP	Pulmonary artery pressure
PTMC	Percutaneous transluminal mitral commissurotomy
RV	Right ventricle
TR	Tricuspid regurgitation

Introduction

The incidence and prevalence of rheumatic heart disease exhibit considerable variation across different age groups and geographical regions. While the prevalence in developed countries is relatively low (0.2–0.3 per 1000), it remains significantly higher in developing countries (5.5–5.7 per 1000) [1]. Despite a steady global decline in prevalence, rheumatic fever persists as the primary cause of MV stenosis worldwide, disproportionately affecting individuals in developing nations, where 79% of cases are reported [2].

Rheumatic fever stands as the primary cause of MS, with rheumatic changes detected in 99% of stenotic MVs at the time of MV replacement. Among rheumatic heart disease patients, around 25% experience isolated MS, while approximately 40% present with combined MS and MR. Multivalve involvement is observed in 38% of MS patients, affecting the aortic valve in roughly 35% and the tricuspid valve in approximately 6% [3]. The pulmonic valve is infrequently affected by rheumatic fever. There is a notable gender predisposition, with female patients accounting for two-thirds of all rheumatic MS cases. The time interval between the initial rheumatic fever episode and the clinical manifestation of MV obstruction can range from a few years to over 20 years. Rheumatic fever induces characteristic alterations in the MV, such as thickened leaflet edges, commissural fusion, and shortened, fused chordae. The symmetric commissural fusion results in a small, central oval orifice during diastole that resembles a fish mouth or buttonhole shape on pathologic specimens due to the atypical open position of the anterior leaflet [1]. The degree of valve opening during diastole, characterized by MVA, serves as the most valuable indicator of MV obstruction severity. In healthy adults, MVA measures between 4 cm^2 and 6 cm^2. With mild MS, defined as an MVA reduction to around

2 cm^2, blood flow from the LA to the LV necessitates an abnormal, albeit minimal, pressure gradient. An MVA decrease to 1 cm^2 signifies severe MS, requiring a left AV pressure gradient of about 20 mm Hg to preserve the normal resting cardiac output [3].

During the early stages of MS, patients typically experience exertional dyspnea, orthopnea, pulmonary edema, hemoptysis, palpitations (atrial fibrillation), and systemic embolic events, with cerebral emboli occasionally presenting as the initial symptom. Prolonged severe MS and increased LA pressure can result in pulmonary hypertension (reactive or obliterative), ultimately causing RV failure and severe TR. At this stage, patients often present with fatigue, peripheral edema, and evidence of low cardiac output.

Diagnostic Workup

In the early stages of the disease, when the valve is still pliable, the S1 heart sound is audible. Nonetheless, in severe, advanced cases with sinus rhythm and rigid, heavily calcified leaflets, the S1 sound and opening snap disappear. Instead, a diastolic rumbling murmur can be heard at the apex, persisting until the end of diastole with presystolic accentuation. Additional systolic murmurs of TR, diastolic murmurs of pulmonary insufficiency, and P2 accentuation may also be audible, accompanied by an RV heave in advanced cases.

ECG in MS often reveals atrial fibrillation, LA enlargement (present in 90% of cases with sinus rhythm), and eventually right atrial enlargement and RV hypertrophy in cases complicated by pulmonary hypertension.

Transthoracic echocardiography serves as the primary diagnostic tool, enabling the assessment of left and RV size and function, MS severity, the Wilkins score, concomitant valvular disease, and estimation of PAP. Transesophageal echocardiography is essential before PTMC, aiding in excluding LA and LAA thrombi, evaluating MR severity, determining commissural fusion symmetry, and characterizing leaflet and subvalvular apparatus calcification.

Management

The mainstay medical treatment for MS includes diuretics, beta-blockers, calcium channel blockers, and anticoagulants. The ultimate decision to proceed with PTMC or surgical MV repair (Fig. 15.1) is influenced by the patient's functional class, MS severity, PAP, and other echocardiographic findings. Although both catheter-based and surgical interventions can yield excellent results by splitting fused commissures to alleviate stenosis, they necessitate a high degree of expertise and should be conducted at centers with extensive experience in these procedures.

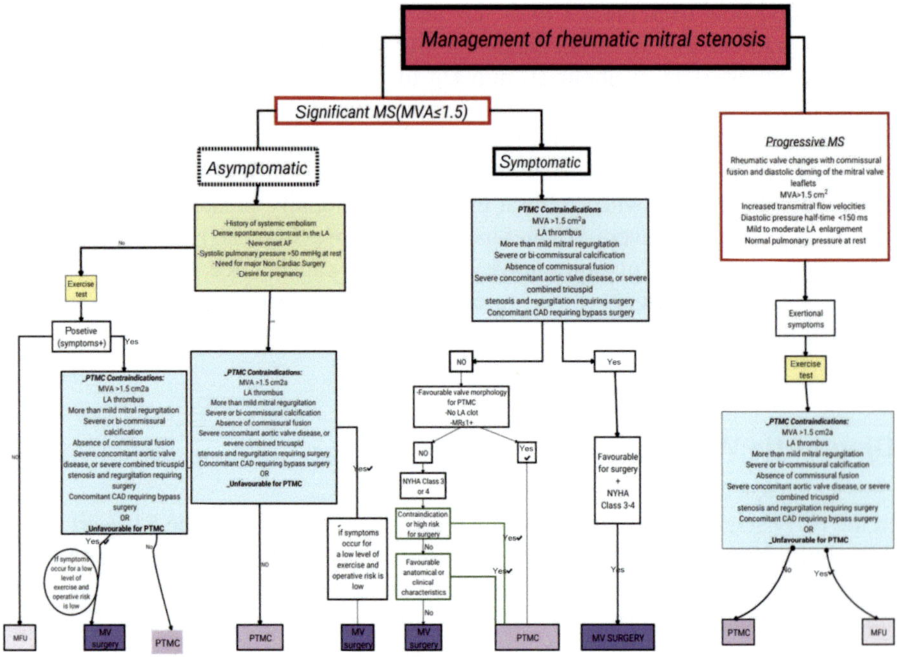

Fig. 15.1 Management of rheumatic mitral stenosis. AF atrial fibrillation, LA left atrium/left atrial, MVA mitral valve area

Challenging Cases

Case 1

The patient was a 45-year-old woman presenting with exertional dyspnea (New York Heart Association functional class III) and orthopnea. On physical examination, the heart rhythm was regular, and hemodynamics were stable. Elevated jugular venous pressure was noted. A loud S1 sound and a diastolic rumbling murmur (grade III) were audible at the apex. Additionally, a loud P2 sound was heard in the left second intercostal space. There were no signs of RV failure. The ECG appeared normal, while the chest X-ray revealed evidence of venous congestion.

Transesophageal echocardiography indicated normal LV size and systolic function. RV size and function were also within normal limits. The anterior MV leaflet appeared thickened, calcified, and dome-shaped, with restricted motion of the posterior MV leaflet. Severe MS was observed, with a mean pressure gradient of 11 mm Hg, a pressure half-time of 250 ms, and an MV area of 0.8 cm^2 by planimetry. Bilateral symmetric commissural fusion was noted (the Wilkins score: 9). No LA or LAA thrombus was detected. The tricuspid and aortic valves showed no evidence of stenosis, though mild aortic insufficiency and moderate TR were present (TR

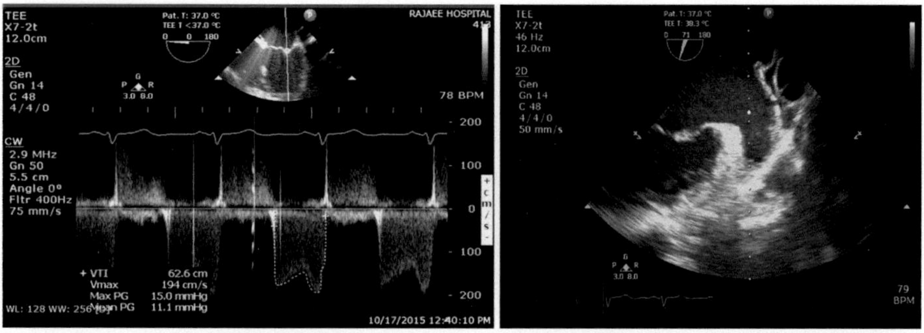

Fig. 15.2 The patient's transesophageal echocardiogram revealed severe MS, with no signs of clot or spontaneous contrast in the LA and LAA

gradient: 45 mm Hg and systolic PAP: 55 mm Hg). No congenital heart abnormalities were identified (Fig. 15.2).

Based on the patient's symptoms and transesophageal echocardiographic findings, a treatment plan involving beta-blockers and diuretics was initiated. She was subsequently deemed a suitable candidate for PTMC.

PTMC [4]

Transesophageal echocardiography must be performed at least 48 h prior to the procedure to exclude contraindications to PTMC. These contraindications include the presence of an LA thrombus, moderate-to-severe MR, severe bi-commissural calcifications, absence of commissural fusion, or any mass on the interatrial septum. Transesophageal echocardiography also confirms the patient's suitability for PTMC. In addition, the international normalized ratio (INR) should be maintained below 1.5 during the procedure.

Under local anesthesia, sheaths (6 Fr and 8 Fr) were inserted into the right common femoral artery and vein, respectively. An LV injection was performed in the right anterior oblique projection to assess the MV, subvalvular apparatus, MR severity, and any regional wall motion abnormalities (Fig. 15.3a).

The critical step in the procedure is septostomy. Using the angiographic method in the anteroposterior projection, a pigtail catheter was placed in the ascending aorta at the right coronary cusp. Simultaneously, right atrial angiography with a delayed phase was performed using another pigtail catheter to delineate the lateral and inferior borders of the LA (Figs. 15.3 and 15.4b).

The upper end of the tricuspid valve during systole (point T, marked with an asterisk) is identified on a stop-frame frontal right atrial angiographic image (a) and then transferred to a stop-frame LA image (b). On the LA image, a horizontal line is drawn from point T to point L, where the line first intersects the lateral border of the atrium (typically the LA). Point L is considered the posterior limit of the

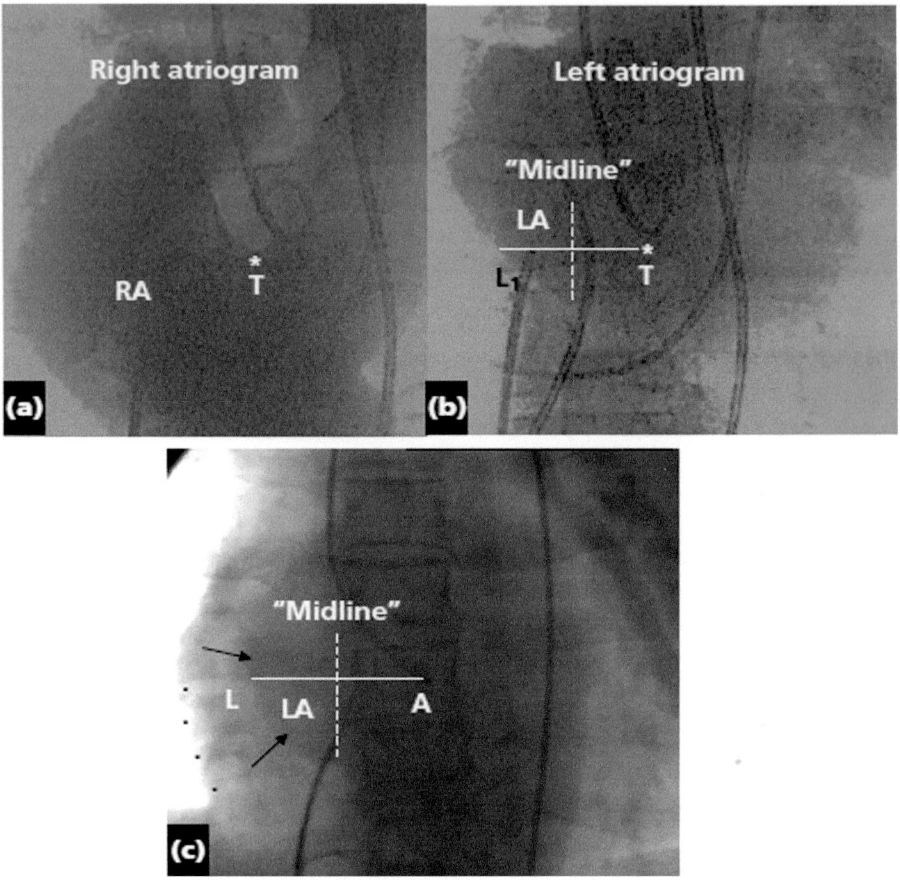

Fig. 15.3 In the angiographic technique, the anteroposterior view shows one pigtail catheter inserted into the ascending aorta at the right coronary cusp. A second pigtail catheter is used to perform right atrial angiography with a delayed phase to delineate the lateral and inferior borders of the LA. The upper edge of the tricuspid valve during systole (point T, indicated by an asterisk) is identified on a stop-frame frontal right atrial angiographic image (**a**) and then transferred to a stop-frame LA image (**b**). In the latter image, an imaginary horizontal line is drawn from point T to point L, where the line first intersects the lateral border of the atrium, typically the LA. Point L is considered the posterior limit of the septum. The dotted vertical line intersecting at the midpoint between points T and L represents the "midline." In the fluoroscopic method (**c**), a horizontal line is drawn from the tip of the pigtail catheter (point A) to L, the LA silhouette (indicated by black arrows), to establish the "midline." The dotted line represents the right atrial silhouette

septum. A vertical dotted line, drawn at the midpoint between T and L, represents the midline.

In the fluoroscopic method (c), a horizontal line is drawn from the tip of the pigtail catheter (point A) to point L, which corresponds to the LA silhouette (indicated by black arrows), to establish the "midline." The right atrial silhouette is marked by a dotted line (2).

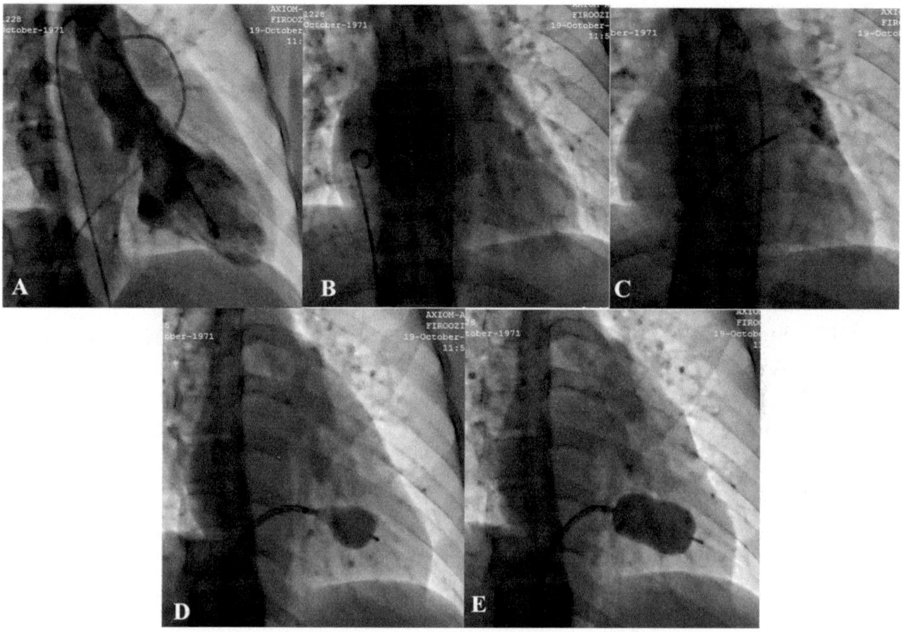

Fig. 15.4 (**a**–**e**) An LV injection in the right anterior oblique projection before the procedure showed a normal LV size, a dome-shaped MV, and mild MR (**a**). Using the angiographic approach, the precise location of the septostomy was assessed following right atrial injection (**b**). After the septostomy, a contrast injection verified the catheter's position in the LA (**c**). Upon entering the LV, the distal portion of the balloon was inflated, and the entire system was retracted toward the MV (**d**). Finally, both the proximal section and waist of the balloon were inflated (**e**). As shown, the central waist of the balloon appeared to be inflated symmetrically

The puncture site is determined based on the size of the LA, typically located two-thirds above the lower border of the LA (with the optimal site being the fossa ovalis) along the vertical line. A Mullins catheter is introduced through the right femoral vein over a 0.032-inch J-tip guidewire into the superior vena cava. After the wire is removed, the catheter is aspirated and flushed.

Next, the Brockenbrough needle, connected to a pressure connector, is inserted into the catheter and carefully advanced under fluoroscopic guidance until its tip reaches the predetermined position. The needle is allowed to rotate freely during advancement. The operator's right-hand stopper finger is firmly placed between the catheter hub and the needle's direction indicator to prevent unintended forward movement of the needle.

In the anteroposterior projection, the needle-fitted catheter, with its direction indicator positioned at approximately 4 o'clock, is slowly withdrawn downward (caudally) from the superior vena cava. A clockwise rotation is applied to the direction indicator to align the catheter/needle along the "midline." The catheter/needle tip is further withdrawn until it reaches the target point, indicated by the sensation of a septal bounce. The position of the catheter-needle is then assessed in the 30°

right anterior oblique projection to confirm the optimal septal puncture site and to avoid puncturing adjacent structures. At this stage, the catheter-needle tip is typically visualized just anterior to the vertebral column and inferior to the ascending aorta.

Once the operator confirms the intended puncture site, the catheter/needle is firmly pressed against the septum. While maintaining steady pressure to prevent the catheter from slipping away from the puncture site, the operator releases the stopper finger and advances the needle forward. The needle is aspirated, and the pressure curve is monitored. A contrast injection is performed to confirm entry into the LA (Fig. 15.4c).

Following this, the Mullins catheter is advanced to the mid-LA, and the needle is removed. A spiral wire (0.025-inch coiled tip) is introduced through the catheter into the LA. After septal dilatation using a dedicated tapered dilator (14 Fr), intravenous heparin is administered [4].

The Inoue balloon catheter is constructed from polyvinyl chloride and features a balloon attached to its distal end. The balloon consists of two latex layers with a polyester micromesh sandwiched between them. The catheter is available in a 12 Fr diameter and measures 70 cm in length, with each balloon measuring 2.5 cm when unstretched. The balloon is divided into three distinct sections, allowing for sequential inflation and stable positioning across the valve. It comes in four sizes (24, 26, 28, and 30 mm), and its diameter can be adjusted by up to 4 mm as needed. Balloon inflation and catheter venting are controlled by two proximally located stopcocks [5].

After pretesting the balloon-syringe matching and meticulously de-airing the catheter, the stretching tube is advanced over the inner tube and secured to the golden tube. Both tubes are then advanced and tightened together to slenderize the balloon.

For this patient (section "Case 1"), a 28 mm Inoue balloon was selected (Table 15.1). The balloon was gently advanced through the subcutaneous tissue and maneuvered across the septum using a combination of clockwise rotation and

Table 15.1 Selection of the Inoue balloon catheter

Reference size (RS) (mm)	
Height (cm) (rounded to the nearest 0) × 1/10 + 10	
For instance,	
Height = 147 cm	
RS = 150 × 1/10 + 10 = 25 mm	
Catheter selection	
Valvular status	**Balloon catheter**
Pliable	RS matched (e.g., PTMC −26 for RS = 25 mm)
Calcified/stenotic lesion	One size smaller than RS matched (e.g., PTMC −24 for RS = 25 mm)

forward pressure under anteroposterior fluoroscopic guidance. At the 12 o'clock position, the stretching tube was pulled back while the balloon was simultaneously pushed forward. At the 3 o'clock position, the golden tube and wire were pulled back as the balloon was advanced further, allowing the distal portion of the balloon to inflate.

In the right anterior oblique projection, the stylet was inserted into the catheter tip. The partially inflated distal balloon was subsequently directed toward the anteriorly positioned mitral orifice by applying a counterclockwise twist (typically 180°) to the stylet using the right hand. The catheter was slowly withdrawn with the left hand until a horizontal bobbing motion of the balloon was observed, indicating its proximity to the MV. Crossing the MV was attempted by gently pulling the stylet back with a counterclockwise rotation while simultaneously advancing the balloon catheter with light pressure.

After successfully crossing the MV and ensuring proper alignment of the balloon with the MV-apex axis, the distal portion of the balloon was inflated using a syringe filled with diluted contrast to achieve a 26 mm diameter (Fig. 15.4d). Thereafter, the balloon catheter was retracted toward the mitral orifice while the stylet was gently pushed toward the end of the balloon. This allowed the proximal and, subsequently, the central parts of the balloon to inflate sequentially, resulting in the symmetrical disappearance of the central waist at full inflation (The total inflation and deflation time should not exceed 6–8 s) (Fig. 15.4e).

If the residual waist is unsatisfactory, a stepwise dilation technique is employed.

Based on the baseline Wilkins score, the degree of commissural and valvular calcification, subvalvular involvement, and the final MVA and MR severity assessed by transthoracic echocardiography, the balloon size was incrementally increased by 1 mm up to the maximum size and inflated in the same manner.

Finally, the stylet was removed, and after measuring the LV end-diastolic pressure, the catheter was withdrawn into the LA and inflated. The distal portion of the balloon was inflated to secure its position within the LA, and LA pressure was measured. An LV injection was performed in the right anterior oblique projection to assess the final degree of MR and to check for any new wall motion abnormalities. Afterward, the balloon was deflated and removed in the reverse order of its insertion. This involved reintroducing the spiral wire, stretching the tube into the inner catheter, and securing it to the golden tube to fully slenderize the balloon before withdrawal. The catheter and wire were withdrawn, ensuring that only the floppy portion of the wire remained external to avoid a cutting effect, which could occur if the stiff part of the wire protruded from the balloon tip.

Following the procedure, the patient exhibited confusion and decreased consciousness. Immediate intubation was performed, and four-vessel cranial angiography revealed complete occlusion of the right anterior and middle cerebral arteries (Fig. 15.5a).

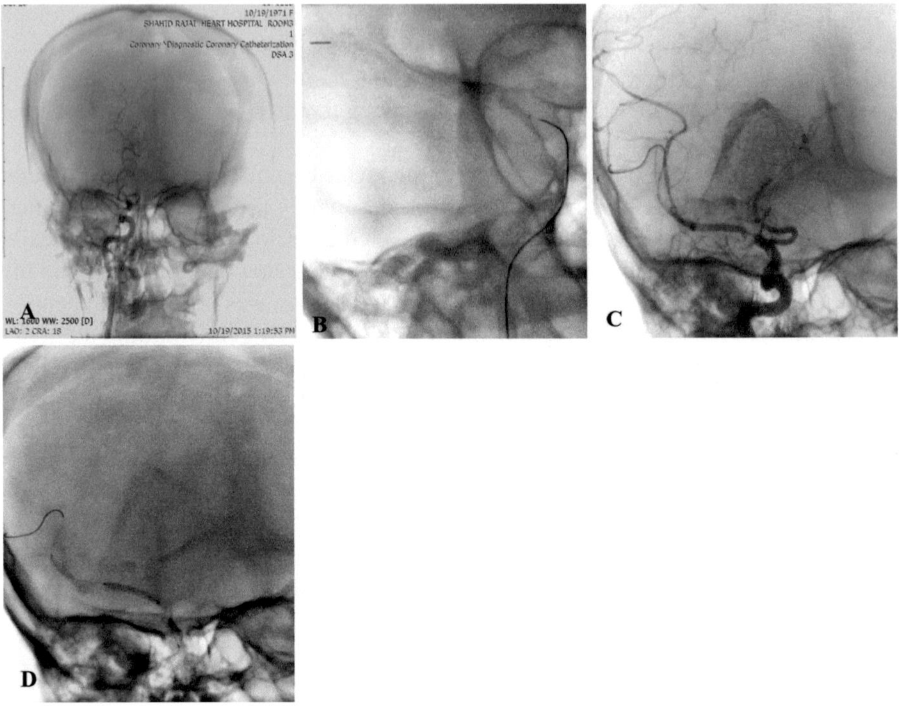

Fig. 15.5 (**a–d**) The initial brain angiography confirmed total occlusion of the right anterior cerebral artery and middle cerebral artery (**a**). After multiple attempts using stent retrievers (SOLITAIRE and ERIC), the lesion remained unresolved (**b**, **c**). Considering the likely cause of the embolization to be calcium rather than thrombus, predilation of the vessel was performed using a coronary balloon at high pressure, followed by the deployment of a coronary stent (**d**)

Initial suspicion pointed toward a thrombotic event, prompting the selection of a stent retriever strategy (Fig. 15.5b). Despite three attempts using SOLITAIRE and ERIC stent retrievers, the occlusion remained unresolved (Fig. 15.5c). Post-retrieval imaging consistently showed a dense lesion resembling calcification rather than a thrombus. Ultimately, it was concluded that a calcified particle from the MV had embolized.

The lesion was predilated using a 2.5 × 12 mm Maverick balloon. Due to the firm nature of the lesion, high-pressure balloon predilation was required. Finally, a 3 × 18 mm XIENCE Xpedition stent was successfully deployed.

The patient was transferred to the ICU and was successfully extubated the following day. Fortunately, her neurocognitive function recovered rapidly. Postprocedural echocardiography revealed an MVA of 1.6 cm^2 at discharge. The patient exhibited only mild left upper limb paresis, which resolved completely during a follow-up visit 2 months later.

Case 2

The patient was a 32-year-old pregnant woman (G1P0, gestational age: 26 weeks) referred with a diagnosis of rheumatic MV stenosis. She was hemodynamically stable but reported recent-onset exertional dyspnea (New York Heart Association functional class III) and palpitations. The ECG was normal.

Transesophageal echocardiography revealed normal LV size and function, with no evidence of LV hypertrophy. RV size and function were also normal. The anterior MV leaflet appeared thickened, calcified, and dome-shaped, with restricted motion of the posterior leaflet. Severe MS was noted, with a mean pressure gradient of 12 mm Hg, a pressure half-time of 260 ms, and an MV area of 0.7 cm^2 by planimetry and continuity equation (the Wilkins score: 9). Bilateral symmetric commissural fusion was observed, along with mild-to-moderate MR. No LA or LAA thrombus was detected, although a smoky pattern was noted in the LAA. The tricuspid and aortic valves showed no stenosis, but mild-to-moderate aortic insufficiency and moderate TR were present (TR gradient: 55 mm Hg and systolic PAP: 65 mm Hg). No pericardial effusion was observed.

Given the patient's condition (very severe MS, gestational age of 26 weeks, and moderate pulmonary hypertension), beta-blockers and diuretics were initiated. Following a heart team discussion, she was scheduled for PTMC. After a thorough discussion with her family regarding the risks and benefits of the procedure, including the potential risks of maternal and fetal radiation exposure, she was ultimately deemed a suitable candidate for PTMC.

PTMC

Under general anesthesia and with transesophageal echocardiography guidance, sheaths (6 Fr and 8 Fr) were inserted into the right common femoral artery and vein, respectively. Removable lead shielding (from a catheterization laboratory table shield) was placed on the procedure table beneath the patient, covering the area from above the umbilicus to below the mid-thigh. This removable lead shielding allowed for emergency imaging of the abdomen and pelvis if needed. Standard gonadal shielding was also positioned to further protect the uterus and fetus.

The procedure was performed under general anesthesia with a cardiac anesthesiologist, cardiac surgeon, gynecologist, and echocardiographer in attendance. A pigtail catheter was positioned in the right coronary cusp as a marker. Subsequently, a Mullins catheter was advanced over a 0.032-inch guidewire into the superior vena cava without the use of radiation.

Using a minimal frame rate, the Brockenbrough needle, attached to a pressure connector, was inserted into the catheter and carefully advanced until its tip reached the predetermined position. The needle was allowed to rotate freely during advancement. The operator's right-hand stopper finger was firmly placed between the catheter hub and the needle's direction indicator to prevent unintended forward movement of the needle.

In the anteroposterior view, the needle-fitted catheter, with its direction indicator positioned at approximately 4 o'clock, was slowly withdrawn (caudally) from the superior vena cava. The catheter/needle tip was further withdrawn until it reached the target point, indicated by the sensation of a septal bounce. At this stage, septal tenting observed at 45° and 0° degrees on transesophageal echocardiography confirmed the most reliable site for proper septostomy.

After confirming the correct position of the fossa ovalis and the needle, the needle was advanced, and immediate back-bleed and pressure curve assessments were performed. Following confirmation by the echocardiographer, the Mullins catheter was advanced into the mid-LA.

A spiral wire was then advanced into the mid-LA, and septal dilatation was performed using a dedicated tapered-tip catheter (14 Fr). Notably, all steps up to this point were carried out without the use of radiation. Intravenous heparin was administered, and the de-aired Inoue balloon (selected size: 28 mm for this patient) was gently introduced through the femoral vein under anteroposterior projection.

Using fluoroscopic guidance, the balloon was advanced into the LA (as described in the previous case), and the distal portion of the balloon was inflated.

A counterclockwise rotation to the stylet catheter aligned the tip of the balloon with the mitral inflow. At this stage, the balloon was advanced into the LV by pulling back the stylet while simultaneously pushing the balloon forward. To ensure optimal balloon alignment and positioning relative to the MV, echocardiographic images helped minimize radiation exposure.

Once the proper balloon position was confirmed using echocardiographic guidance, the distal, proximal, and, finally, the central waist of the balloon were sequentially inflated. Given the critical importance of minimizing maternal and fetal radiation exposure while achieving the best possible outcome without complications (particularly severe MR), the procedure was concluded by inflating the balloon to 1 mm below the reference size (typical in pregnant patients). In this case, the balloon was inflated with 27 mm of diluted contrast (Fig. 15.6).

The patient was discharged 2 days after the procedure. The final MVA was 1.5 cm^2, with up to moderate MR and a systolic PAP of 35 mm Hg. She completed the delivery and postpartum period without complications.

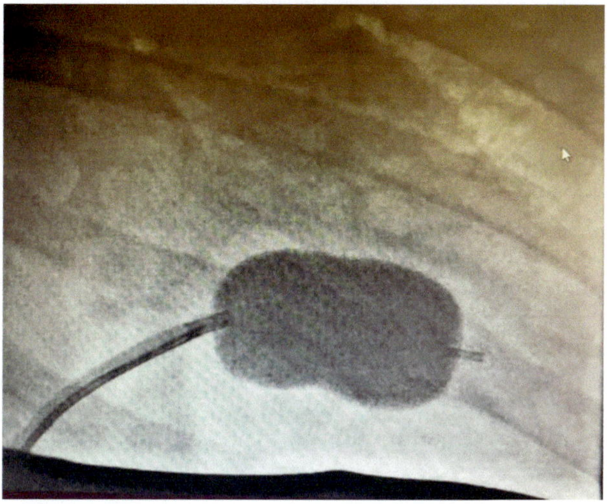

Fig. 15.6 The balloon was inflated with 27 mm of diluted contrast

Discussion

As depicted in Fig. 15.1, for symptomatic patients with severe MS (MVA $\leq$1.5 cm^2) possessing suitable valve morphology and no contraindications to PTMC, this intervention is now the preferred approach, particularly in developing nations. This shift is partly due to the non-negligible complication rates associated with OMVC. Previous studies have shown comparable long-term outcomes (at least 10 years) between PTMC and OMVC. Nonetheless, with the current trend of diagnosing these patients earlier (attributed to enhanced prodromal and primary prevention strategies), more individuals are found to be suitable candidates for PTMC [6].

As with any invasive procedure, PTMC is associated with potential complications. The most frequently reported complications include vascular site complications, tamponade, severe MR, pulmonary edema, systemic embolization (thrombotic or calcified particles; see section "Case 1"), and atrial septal defect.

Twenty-four to 48 h post-procedure, transthoracic echocardiography is conducted to assess the MVA via planimetry, degree of commissural opening, MR severity and underlying mechanism (e.g., commissural MR, leaflet tears, or chordal rupture), presence of residual atrial septal defect, and any pericardial effusion.

Successful Procedure Criteria

1. Final valve area > 1.5 cm^2 and a minimum 25% increase in valve area; and
2. Final valve area > 1.5 cm^2 without MR exceeding grade 2/4.

Failure Rate and Complications

Procedural success rates for PTMC, largely influenced by operator experience, range from 1% to 17% [7]. The most common reasons for procedural failure include difficulties in trans-septal puncture and balloon crossing. Reported complications encompass tamponade (0.5%–12%), severe MR (2%–19%), systemic embolism (coronary or cerebral), left-to-right shunting, and vascular site complications. In section "Case 1", we present an instance of a rare but potentially fatal PTMC complication: calcium particle embolization to the cerebral arterial system.

Furthermore, we encountered an unreported case where calcium embolization to the left circumflex artery resulted in ventricular fibrillation upon coronary angiography, revealing complete left circumflex artery occlusion. In both this case and section "Case 1", the absence of LA or LAA thrombi suggests that calcified particle embolization was the most likely underlying cause.

PTMC During Pregnancy

Valvular heart disease affects an estimated 0.5%–1.5% of pregnancies, with rheumatic heart disease and isolated MS being the most common cardiac valvular complications during pregnancy. Significant MS often worsens during pregnancy due to increased cardiac output and heart rate. Management for symptomatic MS in pregnancy primarily focuses on optimal medical therapy, including beta-blockers, diuretics, and bed rest. For patients with suitable anatomy who do not respond to medical management, balloon valvuloplasty may be a suitable alternative. Given the high fetal and maternal mortality rates associated with surgery (range: 1.8%–33%) and the favorable procedural success rates of PTMC, the latter has emerged as a reliable therapeutic option for pregnant women.

PTMC during pregnancy undoubtedly carries high risks. In addition to the potential procedural stress on the fetus and hemodynamic instability during the intervention, radiation exposure poses another significant concern. Ionizing radiation's nonmalignant effects on the fetus or neonate encompass growth retardation, mental retardation, structural anomalies, and seizures. These adverse effects are not typically observed at doses below 50 mGy but become more likely at doses exceeding 100 mGy. In the event of radiation exposure surpassing 10 rads, pregnancy termination should be strongly considered [8].

Over the past decade, hospital mortality associated with PTMC in pregnant women has decreased to below 1%, with procedural success rates as high as 89%–100% [9, 10]. Firouzi et al. [11] demonstrated remarkably high PTMC success rates and minimal complications during pregnancy. Moreover, their findings revealed no significant correlation between the Apgar score, birth weight, height, and head circumference at birth and radiation exposure duration.

Conclusions

PTMC during pregnancy proves to be a safe and effective intervention. Employing low frame-count techniques to minimize radiation exposure time significantly reduces radiation-associated complications.

References

1. Ranjan R, Pressman GS. Aetiology and epidemiology of mitral stenosis. E-J Cardiol Pract. 2018;16:14. Available at: https://www.escardio.org/Journals/E-Journal-of-Cardiology-Practice/Volume-16/Aetiology-and-epidemiology-of-mitral-stenosis
2. Carapetis JR, Steer AC, Mulholland EK, Weber M. The global burden of group A streptococcal diseases. Lancet Infect Dis. 2005;5(11):685–94. https://doi.org/10.1016/S1473-3099(05)70267-X. PMID: 16253886
3. Tsang W, Freed BH, Lang RM. Three-dimensional anatomy of the aortic and mitral valves. In: Otto CM, Bonow RO, editors. Valvular heart disease: a companion to Braunwald's heart disease. Philadelphia: Saunders; 2013. p. 14–29.
4. Tsiaras S, Poppas A. Mitral valve disease in pregnancy: outcomes and management. Obstet Med. 2009;2(1):6–10. https://doi.org/10.1258/om.2008.080002. PMID: 27582803
5. Firouzi A, Hosseini Z. Percutaneous trans-mitral valve commissurotomy (PTMC). In: Maleki M, Alizadehasl A, editors. Case-based clinical cardiology. London: Springer; 2021. p. 623–30. https://doi.org/10.1007/978-1-4471-7496-7_68.
6. Inoue K, Hung JS, Chen CR, Cheng TO, Lau KW, Owada K, et al. Mitral stenosis: Inoue balloon catheter technique. In: Cheng TO, editor. Percutaneous Balloon Valvuloplasty. New York: Igaku-Shoin Medical Publishers, Inc.; 1992. p. 237–79.
7. Otto CM, Nishimura RA, Bonow RO, Carabello BA, Erwin JP 3rd, Gentile F, et al. 2020 ACC/AHA guideline for the management of patients with valvular heart disease: a report of the American College of Cardiology/American Heart Association Joint Committee on Clinical Practice Guidelines. Circulation. 2021;143(5):e72–e227. https://doi.org/10.1161/CIR.0000000000000923. PMID: 33332150. Erratum in: Circulation 2021;143(5):e229
8. Palacios IF, Sanchez PL, Harrell LC, Weyman AE, Block PC. Which patients benefit from percutaneous mitral balloon valvuloplasty? Prevalvuloplasty and postvalvuloplasty variables that predict long-term outcome. Circulation. 2002;105(12):1465–71. https://doi.org/10.1161/01.CIR.0000012143.95596.F1. PMID: 11914256
9. Schenker JG, Polishuk WZ. Mitral valvotomy during pregnancy. Surg Gynecol Obstet. 1968;127(3):593–7. PMID: 5673222
10. Norrad RS, Salehian O. Management of severe mitral stenosis during pregnancy. Circulation. 2011;124(24):2756–60. https://doi.org/10.1161/CIRCULATIONAHA.111.071779. PMID: 22155994
11. Abdi S, Salehi N, Ghodsi B, Basiri HA, Momtahen M, Firouzi A, et al. Immediate results of percutaneous trans-luminal mitral commissurotomy in pregnant women with severe mitral stenosis. Clin Med Insights Cardiol. 2012;6:35–9. https://doi.org/10.4137/CMC.S8508. PMID: 22442698

Chapter 16
Percutaneous Balloon Aortic Valvuloplasty

Ata Firouzi, Zahra Hosseini, and Ehsan Khalilipur

Abstract Calcific AS is the most prevalent acquired valve disorder in the Western world. For patients with severe symptomatic AS who do not undergo definitive intervention, the prognosis is poor, with a mean survival of 3 years following the onset of angina or syncope and 2 years after the development of heart failure. ABV serves as a palliative treatment to alleviate symptoms and can act as a bridge to more definitive interventions, such as surgical valve replacement or TAVI, particularly in hemodynamically unstable patients. Additionally, any form of LVOTO during pregnancy is considered high risk. In such cases, ABV can be a life-saving intervention for pregnant individuals with symptomatic severe AV stenosis.

Keywords Aortic stenosis · High mortality rate · Aortic balloon valvuloplasty · Palliative therapy · LVOTO · Pregnancy

Abbreviations

ABV	Aortic balloon valvuloplasty
AI	Aortic insufficiency
AR	Aortic regurgitation
AS	Aortic valve stenosis
AVA	Aortic valve area
AVR	Aortic valve replacement
LV	Left ventricle
LVOT	Left ventricular outflow obstruction
NYHA	New York Heart Association
SAVR	Surgical aortic valve replacement
TAVI	Transcatheter aortic valve implantation
TAVR	Transcatheter aortic valve replacement

A. Firouzi · Z. Hosseini (✉) · E. Khalilipur
Cardiovascular Intervention Research Center, Rajaie Cardiovascular Institute, Iran University of Medical Sciences, Tehran, Iran

A. Firouzi et al. (eds.), *Mastering Skills in Complex Cardiovascular Interventions*, https://doi.org/10.1007/978-3-031-97321-5_16

Introduction

Calcific AS is the most frequently acquired valve disorder in the Western world. Moderate AS affects 2%–7% of individuals over 65, while severe AS is found in 5%–7% of those over 80 and 10%–15% of those over 90 [1]. Surgical AVR remains the preferred treatment for symptomatic calcific AS. Without surgical intervention, the prognosis for patients with severe symptomatic AS is poor, with an average survival of 3 years after the onset of angina or syncope and 2 years following the development of heart failure [2]. Despite this, many symptomatic patients are not candidates for surgery due to high operative risks associated with multiple comorbidities. According to the Euro Heart Survey, 31.8% of patients with clear indications did not undergo surgical valve replacement [3]. The narrowing of the valve area leads to progressively increasing pressure overload on the LV. This pressure overload triggers LV remodeling, followed by hypertrophy—which reduces coronary microvascular blood flow—and eventually results in cavity dilation and impaired contractile function. This progression can cause LV dilation, accompanied by systolic dysfunction, as energy is diverted to overcome the resistance posed by the stenotic valve (a phenomenon known as "afterload mismatch") [4].

TAVI has emerged as an alternative treatment option for severe, symptomatic AS, applicable to patients at high, moderate, and even low surgical risk. Many patients with severe, symptomatic AS who are evaluated for TAVI trials but deemed ineligible due to strict inclusion and exclusion criteria are referred for ABV. First introduced by Cribier et al. [5] in 1986, ABV serves as a valuable palliative treatment to alleviate symptoms and can act as a bridge to more definitive interventions, such as surgical valve replacement or TAVI, particularly in hemodynamically unstable patients [6]. ABV leads to significant and immediate reductions in AV gradient and increases in AVA, accompanied by rapid symptomatic improvement. While the precise mechanism of ABV remains unclear, it is thought to involve the fracturing of calcium deposits to create "hinge-points" in the valve cusps or the separation of fused leaflets [7].

Overall, ABV typically results in a final AVA of between 0.7 cm^2 and 1.2 cm^2 [8]. This AVA is smaller than the valve area achieved with a surgical AVR prosthesis, which generally provides a valve area exceeding 1.5 cm^2.

A Challenging Case

Percutaneous ABV During Pregnancy

The patient was a 26-year-old woman, G1P0, at 28 weeks of gestation, who was referred to our center with complaints of exertional dyspnea (NYHA functional class III-IV). Her hemodynamic status remained stable. On physical examination, a harsh ejection systolic murmur (grade III) was audible in the aortic zone, radiating to the apex. Blood pressure measurements in the upper and lower extremities were symmetric, with no delay observed. ECG revealed LV hypertrophy criteria with a strain pattern in the precordial leads. Transthoracic echocardiography showed normal LV size and systolic function, moderate concentric LV hypertrophy, and no evidence of a ventricular septal defect. Right ventricular size and function were normal. The patient had a bicuspid AV (type I) with severe AS (mean pressure gradient: 80 mm Hg, peak pressure gradient: 130 mm Hg, mitral valve area by planimetry: 0.6 cm^2, and annulus diameter: 22 × 25 mm) and mild AI. The sinuses of Valsalva were normal in diameter, with mild aortic root dilatation. No pathologic findings were noted in the other valves, and there was no evidence of patent ductus arteriosus or coarctation.

Given the patient's persistent symptoms despite guideline-directed medical therapy, the heart team decided to proceed with ABV. After a thorough discussion with her family regarding the risks and benefits of the procedure, as well as the potential risks of maternal and fetal radiation exposure, she was deemed a candidate for percutaneous ABV.

Procedural Technique

Removable lead sheeting, typically used as part of a catheterization laboratory table shield, was placed on the procedure table beneath the patient. This shielding extended from above the umbilicus to below the mid-thigh, providing protection while allowing for emergency imaging of the abdomen and pelvis if needed. Standard gonadal shielding was also positioned to further safeguard the uterus and fetus.

The procedure was performed under conscious sedation, with a cardiac anesthesiologist, cardiac surgeon, gynecologist, and echocardiographist in attendance. Access was achieved through the left common femoral artery, where a pigtail catheter was inserted into the aortic root. Through the right common femoral artery (using a 10 Fr sheath), an AL1 catheter was used to guide a 260 cm straight 0.35-inch guidewire across the AV. The catheter was then advanced over the wire, and the guidewire was exchanged for a 0.35-inch Super Stiff guidewire (Fig. 16.1a).

The chosen balloon was a 22 × 40 mm BIB, selected based on the smallest annulus diameter (22 mm). Given the patient's pregnancy, the largest available balloon was used to maximize AV dilatation and avoid additional balloon exchanges, thereby minimizing total radiation exposure. This approach was deemed appropriate since even severe AI is generally well tolerated during pregnancy. Fluoroscopy was used sparingly to further reduce radiation exposure, with minimal frame rates and the lowest possible contrast injection. A temporary pacemaker was not employed to avoid additional radiation and reduce the risk of arrhythmias during rapid pacing.

Using the pigtail catheter positioned in the non-coronary cusp, the balloon's proper placement was confirmed. Under fluoroscopic guidance, the balloon was inflated until the waist disappeared (Fig. 16.1b, c). Only 3D transesophageal

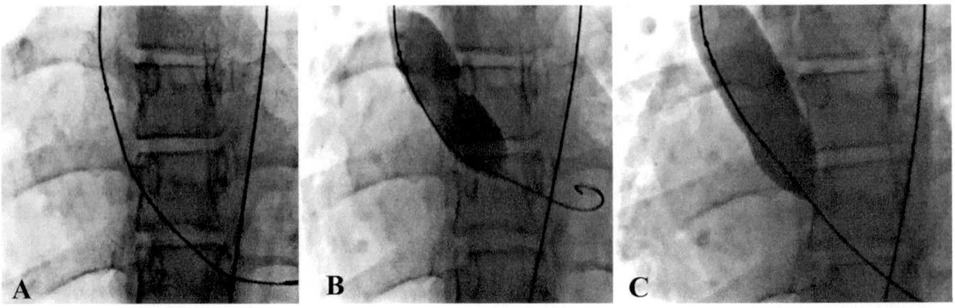

Fig. 16.1 (**a**–**c**) Using a retrograde approach, the AL1 catheter was advanced over the wire toward the LV after wiring the AV. The wire was then exchanged for a 0.035-inch Super Stiff guidewire (**a**). A BIB balloon (22 × 40) was selected based on the smallest annulus diameter. During a detailed hemodynamic assessment, the inner balloon was initially inflated at a low atmospheric pressure. Once the balloon's position was stabilized, the outer balloon was inflated to the rated burst pressure (**b**). Final angiography revealed symmetrical expansion of the commissures (**c**). The aortic root injection showed no signs of annulus perforation or coronary artery involvement, with only mild-to-moderate AI. Transthoracic echocardiography demonstrated a significant reduction in mean pressure gradient and peak pressure gradient to 15 mm Hg and 25 mm Hg, respectively, while AVA by planimetry increased to 1.5 cm^2

echocardiography, fluoroscopy, and fluoro saves (without cineangiography) were utilized to assist in balloon positioning.

- **Final injection findings**
  - Mild-to-moderate AI; and
  - No evidence of annulus perforation.
- **Final echocardiographic results**:
  - Significant reduction in trans-AV gradient:
    - Mean pressure gradient: 15 mm Hg;
    - Peak pressure gradient: 25 mm Hg;
    - AVA: 1.5 cm^2; and
  - Up to moderate AI.

In the catheterization laboratory, critical consideration for high-risk patients includes the crucial availability of transcatheter heart valve systems, such as the SAPIEN families or Medtronic CoreValve. Furthermore, it is essential to ensure readiness for emergent TAVR if needed.

During follow-up, the patient remained symptom-free and successfully underwent a cesarean section delivery at 38 weeks of gestation without any complications before, during, or after the procedure. Ten years later, the patient underwent mechanical AVR.

Discussion

AVS represents around 3%–6% of congenital heart disease, with an estimated incidence of 1–4 per 10,000 live births. AS accounts for two-thirds of the lesions that obstruct the LV outflow tract. The ABV procedure can be performed via retrograde or antegrade approach [9, 10]. The retrograde approach is more common and involves access via the femoral artery with a sheath (10–14 Fr). An Extra Stiff guidewire (0.035 inch) is required to support the balloon during the procedure. The wire should be gently positioned in the LV apex to avoid the risk of ventricular perforation. Typically, undersized balloons (about 1 mm < the smallest size of the annulus diameter or balloon size/annulus diameter < 80%) are used to minimize the risk of annular rupture or severe AI (Table 16.2). If the result is unacceptable, a larger balloon sized to the annulus may be used. Rapid ventricular pacing (160–180 bpm) via a temporary trans-venous pacemaker can stabilize the balloon during inflation by reducing forward cardiac output. The antegrade transvenous approach is done via the femoral vein to the ascending aorta after trans-septal puncture. An Inoue balloon or a traditional valvuloplasty balloon can be used. With one 26-mm balloon, multiple inflations can be performed at sizes ranging from 20 to 26 mm during hemodynamic monitoring. The antegrade approach has diminished vascular complications, reduced stroke, and resulted in a greater increase in the post-ABV area. Nonetheless, it is more technically demanding given the need for a trans-septal puncture and the subsequent looping of the guidewire in the LV apex. The immediate hemodynamic results include an increase in AVA (although rarely >1.2 cm^2) and a reduction of the trans-valvular gradient. Despite only modest changes in the valve parameters, the procedure can lead to symptomatic improvement dramatically in such high-risk patients.

Previous studies have confirmed a high risk of mortality in patients undergoing ABV. Liberman et al. [11] found survival rates of 52%, 31%, and 18% at 1, 2, and 3 years after ABV, respectively. Otto et al. [12] reported 55% survival at 1 year, 35% at 2 year, and 23% at 3 years. Initial enthusiasm for ABV in adults with AS was subsequently mitigated by studies demonstrating that although there was initial symptomatic improvement, it was associated with high complication and recurrence rates (>80% in 1 year) [13], and negligible impact on long-term survival [14]. Several factors explain why ABV does not significantly impact long-term outcomes. Firstly, the patients undergoing this procedure are often frail. Secondly, a substantial proportion (up to 50%) of these patients have concurrent significant coronary artery disease. Thirdly, even after ABV, the average AVA typically remains between 0.7 cm^2 and 1.2 cm^2, indicating that the stenosis persists at a relatively severe level. Therefore, BAV should be considered a bridge to more definitive treatment in all suitable cases.

Symptomatic valvular disease during pregnancy is a rare but challenging clinical scenario in North America. Cardiovascular diseases complicate 0.2%–4% of all pregnancies, and untreated cardiac conditions are responsible for 10%–15% of maternal deaths. Among congenital heart defects, bicuspid AV disease is relatively common, affecting approximately 2% of the general population. Historically, bicuspid AV disease was regarded as a relative contraindication for TAVR. Nevertheless, TAVR procedures are now being performed with increasing success to treat this condition.

Table 16.1 Valvular heart lesion classifications by maternal and fetal risks

Risk category	Conditions
High maternal risks	Ejection fraction <40%
	Previous heart failure
	Previous stroke or transient ischemic attack
High maternal and fetal risks	Severe AS, with or without symptoms
	AR, NYHA functional class II, III, or IV
	Mitral stenosis, NYHA functional class II, III, or IV
	Mitral regurgitation, NYHA functional class III or IV
	Aortic or mitral valve disease with pulmonary hypertension
	Aortic or mitral valve disease with LV dysfunction
	Maternal cyanosis
	Reduced functional status, NYHA functional class III or IV
Low maternal and fetal risks	Asymptomatic AS with a low mean outflow gradient (<50 mm Hg) and normal LV systolic function
	AR, NYHA functional class I or II, with normal LV function
	Mitral regurgitation, NYHA functional class I or II, with normal LV function
	Mitral valve prolapse with none-to-moderate mitral regurgitation and normal LV function
	Mild-to-moderate mitral stenosis, no pulmonary hypertension
	Mild-to-moderate pulmonary valve stenosis

Current guidelines provide specific recommendations for managing valvular heart disease before pregnancy, including counseling and medical treatment. Still, for pregnant patients with class IIb cardiovascular disease and symptomatic AS, where surgical intervention is indicated, the guidelines offer limited direction beyond monitoring and echocardiography requirements. Pregnancy induces significant physiological changes, such as increased cardiac output and hypervolemia. In patients with severe AS, however, the stenotic valve restricts this increase in cardiac output, resulting in a relatively fixed cardiac output. The ability to increase cardiac output in response to exercise is severely restricted in these patients. A significant drop in venous return can lead to a dramatic decline in cardiac output, resulting in syncope or even sudden death. It is well-established that symptomatic stenotic valvular lesions are poorly tolerated during pregnancy and are associated with a high risk of complications and mortality during delivery.

SAVR during pregnancy carries an increased, but acceptable, risk for the mother. Nonetheless, it poses a significantly higher risk to the fetus, with an estimated fetal demise rate of approximately 20%. Another concern during ABV is radiation exposure. At 22 weeks' gestation, acute radiation doses of ≤50 mGy are not associated with noncancerous health effects for the pregnancy or the fetus. Moreover, radiation doses below 50 mGy are estimated to result in a childhood cancer incidence rate of <6% across all gestational ages.

Computed tomography angiography is generally not recommended during pregnancy due to heightened awareness and concerns about radiation exposure and the associated lifetime risk of breast cancer in women with metabolically active breast

Table 16.2 Common types of valvuloplasty balloons

Balloons	Z-Med/Z-Med II/NuCLUES/ NUCLEUS-X/TYSHAK/TYSHAK II	V8	True balloon
Advantages	Quick inflation and deflation time	Designed to lock into valve, limiting movement	True reliable sizing
	Short flexible distal tip with short balloon taper aids maneuverability through tortuous anatomy	Shape maintained, reducing likelihood of annular rupture	Fast
	Coaxial haft design provides enhanced column strength and pushability	Quick inflation and deflation time	Rupture resistant
	Z-Med II provides higher strength and, in some instances, a larger introducer		

Specifications	Z-Med	Z-Med II	Z-Med II TAVR	NUCLEUS-X	TYSHAK II	True balloon
Balloon diameter (mm)	10–25	5–25	20–25	18–30	Waist: 17–23 mm/ Bulbous segment: 22–27.5 mm	20–26 mm
Balloon length (cm)	2–4	2–6	4.5	4–6		4.5 cm
Introducer size (Fr)	7–12	6–14	11–12	10–14	12Fr	11–13 Fr
Shaft size (Fr)	6–9	5–9	10	9		
Usable length (cm)	100	100	110	110		110 cm
Guide wire (inches)	0.035	0.025–0.035	0.035	0.035	0.035	
Rated burst (ATM)	9–3	15–4	5–4	4–2		
Nominal pressure (ATM)	4–2	6–2	2		Inflation volume: 17 mm/16 mL, 19 mm/20 mL, 21 mm/23 mL, 23 mm/27 mL	

tissue during and after pregnancy. Parker et al. demonstrated that computed tomographic pulmonary angiography exposed the breasts of an average-sized woman to a radiation dose of 2.0–5.0 rad (20–50 mGy). This dose is approximately equivalent to 10–25 two-view mammograms or as many as 100–400 chest radiographs. If the imaging procedure must be performed during pregnancy, radiation exposure should be minimized using high-speed scanning techniques and appropriate shielding [15] (Table 16.1).

Nowadays, TAVR is the preferred treatment for intermediate- to high-risk patients unsuitable for SAVR. On the other hand, ABV has limited indications (Class IIB). It is primarily considered a bridge to SAVR or TAVR in hemodynamically unstable patients who are at high surgical risk or in symptomatic AS patients requiring urgent major noncardiac surgery (Table 16.2).

Acute and Long-Term Complications

The most serious complications associated with ABV include death, athero/thromboembolic events, disruption of the aortic root or leaflets leading to massive AR, myocardial infarction, and peripheral vascular injury due to the use of large-sized sheaths. A significant long-term complication of ABV is AI, which often necessitates surgical valve repair or replacement. Most studies indicate a trend toward worsening AI over time, with the degree of AI increasing as follow-up duration lengthens. Factors contributing to progressive AI include greater gradient relief immediately post-ABV, Doppler-quantified AI before and after the procedure, unicommissural AVs, AV prolapse, poor valve morphology, and a large balloon-to-annulus ratio. While some argue that the balloon-to-annulus ratio may not directly influence AI, the degree of AI immediately following ABV is a predictor of significant late AI development.

Another notable complication is the high recurrence rate of stenosis, as mentioned earlier. Ventricular perforation has also been reported. Conduction system damage is another potential risk, given the aortic annulus's proximity to the conduction system. In one study, the incidence of new conduction defects was 8.5%, with only 1.5% of patients requiring permanent pacemaker implantation. The risk of new conduction defects was associated with the ratio of balloon size to the LVOT (1.2 in patients with new defects vs. 1.15 in those without).

During pregnancy, the primary concern is radiation exposure to both the mother and fetus. By implementing prophylactic measures (as outlined earlier), the radiation dose can be minimized.

Conclusions

Although the intermediate and long-term outcomes of ABV are generally disappointing, with a high recurrence rate observed within the early months post-procedure, ABV can serve as a life-saving bridge therapy in selected patients. For instance, during pregnancy, ABV can delay elective valve replacement until after delivery. The most critical factor in maximizing the acute success rate and minimizing complications during ABV is the balloon-to-annulus diameter ratio.

References

1. Pedersen WR, Klaassen PJ, Pedersen CW, Wilson JA, Harris KM, Goldenberg IF, et al. Comparison of outcomes in high-risk patients >70 years of age with aortic valvuloplasty and percutaneous coronary intervention versus aortic valvuloplasty alone. Am J Cardiol. 2008;101(9):1309–14. https://doi.org/10.1016/j.amjcard.2007.12.033. PMID: 18435963

2. Bach DS, Cimino N, Deeb GM. Unoperated patients with severe aortic stenosis. J Am Coll Cardiol. 2007;50(20):2018–9. https://doi.org/10.1016/j.jacc.2007.08.011. PMID: 17996572
3. Iung B, Baron G, Butchart EG, Delahaye F, Gohlke-Bärwolf C, Levang OW, et al. A prospective survey of patients with valvular heart disease in Europe: the Euro Heart Survey on Valvular Heart Disease. Eur Heart J. 2003;24(13):1231–43. https://doi.org/10.1016/S0195-668X(03)00201-X. PMID: 12831818
4. Otto CM, Bonow RO. Valvular heart disease. In: Zipes DP, Libby P, Bonow RO, Braunwald E, editors. Braunwald's Heart Disease: A Textbook of Cardiovascular Medicine. 8th ed. St. Louis, MO: WB Saunders; 2007. p. 1625–34.
5. Cribier A, Savin T, Saoudi N, Rocha P, Berland J, Letac B. Percutaneous transluminal valvuloplasty of acquired aortic stenosis in elderly patients: an alternative to valve replacement? Lancet. 1986;1(8472):63–7. https://doi.org/10.1016/S0140-6736(86)90716-6. PMID: 2867315
6. Moreno PR, Jang IK, Newell JB, Block PC, Palacios IF. The role of percutaneous aortic balloon valvuloplasty in patients with cardiogenic shock and critical aortic stenosis. J Am Coll Cardiol. 1994;23(5):1071–5. https://doi.org/10.1016/0735-1097(94)90592-4. PMID: 8144770
7. Hara H, Pedersen WR, Ladich E, Mooney M, Virmani R, Nakamura M, et al. Percutaneous balloon aortic valvuloplasty revisited: time for a renaissance? Circulation. 2007;115(14):e334–8. https://doi.org/10.1161/CIRCULATIONAHA.106.657405. PMID: 17420351
8. Jabbour RJ, Dick R, Walton AS. Aortic balloon valvuloplasty—review and case series. Heart Lung Circ. 2008;17(Suppl 4):S73–81. https://doi.org/10.1016/j.hlc.2008.08.006. PMID: 18964248
9. Cubeddu RJ, Jneid H, Don CW, Witzke CF, Cruz-Gonzalez I, Gupta R, et al. Retrograde versus antegrade percutaneous aortic balloon valvuloplasty: immediate, short- and long-term outcome at 2 years. Catheter Cardiovasc Interv. 2009;74(2):225–31. https://doi.org/10.1002/ccd.22009. PMID: 19405163
10. Sakata Y, Syed Z, Salinger MH, Feldman T. Percutaneous balloon aortic valvuloplasty: antegrade transseptal vs. conventional retrograde transarterial approach. Catheter Cardiovasc Interv. 2005;64(3):314–21. https://doi.org/10.1002/ccd.20287. PMID: 15736248
11. Lieberman EB, Bashore TM, Hermiller JB, Wilson JS, Pieper KS, Keeler GP, et al. Balloon aortic valvuloplasty in adults: failure of procedure to improve long-term survival. J Am Coll Cardiol. 1995;26(6):1522–8. https://doi.org/10.1016/0735-1097(95)00363-0. PMID: 7594081
12. Otto CM, Mickel MC, Kennedy JW, Alderman EL, Bashore TM, Block PC, et al. Three-year outcome after balloon aortic valvuloplasty: insights into prognosis of valvular aortic stenosis. Circulation. 1994;89(2):642–50. https://doi.org/10.1161/01.CIR.89.2.642. PMID: 8313554
13. Safian RD, Berman AD, Diver DJ, McKay LL, Come PC, Riley MF, et al. Balloon aortic valvuloplasty in 170 consecutive patients. N Engl J Med. 1988;319(3):125–30. https://doi.org/10.1056/NEJM198807213190301. PMID: 3386698
14. Kuntz RE, Tosteson AN, Maitland LA, Goldman L, Baim DS, Gordon PC, et al. Immediate results and long-term follow-up after repeat balloon aortic valvuloplasty. Catheter Cardiovasc Diagn. 1992;25(1):4–9. https://doi.org/10.1002/ccd.1810250103. PMID: 1555248
15. Parker MS, Hui FK, Camacho MA, Chung JK, Broderick DJ, LeBerge JM. Female breast radiation exposure during CT pulmonary angiography: implications for pregnancy. Radiology. 2005;234(2):546–51. https://doi.org/10.1148/radiol.2342031428. PMID: 15671005

Chapter 17
Transcatheter Aortic Valve Implantation (Replacement)

Ata Firouzi, Zahra Hosseini, and Ehsan Khalilipur

Abstract AVR remains the only effective treatment to reduce mortality in symptomatic patients with severe AS. Until recently, SAVR was considered the gold standard for most patients, with TAVI reserved primarily for those at increased surgical risk. Nonetheless, it is now evident that TAVI is non-inferior to SAVR, even in intermediate- and low-risk patients with a life expectancy of at least 8 years. TAVI is associated with lower mortality rates, fewer strokes, reduced major bleeding, decreased incidence of atrial fibrillation, and shorter hospital stays and recovery times. When the appropriate device is selected for the right patient by an experienced interventionist, the intermediate outcomes of TAVI appear comparable to those of SAVR, despite higher rates of regurgitation and conduction disturbances in TAVI patients.

Keywords SAVR · TAVI · Intermediate-risk patients · Low-risk patients · Right patient and right device · Outcomes

Abbreviations

AI	Aortic insufficiency
AS	Aortic stenosis
AV	Aortic valve
AVA	Aortic valve area
AVR	Aortic valve replacement
CT	Computed tomography
EF	Ejection fraction
LA	Left atrium
LBBB	Left bundle branch block

A. Firouzi · Z. Hosseini · E. Khalilipur (✉)
Cardiovascular Intervention Research Center, Rajaie Cardiovascular Institute, Iran University of Medical Sciences, Tehran, Iran

A. Firouzi et al. (eds.), *Mastering Skills in Complex Cardiovascular Interventions*, https://doi.org/10.1007/978-3-031-97321-5_17

LV	Left ventricle
MR	Mitral regurgitation
NYHA	New York Heart Association
RA	Right atrium
RV	Right ventricle
SAVR	Surgical aortic valve replacement
TAVI	Transcatheter aortic valve implantation
TEE	Transesophageal echocardiography
TR	Tricuspid regurgitation
TTE	Transthoracic echocardiography

Introduction

AVR is the only effective treatment to reduce mortality in symptomatic patients with severe AS [1]. Until recently, SAVR was the gold standard for most patients, with TAVI primarily reserved for those at increased surgical risk. Nevertheless, it is now evident that TAVI is non-inferior to SAVR, even in low-risk patients, offering lower mortality rates, fewer strokes, reduced major bleeding, decreased incidence of atrial fibrillation, and shorter hospital stays and recovery times [1, 2]. These benefits have significantly increased the appeal of TAVI, to the extent that, over the past 5 years in the United States, the number of TAVI procedures has surpassed both isolated SAVR and all SAVR cases combined [3]. This dramatic shift in clinical practice can be attributed to a virtuous cycle driven by growing operator experience on the one hand and advancements in technology on the other.

The steadily increasing number of TAVI procedures has benefited from enhanced preprocedural assessment, which now features dedicated protocols for CT image acquisition and reconstruction. These protocols incorporate advanced, albeit still experimental, 3D software designed to simulate procedural outcomes, such as the severity of aortic regurgitation and the likelihood of pacemaker implantation (e.g., FEops HEART Guide, Gent, Belgium) [4].

The procedural technique itself has been refined, aiming for a minimalistic approach in selected patients. This approach includes a primary percutaneous femoral access route—now employed in over 95% of cases [3]—and secondary radial access for supravalvular aortic angiography to guide valve deployment and confirm optimal closure of the primary access [5]. Additional features include direct implantation without pre-dilatation [6], vascular and TTE (without TEE), and the growing adoption of conscious sedation and local anesthesia [7].

Engineering advancements in transcatheter valve design have resulted in catheter miniaturization and reduced paravalvular leaks, achieved through optimized sealing "skirts" incorporated into both balloon-expandable and self-expanding valves. Major vascular complications have also been significantly reduced [3], thanks to improvements in prosthesis delivery systems, which now feature thinner profiles of 14–16 Fr, and the use of advanced vascular closure devices, such as the two Perclose Proglide systems (Abbott Medical, Santa Clara, CA, USA).

Challenging Cases

Case 1

A 74-year-old man with a history of coronary artery bypass surgery (10 years prior) and chronic obstructive pulmonary disease was referred due to progressive dyspnea (NYHA functional class III) and orthopnea. On physical examination, his hemodynamic status was stable. A cardiac examination revealed elevated jugular venous pressure, lateral deviation of the LV apex, and a harsh pansystolic murmur with a palpable thrill at the right second sternal border. Lung auscultation detected bilateral fine rales extending to the lower third of the lung fields, along with decreased tactile fremitus. Symmetrical edema was noted in both lower extremities. ECG showed sinus rhythm with first-degree atrioventricular block, left anterior hemiblock, evidence of a previous anterior myocardial infarction, LBBB, and LV hypertrophy with a strain pattern. Chest X-ray findings were notable for prominent venous congestion and pulmonary edema.

TEE Findings

- **LV:**
  - Moderate enlargement (95 mL/m^2);
  - Severe systolic dysfunction (LVEF: 15%); and
  - Severe diastolic dysfunction (E/E': 25).
- **RV:**
  - Mild enlargement and
  - Severe systolic dysfunction.
- **LA:**
  - Severe enlargement.
- **RA:**
  - Moderate enlargement.
- **AV:**
  - Heavily calcified;
  - Calcification extending into the LVOT;
  - Severe AS:
    - Peak pressure gradient: 110 mm Hg;
    - Mean pressure gradient: 75 mm Hg;
    - AV annulus diameter: 22 × 27 mm;
    - AVA: 0.45 cm^2;

    - Sinus of Valsalva diameter: 3.6 cm;
    - Sinotubular junction diameter: 3 cm; and
    - Ascending aorta diameter: 38 mm.
  - Up to moderate AI.
- **Mitral Valve:**
  - Severe functional MR.
- **Tricuspid Valve:**
  - Moderate TR with a TR gradient of 70 mm Hg.
- **Pulmonary hypertension:**
  - Severe.
- **Other:**
  - No pericardial effusion.

Aortic CT Angiography Report

- **AV:**
  - Highly calcified tricuspid AV;
  - Calcium score: 4300;
  - Area: 505 mm^2;
  - Perimeter: 82 mm; and
  - Diameter: 27.5 × 21 mm.
- **Coronary ostial heights:**
  - Right coronary ostial height: 15 mm and
  - Left coronary ostial height: 6.7 mm.
- **Sinus of Valsalva measurements:**
  - Right sinus width: 32.2 mm;
  - Left sinus width: 34 mm;
  - Right sinus height: 20.5 mm; and
  - Left sinus height: 15 mm.
- **Other measurements:**
  - Sinotubular junction: 31 mm and
  - Ascending aorta: 39 mm.
- **Femoral and iliac arteries:**
  - Right common femoral artery diameter: 6.5 mm;
  - Right external iliac artery diameter: 7.3 mm; and
  - Right common iliac artery diameter: 7.5 mm.

Given the patient's very high risk for redo cardiac surgery (AVR), the estimated Society of Thoracic Surgeons (STS) score was 22% for mortality and morbidity. Furthermore, despite the patient's post-coronary artery bypass grafting status, the distance between the left coronary ostium and the annulus was notably short (6.7 mm). Based on the heart team's decision-making process, the patient was deemed suitable for TAVI.

The Procedure

Under conscious sedation, a 6 Fr sheath was inserted into the right common femoral artery, followed by the insertion of 6 Fr sheaths into the left common femoral artery and left common femoral vein. Using two Perclose ProGlide devices (Abbott Medical, Santa Clara, CA, USA) at the right common femoral artery, a 10 Fr sheath was introduced, and intravenous heparin was administered to achieve an activated clotting time of >250 s.

Under meticulous hemodynamic monitoring by a cardiac anesthesiologist, a pigtail catheter was advanced through left arterial access and positioned in the noncoronary cusp. In the working view (suggested by CT angiography), the aortic root was injected, and a coplanar view was established (Fig. 17.1a).

Afterward, an AL1 catheter was advanced to the aortic root via right arterial access. With careful manipulation, a straight-tip 0.035-inch guidewire was passed through the AV, and the catheter was advanced over it into the LV apex. The wire was then exchanged for a Super Stiff guidewire with a well-formed distal loop, positioned under right anterior oblique projection. A temporary pacemaker lead was subsequently inserted into the RV apex.

Given the short distance between the left main coronary artery and the annulus, a prophylactic measure was taken: using an XB guiding catheter, the ramus branch was wired, and an XIENCE stent (3.0 × 28 mm) was positioned there. Due to the extensive calcification, predilation (balloon aortic valvuloplasty) was performed using a 20 × 40 mm balloon.

Based on CT angiography and TTE data, an Edwards SAPIEN 3–26 valve was selected. A long delivery sheath (14 Fr) was introduced into the descending aorta, and the chosen valve was mounted onto the delivery catheter for retrograde delivery. The valve delivery system was inserted with the Edwards E logo facing upward, and the assembly and alignment of the valve were completed in the descending aorta. The entire system was then advanced toward the AV, and the precise positioning of the valve was repeatedly confirmed through root injections. This ensured that the central marker of the valve was aligned with the nadir of the native valve leaflets.

During ventricular over-pacing (rate: 180 bpm), the balloon was inflated once the blood pressure dropped. After the balloon was fully deflated, the temporary pacemaker was turned off. Aortic root injection confirmed the proper positioning of the valve with no paravalvular leaks or conduction blocks (Fig. 17.1b). Nonetheless, the flow in the left main coronary artery was compromised (Fig. 17.1c), prompting chimney stenting of the left main-ramus (Fig. 17.1d). The final result was satisfactory (Fig. 17.1e).

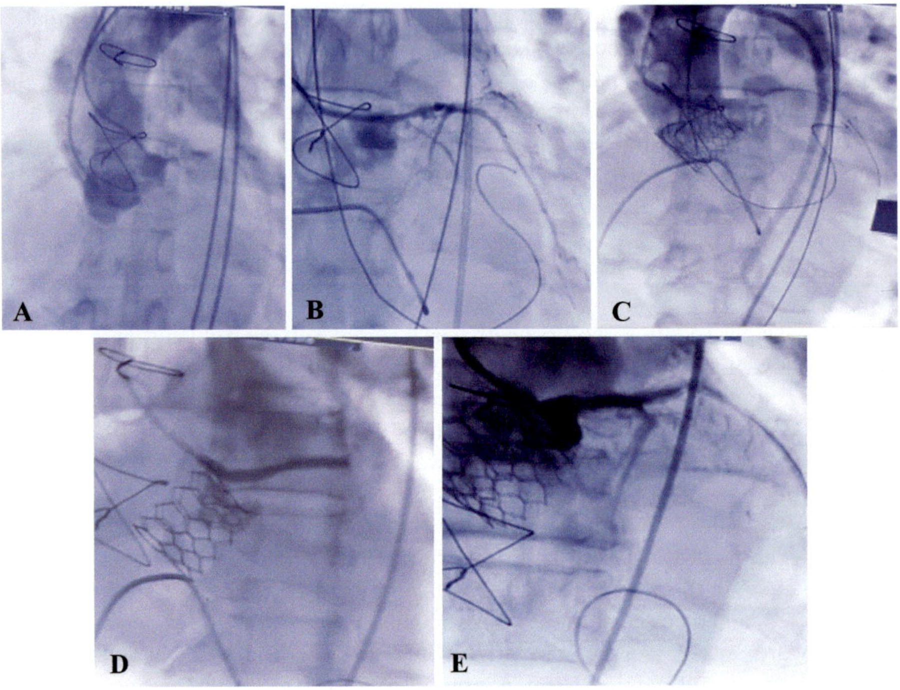

Fig. 17.1 (**a–e**) The aortic root was injected, and a coplanar view was established (**a**). The left main injection revealed a significant ostial lesion extending to the well-developed ramus branch, with a short distance between the left main origin and the annulus (**b**). Consequently, prophylactic wiring of the ramus branch was performed, and an XIENCE stent measuring 3 × 28 mm was positioned there. Following the implantation of the SAPIEN 3 valve, the aortic root injection confirmed the proper positioning of the valve without paravalvular leaks, although the ramus branch was compromised (**c**). Therefore, chimney stenting was performed on the left main to ramus branch, followed by post-dilatation using a 5 × 12 mm balloon (**d**), resulting in an acceptable final outcome (**e**)

The patient was admitted to the ICU, and TTE performed 24 h later revealed an LVEF of 45%, moderate MR, and an appropriately positioned transcatheter heart valve with no paravalvular leaks and good hemodynamic function.

Case 2

An 87-year-old man was referred due to complaints of resting dyspnea and orthopnea. During the physical examination, his hemodynamic status was found to be stable. A cardiac examination revealed elevated jugular venous pressure and a loud systolic murmur (grade III) audible at the left sternal border and the LV apex. ECG displayed an LV hypertrophy pattern.

TTE Findings

- **LV:**
  - Normal size;
  - Normal systolic function (EF: 55%);
  - Moderate concentric LV hypertrophy; and
  - Diastolic dysfunction (grade II).
- **RV:**
  - Normal size and
  - Normal systolic function.
- **Mitral valve:**
  - Mildly thickened and calcified;
  - No mitral stenosis; and
  - Moderate functional MR.
- **AV:**
  - Thickened and heavily calcified;
  - Mean pressure gradient: 55 mm Hg;
  - Peak pressure gradient: 85 mm Hg;
  - AVA by continuity equation and planimetry: 0.6 cm^2;
  - At least moderate eccentric jet AI;
  - AV annulus diameter: 25 × 23 mm; and
  - AV area: 520 mm^2.
- **Aorta:**
  - Ascending aorta diameter: 3.5 cm and
  - Sinotubular junction diameter: 3 cm.
- **Tricuspid valve:**
  - Mild TR.
- **Pulmonary hypertension:**
  - Mild (systolic pulmonary artery pressure: 40 mm Hg).
- **Other:**
  - No pericardial effusion.

Aortic CT Angiography Report

- **AV:**
  - Highly calcified tricuspid AV;

- Calcium score: 5500;
- Area: 515 mm^2;
- Perimeter: 78 mm; and
- Diameter: 25 × 24 mm (Fig. 17.2a, b).

- **Coronary ostial heights:**
  - Right coronary ostial height: 15.7 mm and
  - Left coronary ostial height: 13.5 mm.
- **Sinus of Valsalva measurements:**
  - Right sinus width: 31 mm;
  - Left sinus width: 32 mm;
  - Right sinus height: 22 mm; and
  - Left sinus height: 17 mm.
- **Other measurements:**
  - Sinotubular junction: 31 mm and
  - Ascending aorta: 37 mm.
- **Femoral and iliac arteries:**
  - Right common femoral artery diameter: 6.5 mm;
  - Right external iliac artery: Symmetric focal calcification, diameter: 7 mm; and
  - Right common iliac artery (CIA) diameter: 7.5 mm.

The Procedure

Under conscious sedation, a 6 Fr sheath was inserted into the right common femoral artery, followed by the placement of 6 Fr sheaths in the left common femoral artery and left common femoral vein. Two Perclose ProGlide devices (Abbott Medical, Santa Clara, CA, USA) were deployed at the right common femoral artery, and a 10

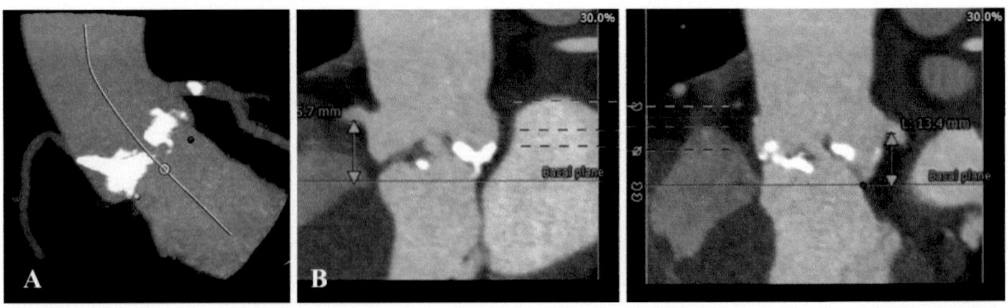

Fig. 17.2 (**a**, **b**) The tricuspid AV exhibited highly calcified leaflets (calcium score: 5500) (**a**), with an acceptable distance between the origins of the coronary arteries and the annulus (**b**)

Fr sheath was introduced. Intravenous heparin was administered to achieve an activated clotting time of >250 s.

Under meticulous hemodynamic monitoring by a cardiac anesthesiologist, a pigtail catheter was advanced through left arterial access and positioned in the noncoronary cusp. In the working view (as suggested by CT angiography), the aortic root was injected, and a coplanar view was established (Fig. 17.3a).

Afterward, via the right arterial access, an AL1 catheter was advanced to the aortic root, and a straight-tip 0.035-inch guidewire was passed through the AV. The catheter was then advanced over the wire into the LV apex. The wire was exchanged for a 260 cm Super Stiff guidewire (0.035-inch Amplatzer) with a well-formed distal loop, positioned under right anterior oblique projection. A temporary pacemaker lead was subsequently inserted into the RV apex. Given the extensive calcification, predilation (balloon aortic valvuloplasty) was performed using a 22 × 40 mm balloon (Fig. 17.3b).

Based on the CT angiography and TTE data, an Edwards SAPIEN 3–26 valve was selected. A long delivery sheath (14 Fr) was introduced into the descending aorta, and the chosen valve was mounted onto the delivery catheter for retrograde

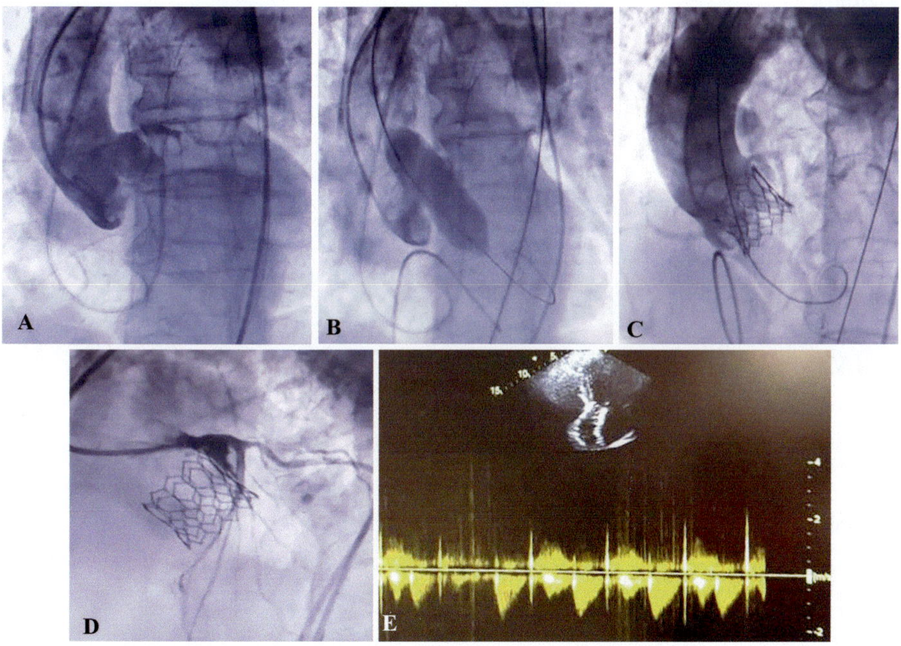

Fig. 17.3 (**a**–**e**) The aortic root was injected, and a coplanar view was established (**a**). Predilation was performed using a 22 × 40 balloon for balloon aortic valvuloplasty (**b**). The final aortic root injection confirmed the proper positioning of the valve without paravalvular leaks (**c**). The majority of the calcification was directed toward the left main, prompting a selective left main injection; fortunately, no flow-limiting events occurred (**d**). The final hemodynamic study via echocardiography indicated an aortic peak-to-peak gradient of 20 mm Hg

delivery. The valve delivery system was inserted with the Edwards E logo facing upward, and the assembly and alignment of the valve were completed in the descending aorta.

Subsequently, the entire system was advanced toward the AV, and the precise positioning of the valve was repeatedly confirmed through root injections, ensuring that the central marker of the valve was aligned with the nadir of the native valve leaflets. During ventricular over-pacing (rate: 180 bpm), the balloon was inflated once the blood pressure dropped. After the balloon was fully deflated, the temporary pacemaker was turned off. Aortic root injection confirmed the proper positioning of the valve with no paravalvular leaks or conduction blocks (Fig. 17.3c). Still, the bulk of the calcification had shifted toward the left main coronary artery, prompting a selective left main injection. Fortunately, no flow-limiting complications were observed (Fig. 17.3d).

The patient was admitted for 3 days, during which TTE demonstrated favorable hemodynamic results (Fig. 17.3e). He was subsequently discharged in good clinical condition.

Case 3

An 83-year-old woman with a history of coronary artery bypass graft surgery was referred due to recurrent syncope. On physical examination, her hemodynamic status was stable. A loud mid-systolic murmur (grade III) was audible at the left sternal border and the LV apex. ECG revealed an LV hypertrophy pattern with a strain pattern in the precordial leads.

TTE Findings

- **LV:**
  - Normal size;
  - Moderate systolic dysfunction (LVEF: 35%); and
  - Moderate concentric LV hypertrophy.
- **RV:**
  - Normal size and
  - Moderate systolic dysfunction.
- **AV:**
  - Tricuspid AV with severely calcified leaflets;
  - Severe AS:
    - Mean pressure gradient: 60 mm Hg;
    - Peak pressure gradient: 90 mm Hg;

    - AVA by planimetry: 0.6 cm^2;
    - Annulus diameter: 23 × 20 mm;
    - Area: 370 mm^2; and
    - Perimeter: 40 mm.
  - Up to moderate AI.
- **Mitral valve:**
  - Mild-to-moderate MR.
- **Tricuspid valve:**
  - Mild TR with a TR gradient of 25 mm Hg.

Aortic CT Angiography Report

- **AV:**
  - Tricuspid AV with severe calcification;
  - Calcium score: 2000;
  - Asymmetric calcification of the noncoronary cusp;
  - Annulus diameter: 23 × 20 mm;
  - Area: 375 mm^2; and
  - Perimeter: 40 mm.
- **Coronary ostial heights:**
  - Right coronary ostial height: 13 mm and
  - Left coronary ostial height: 13.5 mm.
- **Sinus of Valsalva measurements:**
  - Right sinus width: 14.5 mm;
  - Left sinus width: 27 mm;
  - Noncoronary sinus width: 25 mm;
  - Left sinus height: 18 mm; and
  - Right sinus height: 16.5 mm.
- **Other measurements:**
  - Sinotubular junction: 22 mm and
  - Ascending aorta: 26 mm.

The right common femoral artery exhibited moderate stenosis and was deemed unsuitable for access. Considering the patient's symptoms and high risk for SVR, she was considered a candidate for TAVI.

Procedural Technique

Under conscious sedation, 6 Fr sheaths were inserted into the right common femoral artery and common femoral vein, as well as the left common femoral artery. Two Perclose ProGlide devices (Abbott Medical, Santa Clara, CA, USA) were deployed at the left common femoral artery, and a 10 Fr sheath was introduced. Intravenous heparin was administered to achieve an activated clotting time of >250 s.

Under meticulous hemodynamic monitoring by a cardiac anesthesiologist, a pigtail catheter was advanced through the right arterial access and positioned in the noncoronary cusp. In the working view (as suggested by CT angiography), the aortic root was injected, and a coplanar view was established (Fig. 17.4a).

Subsequently, via the left arterial access, an AL1 catheter was advanced to the aortic root, and a straight-tip 0.035-inch guidewire was passed through the AV. The catheter was then advanced over the wire into the LV apex. The wire was exchanged for a 260 cm Super Stiff guidewire (0.035-inch Amplatzer) with a well-formed distal loop, positioned under right anterior oblique projection. A temporary pacemaker lead was inserted into the RV apex. Based on the CT angiography data, a Medtronic Evolut R-23 valve was selected. After introducing the EnVeo R delivery catheter system (14 Fr), the valve was advanced toward the AV. The position of the inflow portion was adjusted based on the working view and overlap view, as recommended by the CT angiography. The valve was then carefully released during repetitive root injections. During deployment, it appeared that the depth of the valve implantation in the LVOT exceeded 10 mm (Fig. 17.4b). As a result, the valve was recaptured, and another attempt was made to deploy it. This time, the depth of the inflow portion in the LVOT was acceptable (Fig. 17.4c). Following ventricular over-pacing (rate: 150 bpm), the remaining part of the transcatheter heart valve was deployed. Nonetheless, the valve suddenly embolized into the aortic root (Fig. 17.4d). It was immediately snared and retrieved (Fig. 17.4e), and its position was stabilized in the ascending aorta just above the vein grafts (Fig. 17.4f). Subsequently, another valve of the same size was advanced and successfully deployed, achieving a good final result (Fig. 17.4g). The final hemodynamic assessment by TTE confirmed the success of the procedure (Fig. 17.4h).

Case 4

A 72-year-old woman was referred due to complaints of exertional dyspnea, classified as NYHA functional class III. During the physical examination, her hemodynamic status was stable. A harsh mid-systolic murmur was audible in the right second intercostal space, radiating to the left sternal border. ECG was normal.

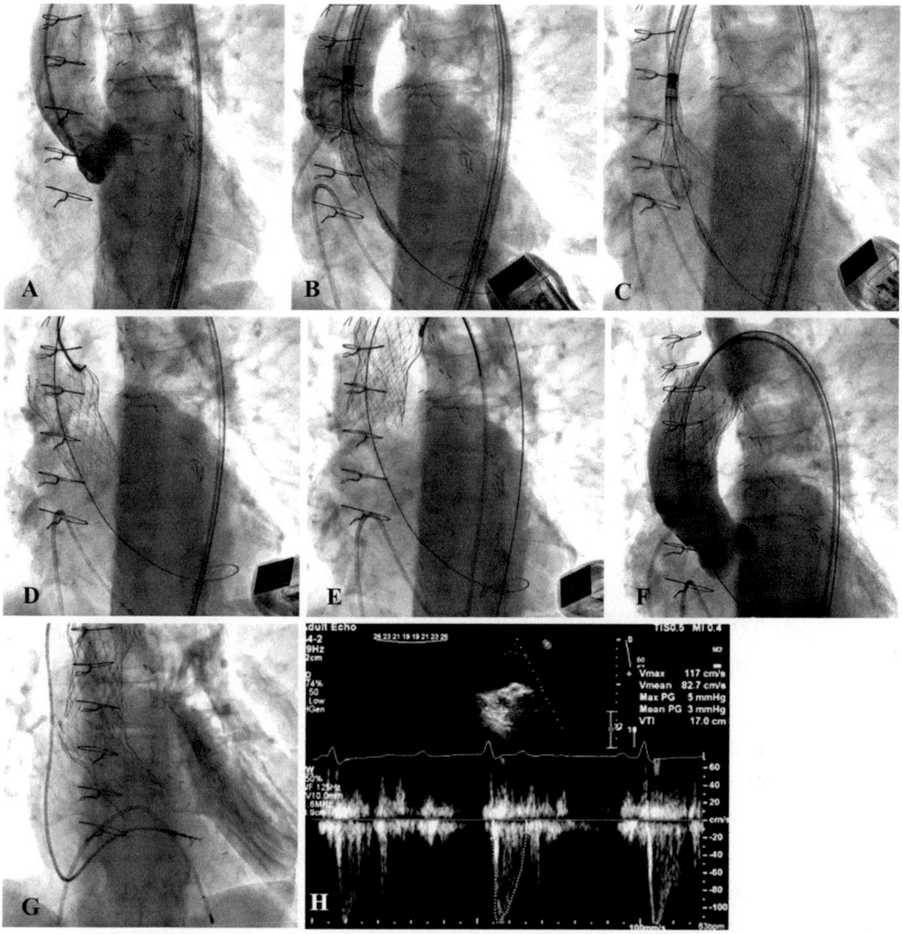

Fig. 17.4 (**a**–**h**). The aortic root was injected, and a coplanar view was established (**a**). During deployment, it appeared that the depth of the valve implantation in the LVOT exceeded 10 mm (**b**). Consequently, the valve was recaptured, and another attempt at deployment was made. This time, the depth of the valve inflow in the LVOT was deemed acceptable (**c**). After complete valve deployment, the valve unexpectedly embolized to the aortic root (**d**). The valve was promptly snared and retrieved (**e**). Its position was stabilized in the ascending aorta, just above the vein grafts (**f**). A second valve of the same size was then advanced and deployed successfully, resulting in a favorable final outcome (**g**). The final hemodynamic data obtained via TTE confirmed the success of the procedure (**h**)

TTE Findings

- **LV:**
  - Small cavity size;
  - Moderate systolic dysfunction (LVEF: 40%); and
  - Mild LV hypertrophy.
- **RV:**
  - Normal size and
  - Normal function.
- **AV:**
  - Tricuspid AV with severe calcification and stenosis;
    - AVA by 2D planimetry: 0.7 cm^2;
    - Mean pressure gradient: 46 mm Hg;
    - Peak pressure gradient: 80 mm Hg;
    - Annulus diameter: 25 × 26 mm;
    - Annular perimeter: 72 mm; and
    - Annular area: 393 mm^2 (Fig. 17.5a, b).
- **Other findings:**
  - No evidence of LVOT obstruction;
  - No systolic anterior motion of the mitral valve;
  - No asymmetric septal hypertrophy; and
  - No intraventricular gradient (Fig. 17.5a, b).

A cardiothoracic surgeon assessed that the patient was at intermediate risk for SVR (STS score: 4.6%). Consequently, trans-femoral TAV was recommended.

Procedural Technique

Under conscious sedation, a 6 Fr sheath was inserted into the right common femoral artery, followed by the placement of 6 Fr sheaths in the left common femoral artery and left common femoral vein. Two Perclose ProGlide devices (Abbott Medical, Santa Clara, CA, USA) were deployed at the right common femoral artery, and a 10 Fr sheath was introduced. Intravenous heparin was administered to achieve an activated clotting time of >250 s.

Under meticulous hemodynamic monitoring by a cardiac anesthesiologist, a pigtail catheter was advanced through left arterial access and positioned in the noncoronary cusp. In the working view (as suggested by CT angiography), the aortic root was injected, and a coplanar view was established (Fig. 17.5c).

Based on TTE and CT angiography data, a balloon-expandable Myval prosthetic valve (Meril Life Sciences Pvt. Ltd., India) size 23 was selected. Subsequently, via

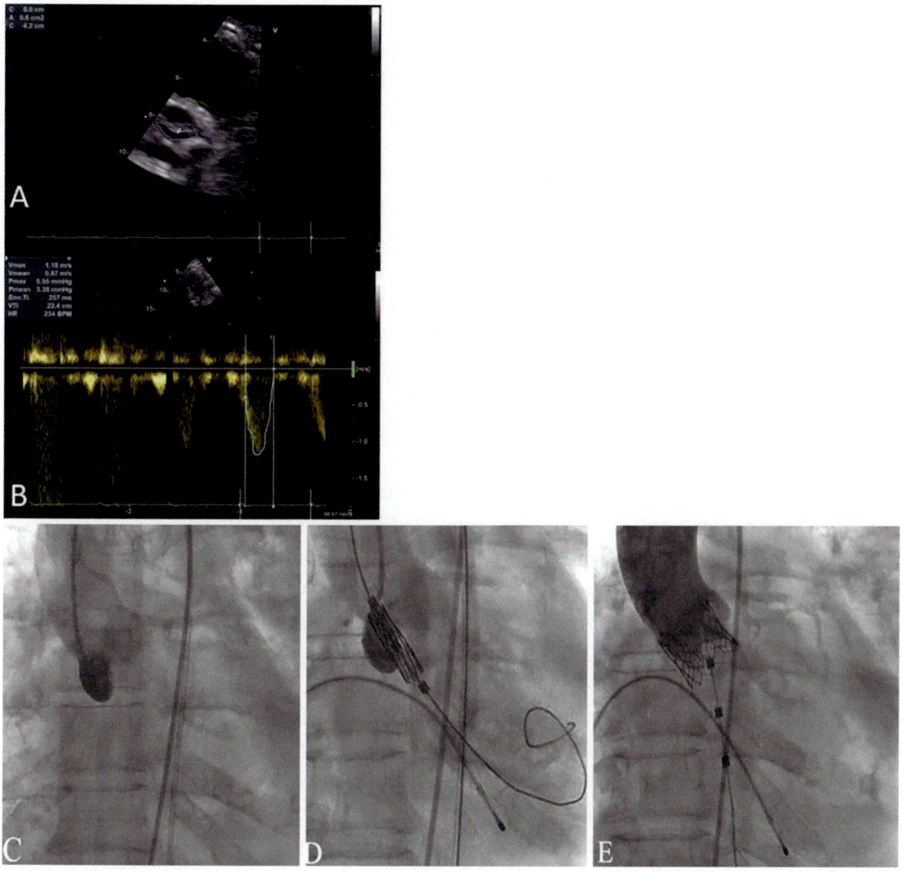

Fig. 17.5 (**a–d**) TTE revealed severe calcification and stenosis of the AV, with AVA measured by 2D planimetry at 0.7 cm^2, mean pressure gradient of 46 mm Hg, and peak-to-peak gradient of 80 mm Hg (**a**, **b**). The aortic root was injected, and a coplanar view was established (**c**). The appropriate positioning of the transcatheter heart valve in the annulus was confirmed through repetitive root injections (**c**). The final root injection after transcatheter heart valve implantation demonstrated the proper positioning of the valve without paravalvular leaks (**d**)

right arterial access, an AL1 catheter was advanced to the aortic root, and a straight-tip 0.035-inch guidewire was passed through the AV. The catheter was then advanced over the wire into the LV apex. The wire was exchanged for a 260 cm Super Stiff guidewire (0.035-inch Amplatzer) with a well-formed distal loop, positioned under right anterior oblique projection. A temporary pacemaker lead was inserted into the RV apex. The proper positioning of the transcatheter heart valve was confirmed by aortic root injection (Fig. 17.5d). The valve was successfully implanted using a deployment method similar to that described in previous cases (Fig. 17.5e). The final root injection demonstrated appropriate valve placement with no paravalvular leaks or conduction disturbances.

The patient was transferred to the ICU, and 4 h later, she experienced severe dyspnea accompanied by sudden hemodynamic deterioration. She developed pulmonary edema and refractory hypotension, prompting immediate intubation. ECG revealed complete heart block, and bedside chest X-ray confirmed pulmonary edema. Bedside echocardiography showed normal LV and RV function, normal transcatheter heart valve function, and satisfactory hemodynamic data without any regurgitation. A "continuous-wave dagger-shaped" Doppler trace was observed, with a peak pressure gradient of 110 mm Hg, indicative of dynamic LVOT obstruction. A temporary pacemaker was immediately implanted, and massive fluid resuscitation was initiated, stabilizing her hemodynamic status relatively.

After 24 h, the patient developed a fever, and a COVID-19 PCR test returned positive. Treatment with remdesivir and dexamethasone was initiated. Follow-up echocardiography confirmed the resolution of LVOT obstruction. Ultimately, the patient was discharged in good condition following a permanent pacemaker implantation.

The development of LVOT obstruction is well-documented in SAVR. Approximately 15% of patients undergoing SAVR for AS have been observed to develop dynamic intraventricular obstruction. Risk factors include asymmetrical hypertrophy (septal-to-ventricular free wall ratio > 1.4), small LV diameter, increased basal septal thickness (>15 mm), narrow LVOT (<18 mm), high valve gradients, elevated EF, and a ratio of anterior-to-posterior mitral valve leaflet length of <1.3.

The mechanisms underlying LVOT obstruction, systolic anterior motion of the mitral valve, and their association with TAVI remain poorly understood. Chronic AS increases intraventricular pressures, which splints the septum and prevents inward motion, thereby averting mid-ventricular and LVOT obstruction. Nonetheless, following AVR, a rapid reduction in LV pressure leads to a smaller LVOT area. Moreover, increased flow across the LVOT creates a Venturi effect, drawing the anterior mitral valve leaflet into the LVOT and causing dynamic obstruction. Managing dynamic LVOT obstruction after AVR remains a significant clinical challenge.

Certain strategies can help minimize obstruction in the acute setting, such as increasing LV preload with fluid boluses and reducing LV contractility by weaning inotropes and increasing beta-blockers. In addition, dual-chamber pacing has been considered for patients in severe shock. Pacing-induced LBBB causes the septum to shift away from the anterior mitral valve leaflet, reducing dynamic outflow obstruction. Maintaining atrioventricular synchronization while increasing LV filling pressure further decreases the gradient. Thus, dual-chamber pacing may be considered when single RV pacing fails to improve hemodynamics. Therefore, in post-TAVI patients who develop hemodynamic disturbances—particularly women with small LV cavities—LVOT obstruction should be considered a potential cause.

Discussion

The recently published American College of Cardiology (ACC)/American Heart Association (AHA) guidelines have undergone significant changes in comparison with the previous edition. The heart team's evaluation and estimated surgical risk remain central to deciding between SAVR and TAVI. The patient's age and life expectancy are also critical factors to consider, provided the decision is made in collaboration with the patient.

In view of the limited data on transcatheter valve durability beyond 5 years (as highlighted in the NOTION trial), TAVI is not yet indicated for younger patients. Nevertheless, valve-in-valve procedures are expected to play a pivotal role, as evidenced by efforts to develop surgical valves specifically designed to enhance valve-in-valve outcomes and reduce patient–prosthesis mismatch (e.g., Edwards Inspiris Resilia [Edwards Lifesciences, Irvine, CA, USA]).

Vascular Complications

Since the first percutaneous heart valve was implanted in 2002 by Cribier et al. [8], the technology has evolved significantly to become minimally invasive while maximizing effectiveness. Numerous advancements, including smaller delivery systems, the introduction of novel percutaneous vascular closure devices, more precise patient and access selection, and increased operator experience, have collectively contributed to reducing vascular complications.

Toggweiler et al. [9], in their early experience with transfemoral TAVI in high-risk patients, identified vascular access as a major source of morbidity and emphasized the importance of meticulous preprocedural screening and improved vascular management in lowering vascular morbidity and mortality risks. According to the recently published STS/ACC registry annual report, the 30-day major vascular access site complications for high-risk patients decreased to 1.8% in 2019, with transfemoral access being the predominant approach [3]. Be that as it may, as reported in large randomized trials and registries, transfemoral access has emerged as the preferred approach for TAVI delivery. The ability to treat 95% of patients via transfemoral access has been associated with lower bleeding rates, shorter hospital stays, faster ambulation, and higher rates of discharge to home. The transfemoral approach is also well-tolerated under local anesthesia and is compatible with the successful use of vascular closure devices.

Over the past decade, various tools have proven effective in preventing periprocedural vascular complications, including ultrasound-guided micro-puncture and peripheral intravascular lithotripsy. Particularly in cases of vascular calcification, careful selection of the cannulation site can significantly reduce risks by avoiding anterior calcification and enabling the successful implantation of percutaneous closure devices. This approach minimizes vascular complications, enhances the likelihood of achieving hemostasis, and improves overall clinical outcomes.

The use of dedicated micro-puncture 21-gauge needles, whose tips are more visible under ultrasound guidance, has been shown to reduce vascular complications, significantly decreasing the incidence of groin hematomas compared with standard large-bore needles [10].

A narrowed luminal diameter with circumferential calcification is particularly concerning, as non-calcified arteries can be stretched, allowing successful insertion even with a lumen as small as 75% of the TAVI sheath's outer diameter. For calcified or tortuous arteries, it is recommended that the lumen be at least 1.25 mm larger than the sheath. For the 14 Fr or 16 Fr inner diameter sheaths used in contemporary delivery systems, this translates to minimum vessel diameters of 6–7 mm for non-calcified and calcified vessels, respectively [11]. In this context, the introduction of peripheral intravascular lithotripsy offers significant advantages in preserving transfemoral access for TAVI procedures.

Non-transfemoral approaches have been associated with a marked increase in early and late mortality following TAVR. Specifically, the transapical approach has been linked to higher early and mid-term mortality in multivariate analyses.

Conduction Disturbances

Another significant issue in TAVI is conduction disturbances. The incidence of conduction disturbances following TAVI varies widely and depends on numerous preexisting and intraprocedural factors. Clinical factors include female sex, preprocedural conduction abnormalities (e.g., prolonged QRS duration, right bundle branch block, first-degree atrioventricular block, or left anterior hemiblock), prior coronary artery bypass graft surgery, diabetes mellitus, AV or LVOT calcification, and a short membranous septum as an anatomical factor. Procedural factors include the use of the CoreValve, transapical access, predilation, valve oversizing, AV overextension, and lower implantation depth.

New-onset LBBB occurs in 4%–30% of cases with the balloon-expandable Edwards SAPIEN XT valve, 12%–22% with the Edwards SAPIEN 3 valve, and 18%–65% with the self-expandable Medtronic CoreValve device. Studies using the self-expandable Portico TAVI device (St. Jude) reported similar incidence rates, while two studies using the mechanically expandable Lotus system (Boston Scientific) documented higher rates of LBBB (55% and 77%). Notably, 85%–94% of TAVI-induced LBBB occurs during the periprocedural period or within 24 h post-procedure, with 62% occurring before actual valve implantation (e.g., during Stiff wire insertion or balloon predilation of the AV).

The most consistently reported procedural factors associated with the occurrence of LBBB are valve type and implantation depth within the LVOT. Regarding valve type, the incidence of both new-onset LBBB and permanent pacemaker implantation is higher after the implantation of the self-expanding CoreValve System

(Medtronic Inc., Minneapolis, MN, USA) compared with the balloon-expandable SAPIEN or SAPIEN XT systems (Edwards Lifesciences, Irvine, CA, USA) [12].

Approximately 50% of new-onset LBBB cases resolve within a few days after TAVI, resulting in a persistent LBBB rate of around 10%. The permanent pacemaker implantation rate for the Medtronic self-expanding CoreValve has decreased with the introduction of the next-generation CoreValve Evolut R, which is repositionable and allows for more precise high implantation. Nevertheless, self-expanding valves still have a significantly higher permanent pacemaker implantation rate than both SAVR and TAVI using SAPIEN valves. The rate of permanent pacemaker implantation ranges from 2% to 51%. During a median follow-up of 4 years, up to 86% of patients with permanent pacemaker implantation after TAVI exhibited ventricular pacing >1% of the time. Nearly all patients with TAVI-induced high-grade atrioventricular block underwent permanent pacemaker implantation within 5 days after the procedure, with a median of 3 days.

Resolution of high-grade atrioventricular block without pacemaker dependency and a low ventricular pacing rate (<1%) can occur over time. Recovery rates observed after 30 days are 59% for acute-onset high-grade atrioventricular block (within 24 h) and 25% for delayed high-grade atrioventricular block (occurring after more than 6 days). Acute-onset high-grade atrioventricular block tends to recover faster than delayed high-grade atrioventricular block. A more recent large American registry reported a higher mortality rate after 1 year in the permanent pacemaker implantation group. Long-term RV pacing may also increase the risk of LV dysfunction and heart failure. Urena et al. [13] reported that the rate of new-onset LBBB after TAVI with Edwards SAPIEN Valves was approximately 20%, with 50% of these cases resolving within a few days, resulting in a persistent LBBB rate of around 10%.

Complete atrioventricular block remains the most commonly reported indication for permanent pacemaker implantation following TAVI. According to a recent systematic review, the overall rate of permanent pacemaker implantation after TAVI with new-generation valves ranges between 2.3% and 36.1% [14]. The early-generation Medtronic CoreValve was associated with a higher risk of permanent pacemaker implantation (16.3%–37.7%), which remained relatively elevated with the newer Medtronic CoreValve/Evolut R valve (14.7%–26.7%). In contrast, the Edwards SAPIEN 3 valve demonstrated a lower risk (4%–24%). More recent studies involving the SAPIEN 3 and SAPIEN 3 Ultra valves have reported new permanent pacemaker rates as low as 4.4%–6.5%, which are broadly comparable to surgical risks. Similarly, recent studies have shown a reduction in the rate of new permanent pacemakers with the self-expanding Evolut R and PRO valves (Medtronic, Minneapolis, MN, USA) to 10%–20% or less.

Key predictors of permanent pacemaker implantation after TAVI include transcatheter heart valve oversizing relative to the annulus or LVOT area, the location and burden of calcification, and the length of the membranous septum. The membranous septum length serves as an anatomical landmark for the distance between the aortic annulus and the exit point of the bundle of His, where the conduction system crosses to the left side of the heart.

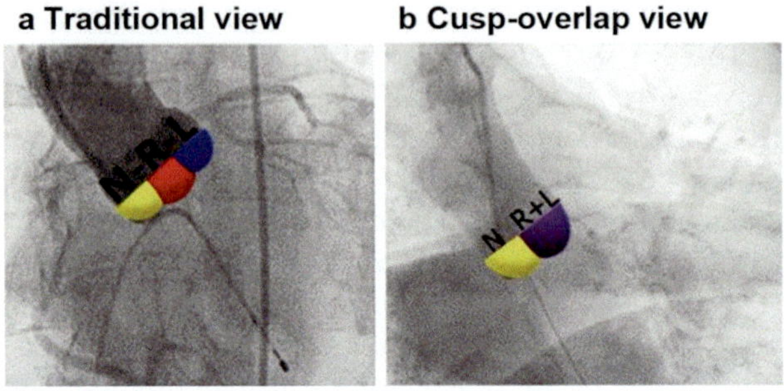

Fig. 17.6 Optimizing fluoroscopic projections for TAVI. Panel (**a**) illustrates the basal aortography of the optimal projection for self-expandable valve implantation, known as the "3 cusp view." In this view, the nadirs of the three cusps (yellow = noncoronary cusp, green = right coronary cusp, and red = left coronary cusp) are perfectly coplanar and equally spaced. Panel (**b**) presents an aortography of a cusp-overlap view, achieved by overlapping the right and left coronary cusps, isolating the noncoronary cusp and minimizing interaction with the conduction system [17]

Hamdan et al. [15] found that the difference between membranous septum length and implantation depth is the single most powerful predictor of atrioventricular block and, along with calcification in the basal septum, the most significant postprocedural predictor of permanent pacemaker implantation with self-expandable valves. Jilaihawi et al. [16] applied a meticulous valve implantation technique in a prospective cohort, considering membranous septum length (referred to as the "MIDAS" [minimizing depth according to the septum] approach) and aiming to avoid mechanical interference with the His bundle. This strategy involved systematically measuring membranous septum length using preprocedural CT scans and targeting a valve depth (measured at the level of the noncoronary cusp) that was less than the membranous septum length.

A strategy aimed at minimizing the risk of interaction between the conduction system and the bioprosthesis, along with a better understanding of the conduction system's location relative to the aortic annulus basal plane, has recently been proposed. Using the cusp overlap technique, the noncoronary cusp, which is the most inferiorly oriented cusp in the LVOT, is isolated from the superimposed left and right coronary cusps (Fig. 17.6) [17].

Aortic Regurgitation

Another significant concern is the degree of AI following TAVI. Post-TAVI leaks can be categorized into transvalvular, paravalvular, and supraskirtal leaks. Transvalvular leaks are relatively rare and typically result from incorrect valve sizing or reduced leaflet motion due to rupture or trauma during post-dilation. Supraskirtal leaks occur when the valve is positioned too low, allowing blood to flow above the valve plane between the metallic struts. These often require emergency correction using a valve-in-valve technique. Paravalvular leaks, the most common type, have a complex etiology and are more challenging to address.

Paravalvular regurgitation is six times more frequent after TAVR than SAVR. While most cases of paravalvular regurgitation are mild, approximately 20% of cases involve moderate-to-severe regurgitation, which has been associated with a worse prognosis. In most cases, paravalvular leaks are caused by one or more factors, such as insufficient prosthesis adherence to the aortic annulus, valve sizing errors, valve malposition, or suboptimal implantation. Multiple anatomical, procedural, and postprocedural risk factors have been identified for the development of paravalvular leaks after TAVI. One anatomical factor is the angle between the axis of the ascending aorta (representing the landing zone for the upper part of the bioprosthesis) and the axis of the LVOT (representing the landing zone for the lower part of the valve). The greater this angle, the higher the incidence of paravalvular regurgitation.

The most notable predictive factors for paravalvular leaks include the severity of the valve and LVOT calcification, the type and size of the valve, and the implantation technique. Proper sizing of the native annulus and LVOT is critical to prevent patient-prosthesis mismatch and paravalvular leaks. While annular eccentricity alone does not predict paravalvular leaks, excessive annular calcification and its asymmetric distribution are significant predictors of aortic regurgitation. In cases of an extremely large annulus, where available transcatheter heart valves require overexpansion, annular eccentricity is associated with paravalvular leaks. Therefore, an accurate preprocedural assessment of the AV annulus is essential for selecting the optimal valve. Studies have shown that paravalvular leaks can be predicted by quantifying LVOT calcification, with a calcium volume $\geq 235\ mm^2$ being a key threshold [18]. Additionally, two other independent predictors of significant paravalvular leaks, identifiable by CT scan, are the valve diameter-to-calculated average annulus diameter (CAAD) ratio and the valve calcification index (VCI), defined as the aortic root calcification volume divided by body surface area.

Although some calcification on the native AV can help secure the anchoring of the prosthesis to the annulus, excessive calcification and its asymmetric distribution can prevent proper positioning against the annular wall, leading to paravalvular leaks and the need for postdilation after valve deployment. Several studies have shown that first-generation self-expanding valves were associated with a higher incidence of paravalvular leaks. Recently, the SOLVE-TAVI trial [19], which randomized 442 patients undergoing TAVI, demonstrated equivalence between self-expanding and second-generation balloon-expandable valves at 30 days in the composite primary endpoint, which included mortality, stroke, moderate-to-severe paravalvular leaks, and pacemaker implantation. Even the incidence of individual endpoints, particularly moderate-to-severe paravalvular leaks, did not significantly differ between the two groups. In a meta-analysis by He et al. [20], which compared three new-generation self-expanding and balloon-expandable valves (SAPIEN 3, Evolut R, and Acurate neo), a significant reduction in paravalvular leaks was observed with the SAPIEN 3 compared with the Acurate neo [21].

Several corrective measures have been proposed to address significant residual paravalvular leaks following TAVI. Severely calcified cusps of the native AV may prevent complete apposition of the prosthesis with the annulus, resulting in a characteristic eccentric paravalvular leak jet. In such cases, balloon postdilation is an option to reduce regurgitation by achieving better expansion of the prosthesis stent frame and improved sealing of the paravalvular space, provided the transcatheter heart valve has been deployed at the correct implantation depth. Postdilation is also the preferred treatment for patients with frame underexpansion as the cause of severe paravalvular leaks, particularly when using self-expanding transcatheter heart valves [22].

Accurate positioning of the transcatheter heart valve relative to the native aortic annulus is critical for a successful procedure. Suboptimal deployment can lead to incomplete apposition of the valve and annulus or, worse, incomplete sealing by the pericardial skirt of the stent frame, resulting in significant diastolic backflow into the LV. For malpositioned transcatheter heart valves with either excessively high (aortic) or excessively low (ventricular) implantation of the prosthesis, valve-in-valve implantation is a viable strategy to reduce significant paravalvular leaks and avoid bailout cardiac surgery. The second valve can be deployed so that the sealing pericardial skirts of both valves overlap, ensuring proper sealing with the native valve annulus. Takagi et al. [23] demonstrated that a low implant (>3 stent meshes below the virtual basal ring) was an independent risk factor for paravalvular leaks (odds ratio: 3.67), while Sherif et al. [24] found that an implant <9.5 mm from the noncoronary cusp predicts severe paravalvular leaks.

Coronary Occlusion

The reported incidence is approximately 2% in recent studies [25]. This complication arises from the displacement of the leaflets toward the ostium of the coronary artery or the sinotubular junction, ultimately leading to hemodynamic instability. Several factors, including anatomical features such as coronary height, the distance between the annulus and the coronary ostia, the dimensions of the sinotubular junction, bioprosthetic valve-in-valve procedures, and technical aspects, can contribute to this issue. New-generation, fully retrievable transcatheter heart valve devices may be preferable when the risk of coronary occlusion is anticipated to be high. If the predicted risk of coronary occlusion is very high, it is recommended to use coronary artery protection with a safety wire, typically accompanied by a pre-loaded stent (Section "Case 1"). In selected cases with nearly 100% probability of occlusion, pre-stenting of the coronary ostia is performed using the "chimney technique" [26–28].

Transcatheter Heart Valve Malpositioning

Bioprosthesis mispositioning occurs when the transcatheter heart valve is placed either too low or too high, leading to suboptimal hemodynamics or embolization. The primary cause is the use of suboptimal fluoroscopic landmarks, which can make optimal transcatheter heart valve positioning challenging. Still, with increased operator expertise and the introduction of repositionable and retrievable transcatheter heart valves, the incidence of malpositioning has decreased from 8% to 4% [29] (Section "Case 3"). Other complications of TAVI include thrombosis, annulus perforation, and patient-prosthesis mismatch.

Conclusion

The challenge of choosing the appropriate device for patients necessitates dedicated trials and their systematic evaluation. Nonetheless, individualized decision-making, leveraging the strengths and weaknesses of each device along with local expertise, will likely remain the cornerstone of clinical practice. Similarly, ancillary devices and techniques will need to undergo formal testing to evaluate their comparative clinical effectiveness and economic viability.

References

1. Otto CM. Informed shared decisions for patients with aortic stenosis. N Engl J Med. 2019;380(18):1769–70. https://doi.org/10.1056/NEJMe1903150. PMID: 31042829
2. Siontis GC, Overtchouk P, Cahill TJ, Modine T, Prendergast B, Praz F, et al. Transcatheter aortic valve implantation vs. surgical aortic valve replacement for treatment of symptomatic severe aortic stenosis: an updated meta-analysis. Eur Heart J. 2019;40(38):3143–53. https://doi.org/10.1093/eurheartj/ehz275. PMID: 31162522
3. Carroll JD, Mack MJ, Vemulapalli S, Herrmann HC, Gleason TG, Hanzel G, et al. STS-ACC TVT Registry of transcatheter aortic valve replacement. J Am Coll Cardiol. 2020;76(21):2492–516. https://doi.org/10.1016/j.jacc.2020.09.595. PMID: 33213729
4. El Faquir N, De Backer O, Bosmans J, Rudolph T, Buzzatti N, Bieliauskas G, et al. Patient-specific computer simulation in TAVR with the self-expanding Evolut R valve. JACC Cardiovasc Interv. 2020;13(15):1803–12. https://doi.org/10.1016/j.jcin.2020.04.036. PMID: 32797906
5. Jhand A, Apala DR, Dhawan R, Katta N, Aronow HD, Daniels MJ, et al. Meta-analysis comparing transradial versus transfemoral secondary access in transcatheter aortic valve implantation. Am J Cardiol. 2020;131:74–81. https://doi.org/10.1016/j.amjcard.2020.06.032. PMID: 32768066
6. Leclercq F, Robert P, Akodad M, Macia JC, Gandet T, Delseny D, et al. Prior balloon valvuloplasty versus direct transcatheter aortic valve replacement: results from the DIRECTAVI trial. JACC Cardiovasc Interv. 2020;13(5):594–602. https://doi.org/10.1016/j.jcin.2019.11.006. PMID: 32113933

7. Thiele H, Kurz T, Feistritzer HJ, Stachel G, Hartung P, Lurz P, et al. General versus local anesthesia with conscious sedation in transcatheter aortic valve implantation: the randomized SOLVE-TAVI trial. Circulation. 2020;142(15):1437–47. https://doi.org/10.1161/CIRCULATIONAHA.120.046451. PMID: 32820667
8. Cribier A, Eltchaninoff H, Bash A, Borenstein N, Tron C, Bauer F, et al. Percutaneous transcatheter implantation of an aortic valve prosthesis for calcific aortic stenosis: first human case description. Circulation. 2002;106(24):3006–8. https://doi.org/10.1161/01.CIR.0000047200.36165.B8. PMID: 12473543
9. Toggweiler S, Leipsic J, Binder RK, Freeman M, Barbanti M, Heijmen RH, et al. Management of vascular access in transcatheter aortic valve replacement: part 1: basic anatomy, imaging, sheaths, wires, and access routes. JACC Cardiovasc Interv. 2013;6(7):643–53. https://doi.org/10.1016/j.jcin.2013.04.003. PMID: 23866177
10. Ben-Dor I, Sharma A, Rogers T, Yerasi C, Case BC, Chezar-Azerrad C, et al. Micropuncture technique for femoral access is associated with lower vascular complications compared to standard needle. Catheter Cardiovasc Interv. 2021;97(6):1379–85. https://doi.org/10.1002/ccd.29216. PMID: 32865860
11. Ramlawi B, Anaya-Ayala JE, Reardon MJ. Transcatheter aortic valve replacement (TAVR): access planning and strategies. Methodist Debakey Cardiovasc J. 2012;8(2):22–5. https://doi.org/10.14797/mdcj-8-2-22. PMID: 22891125
12. Chen S, Chau KH, Nazif TM. The incidence and impact of cardiac conduction disturbances after transcatheter aortic valve replacement. Ann Cardiothorac Surg. 2020;9(6):452–67. https://doi.org/10.21037/acs-2020-av-104. PMID: 33312908
13. Urena M, Webb JG, Cheema A, Serra V, Toggweiler S, Barbanti M, et al. Impact of new-onset persistent left bundle branch block on late clinical outcomes in patients undergoing transcatheter aortic valve implantation with a balloon-expandable valve. JACC Cardiovasc Interv. 2014;7(2):128–36. https://doi.org/10.1016/j.jcin.2013.08.015. PMID: 24556091
14. Van Rosendael PJ, Delgado V, Bax JJ. Pacemaker implantation rate after transcatheter aortic valve implantation with early and new-generation devices: a systematic review. Eur Heart J. 2018;39(21):2003–13. https://doi.org/10.1093/eurheartj/ehx785. PMID: 29300869
15. Hamdan A, Guetta V, Klempfner R, Konen E, Raanani E, Glikson M, et al. Inverse relationship between membranous septal length and the risk of atrioventricular block in patients undergoing transcatheter aortic valve implantation. JACC Cardiovasc Interv. 2015;8(9):1218–28. https://doi.org/10.1016/j.jcin.2015.05.010. PMID: 26315746
16. Jilaihawi H, Zhao Z, Du R, Staniloae C, Saric M, Neuburger PJ, et al. Minimizing permanent pacemaker following repositionable self-expanding transcatheter aortic valve replacement. JACC Cardiovasc Interv. 2019;12(18):1796–807. https://doi.org/10.1016/j.jcin.2019.05.056. PMID: 31537281
17. Rodés-Cabau J. Optimizing valve implantation depth to win the battle against conduction disturbances post-TAVR. JACC Cardiovasc Interv. 2019;12(18):1808–10. https://doi.org/10.1016/j.jcin.2019.06.039. PMID: 31537282
18. Bettinger N, Khalique OK, Krepp JM, Hamid NB, Bae DJ, Pulerwitz TC, et al. Practical determination of aortic valve calcium volume score on contrast-enhanced computed tomography prior to transcatheter aortic valve replacement and impact on paravalvular regurgitation: elucidating optimal threshold cutoffs. J Cardiovasc Comput Tomogr. 2017;11(4):302–8. https://doi.org/10.1016/j.jcct.2017.04.009. PMID: 28595875
19. Kahlert P, Knipp SC, Schlamann M, Thielmann M, Al-Rashid F, Weber M, et al. Silent and apparent cerebral ischemia after percutaneous transfemoral aortic valve implantation: a diffusion-weighted magnetic resonance imaging study. Circulation. 2010;121(7):870–8. https://doi.org/10.1161/CIRCULATIONAHA.109.855866. PMID: 20142456
20. He C, Xiao L, Liu J. Safety and efficacy of self-expandable Evolut R vs. balloon-expandable Sapien 3 valves for transcatheter aortic valve implantation: a systematic review and meta-analysis. Exp Ther Med. 2019;18(5):3893–904. https://doi.org/10.3892/etm.2019.8009. PMID: 31616514

21. Li YM, Tsauo JY, Liao YB, Zhao ZG, Chen M. Comparison of third generation balloon-expandable Edwards Sapien 3 versus self-expandable Evolut R in transcatheter aortic valve implantation: a meta-analysis. Ann Cardiothorac Surg. 2020;9(6):700–8. https://doi.org/10.21037/acs-2020-av-13. PMID: 33312919
22. Masri A, Schoenhagen P, Svensson L, Kapadia SR, Griffin BP, Tuzcu EM, et al. Dynamic characterization of aortic annulus geometry and morphology with multimodality imaging: predictive value for aortic regurgitation after transcatheter aortic valve replacement. J Thorac Cardiovasc Surg. 2014;147(6):1847–54. https://doi.org/10.1016/j.jtcvs.2013.05.047. PMID: 23856190
23. Takagi K, Latib A, Al-Lamee R, Mussardo M, Montorfano M, Maisano F, et al. Predictors of moderate-to-severe paravalvular aortic regurgitation immediately after CoreValve implantation and the impact of postdilatation. Catheter Cardiovasc Interv. 2011;78(3):432–43. https://doi.org/10.1002/ccd.23003. PMID: 21413114
24. Sherif MA, Abdel-Wahab M, Stöcker B, Geist V, Richardt D, Tölg R, et al. Anatomic and procedural predictors of paravalvular aortic regurgitation after implantation of the Medtronic CoreValve bioprosthesis. J Am Coll Cardiol. 2010;56(20):1623–9. https://doi.org/10.1016/j.jacc.2010.06.035. PMID: 21050923
25. Bapat V, Attia R, Redwood S, Hancock J, Wilson K, Young C, et al. Use of transcatheter heart valves for a valve-in-valve implantation in patients with degenerated aortic bioprosthesis: technical considerations and results. J Thorac Cardiovasc Surg. 2012;144(6):1372–80. https://doi.org/10.1016/j.jtcvs.2012.07.104. PMID: 22980066
26. Fede A, Romano M, Buffoli F, Camurri N, Lettieri C. Combination of double chimney technique and prosthesis post-dilation after valve-in-valve implantation. Case Rep. 2020;2(14):2173–5. https://doi.org/10.1016/j.jaccas.2020.08.028. PMID: 34317143
27. Mercanti F, Rosseel L, Neylon A, Bagur R, Sinning JM, Nickenig G, et al. Chimney stenting for coronary occlusion during TAVR: insights from the Chimney Registry. JACC Cardiovasc Interv. 2020;13(6):751–61. https://doi.org/10.1016/j.jcin.2020.01.227. PMID: 32192695
28. Palmerini T, Chakravarty T, Saia F, Bruno AG, Bacchi-Reggiani ML, Marrozzini C, et al. Coronary protection to prevent coronary obstruction during TAVR: a multicenter international registry. JACC Cardiovasc Interv. 2020;13(6):739–47. https://doi.org/10.1016/j.jcin.2019.11.008. PMID: 32192693
29. Noorani A, Radia R, Bapat V. Challenges in valve-in-valve therapy. J Thorac Dis. 2015;7(9):1501–8. https://doi.org/10.3978/j.issn.2072-1439.2015.07.27. PMID: 26543595

Chapter 18
The Transcatheter Tricuspid Valve-in-Valve Technique in Degenerated Bioprostheses

Ata Firouzi, Zahra Hosseini, and Ehsan Khalilipur

Abstract One of the most challenging forms of valvular heart disease is related to the TV. Regardless of the underlying pathology, when surgical intervention is necessary, biologic prosthetic valves appear to be the most suitable option due to the high prevalence of thrombotic events and the requirement for lifelong anticoagulation. Additionally, since most patients undergoing TV surgery are relatively young, the likelihood of reoperation is significant, given the high incidence of failure associated with biologic valves. Conversely, the mortality rate for redo surgeries is notably high, as many of these patients experience advanced right heart failure. In recent years, advancements in TAVR and the management of failed biologic aortic valves have contributed to a growing interest in the TVIV procedure.

Keywords Tricuspid valve · Biologic prosthetic valves · Thrombotic events Degenerative changes · Mortality rate · Re-operation · TAVI · Tricuspid valve-in-valve

Abbreviations

CT	Computed tomography
NYHA	New York Heart Association
SHV	Surgical heart valve
TAVI	Transcatheter aortic valve implantation
TAVR	Transcatheter aortic valve replacement
TEE	Transesophageal echocardiography
THV	Transcatheter heart valve

A. Firouzi · Z. Hosseini · E. Khalilipur (✉)
Cardiovascular Intervention Research Center, Rajaie Cardiovascular Institute, Iran University of Medical Sciences, Tehran, Iran

A. Firouzi et al. (eds.), *Mastering Skills in Complex Cardiovascular Interventions*, https://doi.org/10.1007/978-3-031-97321-5_18

TTE	Transthoracic echocardiography
TV	Tricuspid valve
TVIV	Transcatheter tricuspid valve-in-valve

Introduction

The choice between biological and mechanical prosthetic valves for the tricuspid position remains a topic of debate. In general, bioprosthetic heart valves are often favored over mechanical valves because they eliminate the need for high-level systemic anticoagulation and its associated bleeding risks. In addition, mechanical valves in the tricuspid position have been linked to higher rates of prosthetic valve thrombosis [1, 2]. Despite progress in leaflet design, hemodynamic performance, and anti-calcification treatments, bioprosthetic heart valves in the tricuspid position tend to have a shorter lifespan than those in the systemic circulation. Studies report a reoperation rate of 10% to 22% at 9 years and a freedom-from-reintervention rate of less than 60% at 15 years [3, 4].

Moreover, in cases of redo surgery for a failing bioprosthetic heart valve in the tricuspid position, a bioprosthetic heart valve is again the preferred choice. This suggests that during the decision-making process for reintervention—whether surgical or transcatheter valve-in-valve—the longevity of the valve is not a primary consideration. TVIV procedure has broadened the scope of cardiac surgery by enabling the implantation of stent-valves within degenerated bioprostheses. This approach is applicable to both older adults and younger patients, particularly given that the reported mortality rate for redo surgery in previous studies is at least 37%.

Patients requiring reintervention for the TV often present with advanced right ventricular failure and multiple comorbidities, including advanced age, liver disease or cirrhosis, renal dysfunction, and pulmonary disease [5, 6]. As a result, while surgery remains the gold standard for treatment, TVIV implantation in deteriorated bioprostheses has recently emerged as a viable alternative for patients considered high-risk for redo cardiac surgery. In fact, TVIV is now routinely performed in experienced centers, even for low-risk patients, with consistently favorable clinical outcomes [7, 8].

Decision-making and preprocedural planning by the heart surgical team, utilizing various imaging modalities, are critical to achieving successful outcomes. Fluoroscopic guidance can help the operator position and implant a bioprosthetic valve at the ideal depth and location based on radiopaque markers. Nonetheless, the complete radiolucency of certain bioprosthetic valves poses a significant challenge.

Moreover, two groundbreaking studies have recently advanced the field of transcatheter TV repair. The TRILUMINATE Pivotal Clinical Trials were published in *The New England Journal of Medicine* in 2023. Another study, bRIGHT (An Observational Real-World Study Evaluating Severe Tricuspid Regurgitation Patients Treated With the Abbott TriClip Device), was published in the *Journal of the American College of Cardiology* in the same year.

The bRIGHT study was an open-label trial conducted across 26 centers in Europe. The target population consisted of older adults with multiple comorbidities, torrential tricuspid regurgitation, and NYHA functional class III or IV. Successful device implantation was achieved in 99% of patients, with tricuspid regurgitation reduced to moderate or less in 77% of the population at 30 days. Significant improvements in NYHA functional class were also observed in the remaining participants. This study demonstrated that transcatheter TV repair is a safe and effective approach for reducing TV regurgitation and improving NYHA functional class [9].

The TRILUMINATE Pivotal Clinical Trials were randomized controlled trials conducted across 65 centers in the United States, Canada, and Europe. The target population included patients with symptomatic severe tricuspid regurgitation. Participants were randomly assigned in a 1:1 ratio to undergo either tricuspid transcatheter edge-to-edge repair or receive medical therapy (control group). The study demonstrated that tricuspid transcatheter edge-to-edge repair was a safe and effective procedure for reducing the severity of tricuspid regurgitation, subsequently improving NYHA functional class and quality of life [10].

The purpose of this article is to provide practical tips and strategies for performing TVIV procedures on failing radiolucent bioprostheses.

Challenging Cases

Case 1

A 64-year-old woman with rheumatic mitral and TV disease was referred to our center presenting with symptoms of advanced right ventricular failure, anasarca, and liver congestion. The patient had previously undergone mitral and TV replacement, receiving a St. Jude Cardiac Valve Prosthesis (23 mm) in the mitral position and a radiolucent porcine bioprosthetic valve (31 mm) in the tricuspid position.

Preprocedural Planning

1. Access to the surgical notes from the previous procedure is essential to accurately identify the type and size of the implanted valve. In this case, the note simply stated: "Porcine 31."
2. Echocardiography findings included normal left ventricular size with mild systolic dysfunction, moderate right ventricular enlargement with moderate-to-severe systolic dysfunction, severe biatrial enlargement, a thickened tricuspid aortic valve with moderate aortic regurgitation, and a bileaflet mechanical mitral prosthesis with acceptable leaflet motion and hemodynamic parameters (mean pressure gradient = 4 mm Hg, pressure half-time = 74 ms). The stented biopros-

thetic TV exhibited severe degenerative changes, including highly thickened and calcified leaflets with severe stenosis (mean pressure gradient =12 mm Hg), severe transvalvular leakage, no periprosthetic leakage, and an estimated true internal diameter of 28 mm. The mean pulmonary artery pressure was 20 mm Hg.

3. Cardiac CT scanning revealed normal left ventricular size and function, moderate right ventricular enlargement, severe biatrial enlargement, a calcified tricuspid aortic valve without aortic stenosis, a mechanical prosthetic mitral valve, and a stented bioprosthetic TV (with the annulus aligned parallel to the mitral prosthetic valve) showing severe degenerative changes (calcified leaflets and a true internal diameter = 27 mm). The recommended working views were the right anterior oblique projection (20°) and the caudal projection (5°).

Considering the patient's comorbidities and the estimated risk of redo surgery as calculated by the EuroSCORE risk assessment tool, the heart team reached a consensus to proceed with the TVIV procedure.

The Procedure

At the time of the procedure, no fluoroscopic landmarks were available to provide an anatomic reference for valve deployment (Fig. 18.1a).

Under general anesthesia, sheaths (6 Fr) were inserted into the right and left common femoral veins, and intravenous heparin was administered to achieve an activated clotting time > 250 seconds. To ensure precise valve deployment, the procedure was guided by TEE. Initially, a right atrial injection was performed in the right anterior oblique projection (the working view recommended by cardiac CT imaging) using a Pigtail catheter via the left access to delineate the inflow of the prosthesis (Fig. 18.1b).

Fig. 18.1 In fluoroscopy, just the mechanical mitral valve opacity was seen, and no radiopaque markers of the tricuspid prostheses were visualized (**a**). A right atrial injection in the right anterior oblique projection demonstrated the inflow of the tricuspid SHV (**b**). The SHV was crossed with the support of a steerable Agilis NxT introducer (**c**). Afterward, the hydrophilic wire was exchanged with a Lunderquist Super Stiff wire, the end of which was then positioned in the left pulmonary artery (**d**). Predilation of the SHV was performed with a noncompliant balloon (Bard-24 × 40), with the waist of the balloon as an indicator of the annular plane of the SHV (**e**). The THV was delivered through the SHV. The central marker of the valve (SAPIEN XT-29) and the depth of the implantation were adjusted by repetitive right atrial injections and TEE guidance (**f**). The balloon was slowly inflated; then, before complete valve deployment, the position of the valve was rechecked by TEE and right atrial injection. After the right position of the valve was confirmed, the THV was deployed completely (**g**). The final right atrial injection and echocardiography image illustrate the proper position of the valve without any leakage and with a minimal residual gradient (**h–l**)

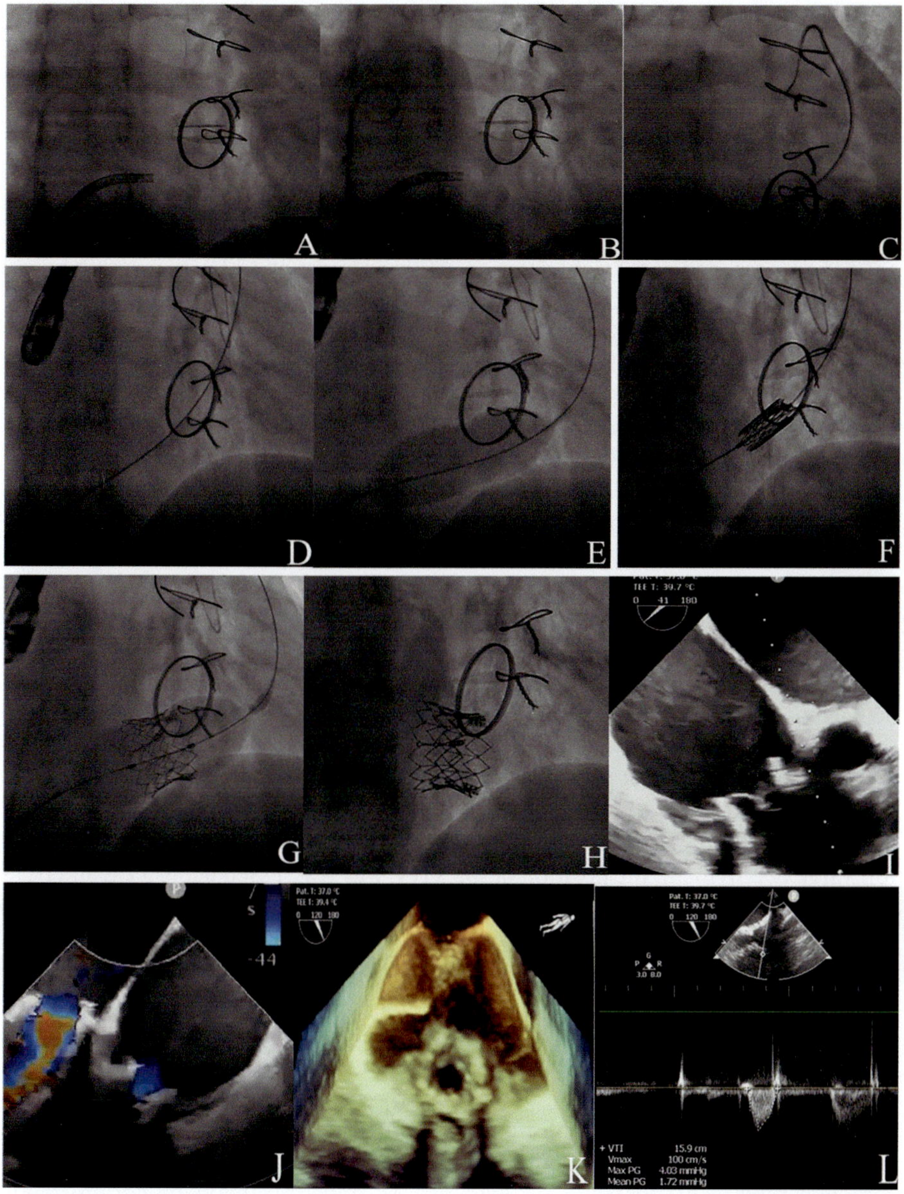
A
B
C
D
E
F
G
H
TEE T: 39.7 °C
0 41 180
I
J
K
Pat. T: 37.0 °C
TEE T: 39.7 °C
0 120 180
+ VTI 15.9 cm
Vmax 100 cm/s
Max PG 4.03 mmHg
Mean PG 1.72 mmHg
L

Next, through the right femoral vein and under fluoroscopic guidance, a diagnostic multipurpose catheter (6 Fr) was advanced into the significantly enlarged right atrium. The catheter was telescoped into an 8.5 Fr large-curve Agilis NxT steerable sheath (St. Jude Medical, St. Paul, Minnesota), and the SHV was crossed using a 0.035-inch hydrophilic guidewire. The distal end of the wire was positioned in the distal portion of the left pulmonary artery (Fig. 18.1c). The wire was then exchanged for a 0.035-inch Lunderquist wire (Cook Medical, Bloomington, Indiana) (Fig. 18.1d). To accurately visualize the annular plane of the SHV, a 24 × 40 Bard balloon was inflated in the right anterior oblique projection (20°), the working view recommended by cardiac CT imaging. The correct positioning of the balloon within the SHV was confirmed by echocardiography.

Following this, the SHV was predilated. This step helped define the annular plane of the SHV (indicated by the proximal waist of the balloon), confirm the appropriate size of the selected THV, and facilitate the delivery of the THV through the SHV (Fig. 18.1e). Subsequently, a long delivery sheath (16 Fr) was introduced. The chosen valve (Edwards SAPIEN XT-29) was mounted onto the delivery catheter for antegrade delivery, with the Edwards E logo oriented downward. The assembly and alignment of the valve were completed within the inferior vena cava.

For optimal functioning of the SAPIEN valve, no more than 20% of the valve should extend into the atrium after deployment. Nevertheless, in this case, due to the radiolucent nature of the SHV, variations in implantation depth of up to 30% were deemed acceptable. After crossing the SHV under fluoroscopic and TEE guidance, the valve was positioned so that the central marker of the balloon aligned precisely with the annular plane. Repeated right atrial injections, combined with TEE guidance, confirmed the proper positioning of the THV (Fig. 18.1f).

The valve was then slowly inflated. During partial balloon inflation and before full valve deployment, the position of the THV was reassessed using TEE and right atrial injection. Once the echocardiographer confirmed proper positioning, the THV was fully deployed (Fig. 18.1g). TEE revealed mild paravalvular leakage, prompting overexpansion with an additional 4 mL of inflation. A final right atrial injection confirmed the correct valve position with no evidence of leakage (Fig. 18.1h). Echocardiography demonstrated optimal valve placement, no residual paravalvular or transvalvular leakage, and a final mean pressure gradient of 2 mm Hg (Fig. 18.1i–l).

The following day, TTE confirmed the appropriate positioning of the THV with acceptable hemodynamic parameters (mean pressure gradient =2 mm Hg and pressure half-time = 60 ms) and no evidence of trans-prosthetic or periprosthetic leakage.

The patient was discharged without complications on the fourth postprocedural day and was prescribed lifelong warfarin plus aspirin.

Case 2

A 65-year-old woman was referred to our center with progressive lower extremity edema over the past six months. The patient had previously undergone mitral valve replacement and TV repair, followed by TV replacement with a 31 mm bioprosthetic porcine valve four years later due to declining TV function. Her medical history included mediastinitis following the last operation and a dual-chamber permanent pacemaker for complete heart block, with one lead in the right atrium and the other in the coronary sinus.

Preprocedural Planning

1. Reviewing the surgical notes from the last operation is critical for determining the type and size of the prosthetic valve to be implanted. In this case, the note stated: "Porcine 31."
2. Echocardiography findings included normal left ventricular size with mild systolic dysfunction, moderate right ventricular enlargement with moderate systolic dysfunction, severe biatrial enlargement, a tricuspid aortic valve with no aortic stenosis and mild aortic insufficiency, a bileaflet mechanical prosthetic mitral valve with normal leaflet motion and hemodynamic parameters (mean pressure gradient =5 mm Hg and pressure half-time = 70 ms), no paravalvular leakage, and a stented bioprosthetic TV with severe degenerative changes (thickened and calcified leaflets, severe stenosis, mean pressure gradient =10 mm Hg and pressure half-time = 300 ms, severe transvalvular leakage, and an estimated true internal diameter of 26 mm). Moderate pulmonary artery hypertension was also noted (mean pulmonary artery pressure = 35 mm Hg).
3. Cardiac CT scanning revealed normal left ventricular size with mild systolic dysfunction (left ventricular ejection fraction =50%), moderate right ventricular enlargement, severe right atrial enlargement with a smoky pattern, a mechanical bileaflet prosthetic mitral valve without evidence of thrombosis, and a stented bioprosthetic TV with severe degenerative changes (thickened and calcified leaflets, estimated true internal diameter = 27 mm). The recommended working views were the right anterior oblique projection (15°) and the caudal projection (5°).

Given the patient's condition and the significantly high risk associated with a third open-heart surgery, she was scheduled for the TVIV procedure.

The Procedure

At the time of the procedure, the radiopacity of the mitral mechanical valve, the right atrial lead, and the coronary sinus lead was prominent. However, no fluoroscopic landmarks of the tricuspid SHV were available to provide an anatomic reference for THV deployment (Fig. 18.2a, b).

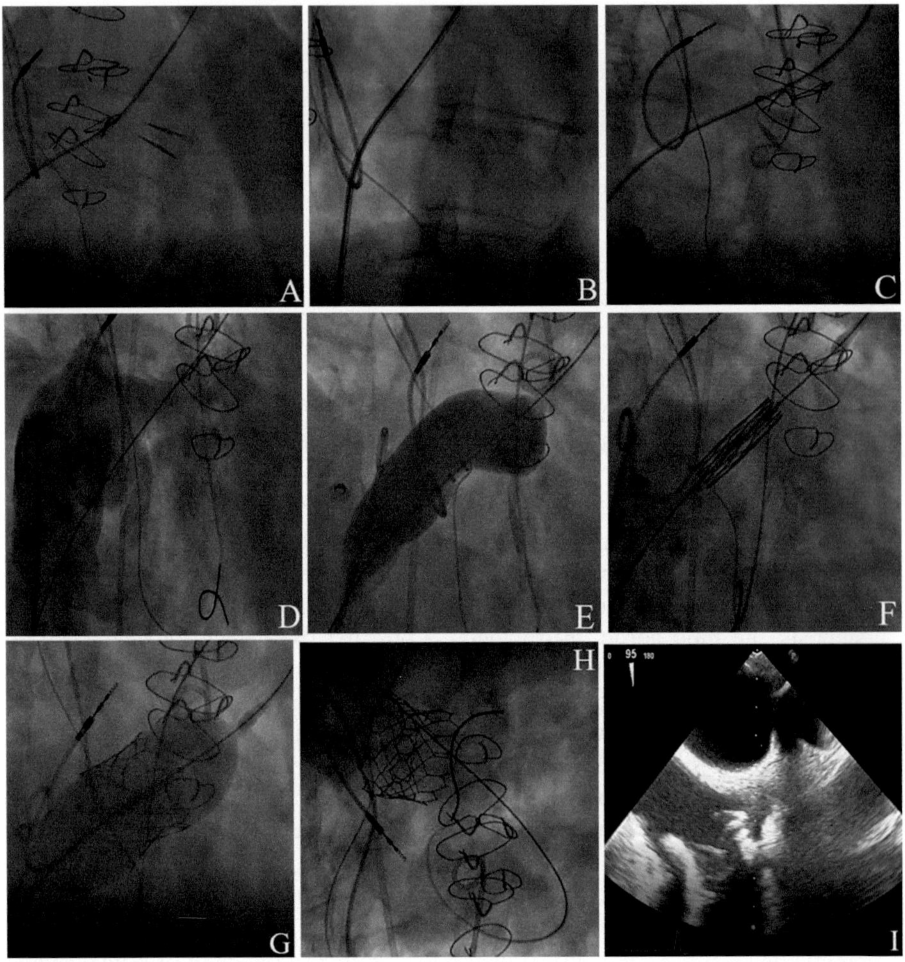

Fig. 18.2 Fluoroscopy visualized only the radio-opacity of the mechanical mitral valve, as well as the right atrial and coronary sinus leads. No radiopaque markers of the tricuspid SHV were shown (**a**, **b**). Under fluoroscopy guidance, the SHV was crossed with a 0.035-inch hydrophilic wire. Subsequently, the wire was exchanged for a 0.035-inch Super Stiff wire, whose end was then positioned in the left pulmonary artery (**c**). Using a right atrial injection in the right anterior oblique projection, the inflow of the tricuspid SHV was demarcated (**d**). Next, the SHV was predilated with a 24 × 50 noncompliant balloon, the central waist of which was an indicator of the annular plane (**e**). The THV was crossed through the SHV. Afterward, the appropriate position of the THV was confirmed both by repetitive right atrial injections and TEE guidance and by considering the location of the right atrial lead and the annular plane (**f**). The balloon was slowly inflated. Prior to complete valve deployment, the right atrium was injected again to rule out any migration. The echo image was then reviewed. As the position of the THV was acceptable (20% of the valve was atrial), the valve was deployed completely (**g**). The final right atrial injection illustrated the suitable position of the THV without any leakage (**h**). TEE showed the appropriate position of the valve (**i**) with a minimal transvalvular pressure gradient (the mean pressure gradient =4 mm Hg) and without any transvalvular or periprosthetic leakage

Under general anesthesia, sheaths (6 Fr) were inserted into the right and left common femoral veins, and intravenous heparin was administered to achieve an activated clotting time exceeding 250 seconds. To ensure precise valve deployment, the procedure was guided by TEE. Initially, through the right femoral vein and under fluoroscopic guidance, the SHV was crossed using a 0.035-inch hydrophilic guidewire. The wire was then advanced into the left pulmonary artery, and a diagnostic multipurpose catheter was passed over it. The wire was subsequently exchanged for a 0.035-inch Super Stiff Lunderquist wire (Fig. 18.2c).

Next, a right atrial injection was performed in the right anterior oblique projection (15°), the recommended working view by CT scanning, to demarcate the inflow of the SHV using a Pigtail catheter through the left access (Fig. 18.2d). The right atrial lead was also highlighted as an anatomical landmark for implantation. To further define the annular plane of the prosthesis, a 24 × 50 balloon was inflated in the shallow right anterior oblique projection. The proximal waist of the balloon served as an indicator of the annulus plane (Fig. 18.2e).

Subsequently, a long delivery sheath (16 Fr) was introduced, and the selected valve (Edwards SAPIEN 3–29) was mounted onto the delivery catheter for antegrade deployment. After crossing the SHV, the valve position was determined using the right atrial lead marker, the central base of the balloon, and TEE guidance, ensuring the central marker of the balloon aligned with the annular plane (Fig. 18.2f).

After multiple right atrial injections and confirmation from the echocardiographer, the appropriate valve position was verified, and the balloon was slowly inflated. Prior to complete valve deployment, another right atrial injection was performed to rule out migration, and the echocardiographic images were reviewed. The position of the THV was deemed acceptable (with 20% of the valve extending into the atrium), and the valve was fully deployed (Fig. 18.2g). A final right atrial injection confirmed precise valve positioning with no evidence of paravalvular or transvalvular leakage (Fig. 18.2h). TEE confirmed the proper positioning of the valve, a significant reduction in the mean valvular gradient (4 mm Hg), and no evidence of leakage (Fig. 18.2i).

The following day, TTE demonstrated appropriate valve placement, acceptable hemodynamic parameters (mean pressure gradient =3 mm Hg and pressure half-time = 55 ms), and no leakage or pericardial effusion.

Discussion

Compared with mechanical valves, tissue valves are less thrombogenic and eliminate the need for long-term anticoagulation. Be that as it may, tissue valves are prone to structural valve deterioration and failure, leading to reduced long-term durability. Structural valve deterioration occurs due to degenerative changes in the valve leaflets, which may involve calcification, pannus or thrombus formation, stenosis, tears, regurgitation, or a combination of these factors [11, 12]. In this context, the TVIV approach has emerged as a safe and effective alternative to redo surgical valve replacement for failing tissue valves [13].

The first TVIV implantation was reported in 2010 [14]. Under fluoroscopic guidance, TVIV involves deploying a THV within a deteriorated SHV. The depth of THV implantation within the damaged tissue valve is critical for achieving optimal hemodynamics [15]. This depth is determined using radiopaque landmarks of the SHV, such as the swing ring, stent frame, or outflow markers. For the SAPIEN 3 valve to function optimally, no more than 15% to 20% of the valve should extend into the atrium after deployment [16]. Another key factor for proper THV functioning is selecting the correct valve size. Undersized valves can lead to embolization and paravalvular leakage, while oversized valves may result in improper THV function or higher residual gradients [17]. Preprocedural multimodality imaging and manufacturers' recommendations assist in selecting the appropriate valve size based on the true internal diameter. Nonetheless, for certain SHVs lacking radiopaque markers under fluoroscopy, balloon sizing during the procedure is essential. This step not only confirms the position and alignment of the valve but also estimates the true internal diameter.

The Melody Valve (Medtronic, Minneapolis, MN, USA) received Conformité Européenne (CE) clearance for pulmonic valve-in-valve procedures in 2006 in Canada and Europe, followed by United States Food and Drug Administration (FDA) approval in 2010. In 2016, the FDA approved the SAPIEN XT Valve (Edwards, Irvine, CA, U.S.A.) for percutaneous implantation in the pulmonic position. Still, neither valve is approved for tricuspid implantation, which is why TVIV remains an off-label catheter-based technique. The decision to proceed with a TVIV procedure should be discussed with patients and their health insurance providers.

The Edwards SAPIEN 3 Valve is a trileaflet bovine pericardial valve manually sutured to a rigid chromium-cobalt stent frame. It is available in three diameter sizes: 23 mm, 26 mm, and 29 mm. When expanded, the stent height ranges from 18 mm for the 23 mm valve to 22.5 mm for the 29 mm valve. The valve is crimped onto a 30 mm long, noncompliant high-pressure balloon using a specialized tool and is delivered through a 14 Fr or 16 Fr sheath via the femoral vein. If the dimensions of the failing bioprosthesis are accurately characterized and the valve size is appropriately selected, the failing bioprosthesis in the TV position generally serves as a suitable landing zone for a TVIV procedure. In recent years, the use of TVIV has increased, with relevant registries now providing longer-term follow-up data [18–20].

Certain valves, such as the Aspire (Vascutek, Inchinnan, U.K.), lack radiopaque markers. As a result, to ensure success and prevent valve embolization, the operator must deploy the THV within the frame of the bioprosthetic SHV. The temptation to position the device lower than necessary can be problematic, especially with the SAPIEN Valve, as it may not fully cover the deteriorated bioprosthetic leaflets. Therefore, both the height of the THV and the design of the SHV are crucial, with optimal positioning providing adequate support while ensuring the valve fully covers the leaflets.

In these two cases, we presented patients with radiolucent failing bioprosthetic valves in the tricuspid position, both of which were addressed using the TVIV procedure.

Our recommendations for achieving optimal results during TVIV procedures in radiolucent prostheses are as follows:

1. Reviewing the surgical notes from the previous operation is essential to accurately identify the type and size of the implanted SHV.
2. Preprocedural planning and multimodality imaging (TTE, TEE, and cardiac CT imaging), combined with the manufacturer-provided stent internal diameter of the SHV, should guide the selection of the most appropriate THV type and size.
3. Similar to the TAVI procedure, preprocedural cardiac CT imaging can help determine the optimal working view.
4. Before the procedure, rotating the fluoroscopy tube can help identify anatomical landmarks, such as other mechanical or radiopaque valves, pacing wires, or implanted devices like septal occluders, and define the most suitable working view (Figs. 18.1a and 18.2a).
5. Most of these patients present with advanced right ventricular failure and are preload-dependent; therefore, patient preparation and sedation should be managed by an expert cardiac anesthesiologist, and hemodynamic monitoring should be conducted by the heart failure team throughout the entire procedure.
6. The use of 3D or 2D TEE images is recommended during the procedure to confirm proper wire crossing of the SHV and to determine the precise depth of THV implantation.
7. The preferred access route is the femoral approach. After inserting 6 Fr sheaths into the right and left common femoral veins, intravenous heparin (100 U/kg) should be administered, and activated clotting time should be monitored every half hour to ensure it exceeds 250 seconds.
8. Routine rapid pacing is not recommended during valve deployment. However, in cases of high cardiac mobility, a pacing wire may be placed in the right atrium or left ventricle. Since many of these patients have permanent pacemakers, rapid pacing can be facilitated using the existing leads.
9. Right atrial injection using a pigtail catheter via the left access in the right anterior oblique projection or the working view recommended by cardiac CT imaging can aid in characterizing the inflow of the SHV.
10. Given that these patients typically have a significantly enlarged right atrium and right ventricle, wire crossing of the SHV is one of the most challenging aspects of the procedure. A low threshold for utilizing a steerable sheath, such as the Agilis NxT Sheath (St. Jude Medical, St. Paul, Minnesota), is recommended. To provide additional support during THV delivery, it is advisable to position the distal end of a Super Stiff wire in the distal portion of the left pulmonary artery.
11. In right anterior oblique caudal or shallow cranial projections, or in the recommended working views provided by cardiac CT imaging, the SHV should be predilated using undersized noncompliant balloons. The accurate positioning of the balloon should be confirmed via echocardiography. Predilation of the SHV serves three purposes: first, to verify the suitability of the selected THV size; second, to facilitate THV delivery; and third, to allow the proximal waist of the balloon to define the annular plane of the SHV.

12. Following the introduction of a long delivery sheath (16 Fr), the chosen THV is mounted for antegrade delivery onto the delivery catheter. The valve delivery system is then inserted with the Edwards E logo facing downward. Assembly and alignment with the valve are performed in the inferior vena cava. Under fluoroscopic and TEE guidance, the prosthesis is advanced. The valve position should be aligned so that the central marker of the balloon is positioned precisely at the level of the annulus plane, ensuring that no more than 30% of the valve is in the atrium during deployment. Right atrial injection and TEE guidance assist in confirming the proper positioning of the THV.
13. The balloon should be inflated slowly. Before complete valve deployment (during partial balloon inflation), the valve position should be verified through right atrial injection and echocardiographic images. The valve should only be fully deployed once the stability of its position is confirmed.
14. Postdilation of the THV with ultra-high-pressure balloons is recommended in cases of high residual gradients or any paravalvular leakage.
15. A final echocardiographic assessment, along with a right ventricular injection, can aid in the early diagnosis of valve malfunction or regurgitation.

The use of a lifelong vitamin K antagonist (e.g., warfarin) and aspirin is recommended to minimize the risk of valve thrombosis, which can lead to a cascade of structural valve deterioration. TVIV is technically feasible, safe, and effective for selected patients experiencing malfunction of surgically implanted bioprostheses in the tricuspid position. The procedure can be performed using either the Melody Valve or the SAPIEN Valve. Patients with congenital heart disease and a failing tricuspid bioprosthesis often have a history of multiple open-heart surgeries and are typically in a compromised clinical state, placing them in a high-risk category for repeated open-heart surgery [21].

TVIV is a relatively straightforward technique with a high success rate and a low incidence of periprocedural complications. Nevertheless, the procedure can be complicated by issues such as embolization of the implanted THV to the right atrium or right ventricle, paravalvular leakage, transvalvular leakage, vascular access problems, and valve thrombosis. These potential complications should be carefully considered when selecting patients for this procedure [22, 23].

The multicenter Valve-in-Valve International Data (VIVID) registry, conducted by McElhinney et al. [24] represents the largest study on TVIV procedures, involving 156 patients with bioprosthetic TV dysfunction. Among these patients, 71% were classified as NYHA functional class III–IV. The valve was successfully implanted in 150 individuals, resulting in a success rate of 99% after 152 attempts (94 using the Melody valve and 58 using the SAPIEN valve). There were no significant differences in outcomes based on valve type; however, the Melody THV was more frequently used in younger patients, those with congenital native valve disease, and patients with smaller surgical bioprostheses. Tricuspid regurgitation was successfully eliminated, and the gradient between the right atrium and right ventricle was significantly reduced. During the follow-up period (median = 13.3 months), 22 patients died, including five within 30 days of the TVIV procedure; all of these patients were in NYHA functional class III or IV, and nine had a history of prior hospitalizations. A total of 10 TV reinterventions were performed, and 3 patients experienced significant recurring TV malfunction. At follow-up, 77% of the patients were classified as NYHA functional class I or II (P = 0.001 vs. their status before TVIV). The authors concluded that TVIV represents a promising therapeutic option for patients with failing tricuspid bioprostheses.

Conclusions

Based on the current evidence, TVIV appears to be a safe, feasible, and effective alternative to redo surgery. This procedure could serve as a first-line treatment option for patients with symptomatic severe dysfunction of bioprosthetic tricuspid heart valves, particularly those who have a reasonable life expectancy and are likely to benefit from the intervention. Careful procedural planning is vital for the success of the TVIV procedure. Selecting the appropriate THV size, paying special attention to the design of the SHV, and recognizing their anatomical landmarks are all critical factors. Additionally, precise fluoroscopy and echocardiography-guided positioning of the THV are essential to prevent malpositioning and embolization of the valve.

The sewing ring serves as the most reliable anchor for positioning the THV. Understanding the procedural steps and technical aspects of the valve-in-valve technique, along with the utilization of fusion imaging (combining CT and fluoroscopy) and echocardiography, can help address the technical challenges associated with radiolucent bioprosthetic valves.

It is important to note that TVIV is an off-label technique, and there is currently limited reliable experience regarding the optimal depth of THV implantation needed to achieve the most effective orifice area and minimal residual gradient. Furthermore, a dedicated THV specifically designed for the tricuspid position has yet to be developed. Moreover, since most patients with failing SHVs are relatively young, the longevity of the implanted THV is crucial, necessitating large multicenter studies with long-term clinical follow-ups.

References

1. Guenther T, Noebauer C, Mazzitelli D, Busch R, Tassani-Prell P, Lange R. Tricuspid valve surgery: a thirty-year assessment of early and late outcome. Eur J Cardiothorac Surg. 2008;34(2):402–9.; discussion 409. PMID: 18554862. https://doi.org/10.1016/j.ejcts.2008.05.006.
2. Marquis-Gravel G, Bouchard D, Perrault LP, Pagé P, Jeanmart H, Demers P, et al. Retrospective cohort analysis of 926 tricuspid valve surgeries: clinical and hemodynamic outcomes with propensity score analysis. Am Heart J. 2012;163(5):851–858.e1. https://doi.org/10.1016/j.ahj.2012.02.008. PMID: 22607864.

3. Vahanian A, Beyersdorf F, Praz F, Milojevic M, Baldus S, Bauersachs J, et al. 2021 ESC/EACTS guidelines for the management of valvular heart disease. Eur J Cardiothorac Surg. 2021;60(4):727–800. https://doi.org/10.1093/ejcts/ezab389. PMID: 34453161
4. Eitz T, Fritzsche D, Kleikamp G, Zittermann A, Horstkotte D, Körfer R. Reoperation of the aortic valve in octogenarians. Ann Thorac Surg. 2006;82(4):1385–90. https://doi.org/10.1016/j.athoracsur.2006.04.086. PMID: 16996938
5. Maganti M, Rao V, Armstrong S, Feindel CM, Scully HE, David TE. Redo valvular surgery in elderly patients. Ann Thorac Surg. 2009;87(2):521–5. https://doi.org/10.1016/j.athoracsur.2008.11.030. PMID: 19161774
6. Russo M, Zilberszac R, Werner P, Kocher A, Wiedemann D, Andreas M. Isolated tricuspid valve regurgitation: old concepts, new insights, and innovation. J Cardiovasc Med (Hagerstown). 2020;21(6):405–13. https://doi.org/10.2459/JCM.0000000000000933. PMID: 32073447
7. Dvir D, Webb JG. Transcatheter aortic valve-in-valve implantation for patients with degenerative surgical bioprosthetic valves. Circ J. 2015;79(4):695–703. https://doi.org/10.1253/circj.CJ-15-0068. PMID: 25746546
8. Walther T, Falk V, Dewey T, Kempfert J, Emrich F, Pfannmüller B, et al. Valve-in-a-valve concept for transcatheter minimally invasive repeat xenograft implantation. J Am Coll Cardiol. 2007;50(1):56–60. https://doi.org/10.1016/j.jacc.2007.03.030. PMID: 17601547
9. Lurz P, Besler C, Schmitz T, Bekeredjian R, Nickenig G, Möllmann H, et al. Short-term outcomes of tricuspid edge-to-edge repair in clinical practice. J Am Coll Cardiol. 2023;82(4):281–91. https://doi.org/10.1016/j.jacc.2023.05.008. PMID: 37207924
10. Sorajja P, Whisenant B, Hamid N, Naik H, Makkar R, Tadros P, et al. Transcatheter repair for patients with tricuspid regurgitation. N Engl J Med. 2023;388(20):1833–42. https://doi.org/10.1056/NEJMoa2300525. PMID: 36876753
11. Ranocchi F, Nicolò F, Russo M, Musumeci F. Minimally invasive implantation of a Sapien 3 Ultra valve in a degenerated tricuspid bioprosthesis. J Card Surg. 2020;35(6):1371–4. https://doi.org/10.1111/jocs.14572. PMID: 32333442
12. Dvir D, Bourguignon T, Otto CM, Hahn RT, Rosenhek R, Webb JG, et al. Standardized definition of structural valve degeneration for surgical and transcatheter bioprosthetic aortic valves. Circulation. 2018;137(4):388–99. https://doi.org/10.1161/CIRCULATIONAHA.117.030729. PMID: 29358344
13. Capodanno D, Petronio AS, Prendergast B, Eltchaninoff H, Vahanian A, Modine T, et al. Standardized definitions of structural deterioration and valve failure in assessing long-term durability of transcatheter and surgical aortic bioprosthetic valves: a consensus statement from the European Association of Percutaneous Cardiovascular Interventions (EAPCI) endorsed by the European Society of Cardiology (ESC) and the European Association for Cardio-Thoracic Surgery (EACTS). Eur Heart J. 2017;38(45):3382–90. https://doi.org/10.1093/eurheartj/ehx303. PMID: 29020341
14. McElhinney DB, Aboulhosn JA, Dvir D, Whisenant B, Zhang Y, Eicken A, et al. Mid-term valve-related outcomes after transcatheter tricuspid valve-in-valve or valve-in-ring replacement. J Am Coll Cardiol. 2019;73(2):148–57. https://doi.org/10.1016/j.jacc.2018.10.051. PMID: 30654887
15. Tam DY, Vo TX, Wijeysundera HC, Dvir D, Friedrich JO, Fremes SE. Transcatheter valve-in-valve versus redo surgical aortic valve replacement for the treatment of degenerated bioprosthetic aortic valve: a systematic review and meta-analysis. Catheter Cardiovasc Interv. 2018;92(7):1404–11. https://doi.org/10.1002/ccd.27686. PMID: 30024111
16. Sanon S, Cabalka AK, Babaliaros V, Rihal CS, Gafoor S, Webb J, et al. Transcatheter tricuspid valve-in-valve and valve-in-ring implantation for degenerated surgical prosthesis. JACC Cardiovasc Interv. 2019;12(15):1403–12. https://doi.org/10.1016/j.jcin.2019.05.029. PMID: 31395211
17. Azadani AN, Reardon M, Simonato M, Aldea G, Nickenig G, Kornowski R, et al. Effect of transcatheter aortic valve size and position on valve-in-valve hemodynamics: an in vitro

study. J Thorac Cardiovasc Surg. 2017;153(6):1303–1315.e1. https://doi.org/10.1016/j.jtcvs.2016.12.032. PMID: 28185685

18. Bonhoeffer P, Boudjemline Y, Saliba Z, Merckx J, Aggoun Y, Bonnet D, et al. Percutaneous replacement of pulmonary valve in a right-ventricle to pulmonary-artery prosthetic conduit with valve dysfunction. Lancet. 2000;356(9239):1403–5. https://doi.org/10.1016/S0140-6736(00)02844-0. PMID: 11052596
19. Tanous D, Nadeem SN, Mason X, Colman JM, Benson LN, Horlick EM. Creation of a functional tricuspid valve: novel use of percutaneously implanted valve in right atrial to right ventricular conduit in a patient with tricuspid atresia. Int J Cardiol. 2010;144(1):e8–e10. https://doi.org/10.1016/j.ijcard.2008.12.070. PMID: 19157602
20. Roberts PA, Boudjemline Y, Cheatham JP, Eicken A, Ewert P, McElhinney DB, et al. Percutaneous tricuspid valve replacement in congenital and acquired heart disease. J Am Coll Cardiol. 2011;58(2):117–22. https://doi.org/10.1016/j.jacc.2011.01.036. PMID: 21718908
21. Eicken A, Schubert S, Hager A, Hörer J, McElhinney DB, Hess J, et al. Percutaneous tricuspid valve implantation: two-center experience with midterm results. Circ Cardiovasc Interv. 2015;8(5):e002155. https://doi.org/10.1161/CIRCINTERVENTIONS.114.002155. PMID: 25940519
22. Eicken A, Ewert P. Percutaneous tricuspid valve implantation in failing bioprosthesis. Cardiovasc Diagn Ther. 2018;8(6):765–70. https://doi.org/10.21037/cdt.2018.09.08. PMID: 30740323
23. Burri M, Vogt MO, Hörer J, Cleuziou J, Kasnar-Samprec J, Kühn A, et al. Durability of bioprostheses for the tricuspid valve in patients with congenital heart disease. Eur J Cardiothorac Surg. 2016;50(5):988–93. https://doi.org/10.1093/ejcts/ezw171. PMID: 27242387
24. McElhinney DB, Cabalka AK, Aboulhosn JA, Eicken A, Boudjemline Y, Schubert S, et al. Transcatheter tricuspid valve-in-valve implantation for the treatment of dysfunctional surgical bioprosthetic valves: an international, multicenter registry study. Circulation. 2016;133(16):1582–93. https://doi.org/10.1161/CIRCULATIONAHA.115.019353. PMID: 26994123

Chapter 19
Transcatheter Mitral Valve in Valve

Ata Firouzi, Zahra Hosseini, and Ehsan Khalilipur

Abstract As the trend toward mitral repair becomes the primary surgical approach for addressing MV pathology, an increasing number of patients may present with failed surgical repairs. Consequently, it is not unexpected that high-risk patients have been evaluated for the mitral ViV procedure instead of undergoing reoperation. When key factors are taken into account during the procedure at expert and well-equipped centers, the success rate appears to be high, with lower mortality and morbidity rates than traditional surgery. Nevertheless, concerns regarding long-term residual gradients, thrombosis, and durability remain.

Keywords Failed bioprosthetic valve · Redo-surgery · Mitral ViV · Success rate Mortality · Residual gradient

Abbreviations

ID	Internal diameter
LA	Left atrium
LV	Left ventricle
LVOT	Left ventricular outflow tract
LVOTO	Left ventricular outflow tract obstruction
MR	Mitral regurgitation
MV	Mitral valve
MVA	Mitral valve area
MVR	Mitral valve replacement

A. Firouzi · Z. Hosseini (✉) · E. Khalilipur
Cardiovascular Intervention Research Center, Rajaie Cardiovascular Institute, Iran University of Medical Sciences, Tehran, Iran

A. Firouzi et al. (eds.), *Mastering Skills in Complex Cardiovascular Interventions*, https://doi.org/10.1007/978-3-031-97321-5_19

PVL	Paravalvular leakage
SHV	Surgical heart valve
STS	Society of Thoracic Surgeons
TEE	Transesophageal echocardiography
THV	Transcatheter heart valve
ViV	Valve-in-valve

Introduction

Several decades ago, rheumatic changes were the most common cause of MV surgery; nonetheless, in recent years, the predominant etiology has become MR. The most important consideration in MV disease is understanding the underlying cause, mechanism, and severity. According to the Carpentier classification, this can range from myxomatous degeneration (including MV prolapse, Barlow disease, and ruptured chordae) to endocarditis, hypertrophic cardiomyopathy, collagen vascular disease, rheumatic disease, and functional or ischemic causes [1].

MV repair is generally preferred over MVR replacement, as it has been shown to offer a better prognosis and delay the need for lifelong anticoagulant therapy in relatively younger patients. Although repair is the most preferred option for those with primary MR, the recent consensus for those with functional MR leans toward MVR [2].

In younger patients, mechanical valves are favored because they can last a lifetime, but they require anticoagulation. In contrast, older patients typically receive a bioprosthetic (pericardial or porcine) valve to avoid the need for anticoagulation. Nevertheless, bioprosthetic valves have a higher rate of failure and structural deterioration over time, with durability ranging from 10 to 15 years in the mitral position. Consequently, re-operation is inevitable for most patients [3].

Outcomes of re-operative MV surgery, based on data from the Society of Thoracic Surgeons (1096 patients, 2002–2016), indicate an operative mortality rate of 11.1%. Further, there has been a 10% annual increase in reoperation numbers, accompanied by a downward trend in observed-to-expected ratios for operative mortality, improving from 1.44 early in the study to 0.72 in the most recent period. This data also identified significant tricuspid regurgitation and the need for revascularization as factors that increase the risk of morbidity and mortality [4].

In recent years, advances in TAVI and aortic bioprosthetic valve-in-valve procedures, along with considerations regarding the risk of redo-surgery, have led to growing interest in mitral ViV and valve-in-ring procedures. As experience in this field increases, our knowledge about case selection and pre-procedure planning continues to expand.

Patient Selection

Patient selection is crucial for the successful execution of transseptal mitral ViV procedures. The heart team should consider this approach for patients with a degenerated MV bioprosthesis who are at high risk for repeat open-heart surgery. Still, several factors may render some individuals unsuitable candidates. For instance, if a patient has previously undergone atrial septal defect repair or MV surgery via a transseptal approach, caution is warranted, and the transseptal method may not be viable. Additionally, patients with a small LVOT and a long, calcified anterior MV leaflet may also be poor candidates, as they face an increased risk of LVOTO, which is the most concerning and potentially life-threatening complication associated with transcatheter MVR [5]. Furthermore, patients exhibiting severe patient-prosthesis mismatch should be selected with care, as placing a transcatheter valve within an already small bioprosthesis can worsen the issue [6]. The presence of endocarditis, severe PVL, thrombosis, or dehiscence of the bioprosthesis may also contraindicate the procedure [7, 8].

Identification of the Bioprosthesis

The first step is to determine the type and size of the bioprosthesis. In the mitral position, unlike in the aortic position, only stented bioprostheses are utilized. Information regarding the bioprosthesis, including its type and size, can be found in the operative notes or obtained from the bioprosthesis manufacturer. If this information is unavailable, the bioprosthesis can be identified based on its unique fluoroscopic appearance, which resembles that of its aortic counterpart [9].

True ID

Once the type of bioprosthesis is identified, the next step is to determine the true ID. Manufacturer labeling provides the labeled size, external stent diameter, and stent ID. Nonetheless, these parameters may be misleading when planning a mitral ViV procedure. The true ID is the most critical dimension, as it dictates the size of the THV to be implanted within a specific bioprosthesis. The true ID essentially represents the mechanical ID after the leaflets are mounted on the stent frame. [10] CT scans can only serve as an adjunct to the sizing information provided in the mitral ViV App.

Differences Between Bioprosthetic Valves ID

- **Porcine SHV**: The porcine valve leaflets are always sutured inside the stent frame, resulting in a true ID that is at least 2 mm below the stent ID. Examples include Hancock II (Medtronic), Mosaic (Medtronic), Aspire (Vascutek, Inchinnan, United Kingdom), CE porcine standard, CE porcine S.A.V. (Edwards Lifesciences), Epic/Biocor (St. Jude Medical), and Epic/Biocor Supra (St. Jude Medical).
- **Pericardial SHV with leaflets sutured inside the stent**: The impact of the pericardial leaflets is less significant than that of the porcine leaflets, leading to a difference of 1 mm between the true ID and the stent ID. Examples include Perimount (Edwards Lifesciences), Perimount 2700 (Edwards Lifesciences), Magna/Magna Ease (Edwards Lifesciences), and Soprano (Sorin, Milan, Italy).
- **Pericardial SHV with leaflets sutured outside the stent**: Since the leaflets are sutured outside the stent, the stent ID and the true ID are similar. Examples include Mitroflow (Sorin) and Trifecta.

Fluoroscopic Appearance

While identifying the type of device is crucial, it is equally important to recognize key fluoroscopic landmarks for optimal positioning of the THV during the procedure. An ideal position is defined as one where the device is secure and provides the best hemodynamic performance. [11] Therefore, the neo-annulus of the bioprosthesis, which represents its narrowest portion, must be located. The neo-annulus corresponds to the sewing ring of the bioprosthesis. Some bioprostheses feature a sewing ring marker; however, in others, where the stent frame or stent post markers (outflow markers) are visible, identifying the neo-annulus can be challenging (Fig. 19.1). Understanding the relationship between the stent frame/stent post

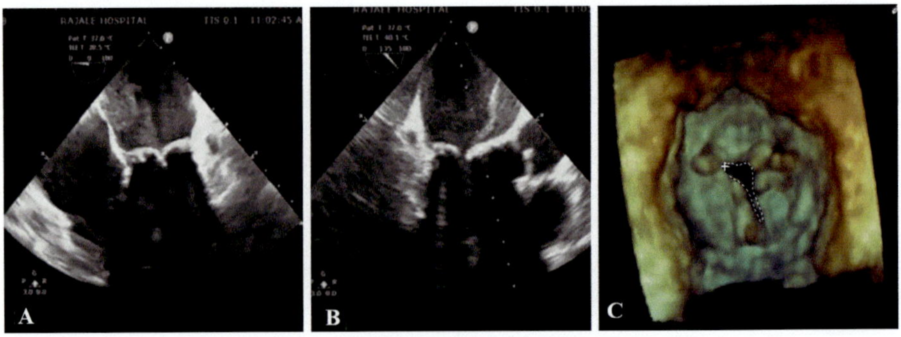

Fig. 19.1 (**a–c**) Highly calcified and thickened biological MV leaflets exhibited restricted motion and a severe smoke pattern in the LA (**a**, **b**). MVA measured by 3D planimetry: 0.5 cm² (**c**)

markers and the sewing ring is essential for achieving optimal positioning, and this information is available in the Mitral ViV app.

THV Type and Size

A THV is selected based on user preference and access route. It is evident that only THV devices with low profiles can be utilized in the mitral position. The most experience has been with the SAPIEN balloon-expandable platform (SAPIEN, SAPIEN XT, and SAPIEN 3, Edwards Lifesciences, Irvine, USA). Other THV devices employed include INNOVRE, MyVal, Lotus, and Direct Flow (only via the transapical approach) [12]. The size of the THV used is determined by the sizing algorithm specific to each device. In the mitral ViV procedure, a minimum oversizing of 2 mm over the true ID is preferred due to higher closing pressures and the associated risk of embolization. Still, excessive oversizing may be detrimental, as it can result in underexpansion of the THV device, increasing the risk of leaflet pin-wheeling, thrombosis, or decreased durability.

Case

A 77-year-old woman with a history of coronary artery bypass graft and biological MVR (size 29) performed 11 years prior was admitted due to pulmonary edema. On physical examination, she exhibited tachypnea (respiratory rate: 28, oxygen saturation: 88%). Her cardiac rhythm was irregular, consistent with atrial fibrillation and a heart rate of 130 bpm. Jugular venous pressure was elevated, and a grade II diastolic murmur as well as a pansystolic murmur were auscultated at the apex and left sternal border, respectively, accompanied by a right ventricular heave. Lung examination revealed bilateral diffuse fine rales and wheezing extending to the apex, along with diminished tactile fremitus. Bilateral lower extremity edema was also noted. An electrocardiogram (ECG) demonstrated rapid AF rhythm, right axis deviation, and incomplete right bundle branch block (RBBB).

High-dose diuretics and vasodilators were administered immediately. Within one hour, the patient showed a dramatic response to the treatment.

- **TEE findings**
  - **LV:**
    - Normal size;
    - Mild systolic dysfunction (LV ejection fraction: 45%–50%);
    - Diastolic dysfunction (Grade III); and
    - No LV hypertrophy.

- **Right ventricle:**
  - Moderate size and
  - Severe systolic dysfunction.
- **MV:**
  - Degenerated, heavily calcified biological MV;
  - Thickened and calcified leaflets;
  - Mean pressure gradient: 15 mm Hg;
  - Pressure half-time: 210 milliseconds;
  - MVA by 3D planimetry: 0.5 cm^2;
  - Mild transvalvular leakage;
  - No PVL; and
  - ID of the biological MV: 27 mm.
- **LA and LA appendage:**
  - Severe smoke pattern
- **Aortic valve:**
  - Thickened and tricuspid;
  - No aortic stenosis; and
  - Mild aortic insufficiency.
- **Tricuspid valve:**
  - No tricuspid stenosis and
  - Severe tricuspid regurgitation.
- **Pulmonary hypertension:**
  - Severe (systolic pulmonary artery pressure: 70 mm Hg).
- **Neo-LVOT:**
  - Estimated area: ~2.8 cm^2.
- **Other findings:**
  - No pulmonary embolism (Fig. 19.1a–c).

The cardiac CT scan estimated the neo-LVOT area to be >1.7 cm^2 (Fig. 19.2).

Considering the patient's clinical condition and high STS score, the heart team decided in favor of a transcatheter mitral ViV procedure. Consequently, the patient was deemed a suitable candidate for the intervention.

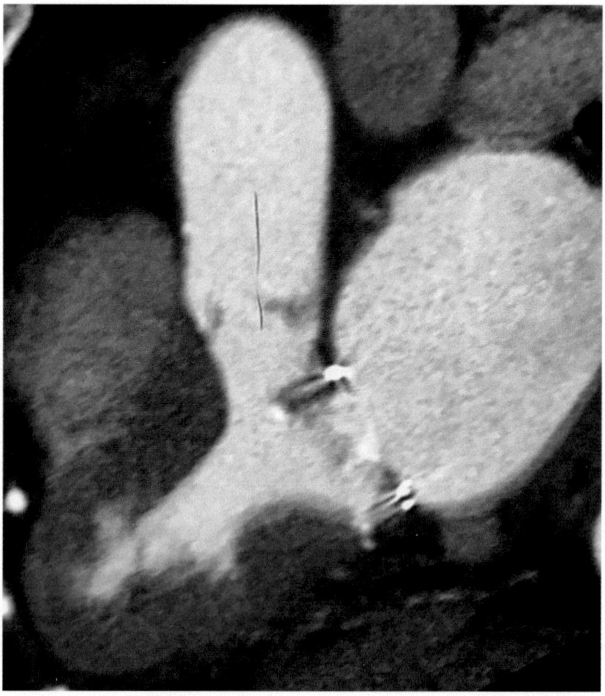

Fig. 19.2 The estimated neo-LVOT on the CT scan was >1.7 cm^2

Procedural Technique

Under general anesthesia, right and left common femoral artery sheaths (6 Fr) were inserted, and a 10 Fr sheath was introduced into the right common femoral artery using two Perclose ProGlide (Abbott Medical, Santa Clara, CA, USA). Under rigorous hemodynamic monitoring by a cardiac anesthesiologist (during cardiac index monitoring) and TEE guidance, a postero-inferior septostomy was performed through the right CFA. Due to the significant enlargement of the LA, the procedure took several minutes, during which TEE revealed impending layered clot formation in the LA (Fig. 19.3a). After the septostomy, intravenous heparin was administered to achieve an activated clotting time of over 250 seconds. Contrast injection confirmed the correct position of the Mullins catheter (Fig. 19.3b). The fluoroscopic appearance of the valve is shown (Fig. 19.3c). After inserting a spiral wire, a diagnostic multipurpose catheter (6 Fr) was telescoped into an 8.5 Fr large curve Agilis NxT steerable sheath (St. Jude Medical, St. Paul, Minnesota). The SHV was crossed

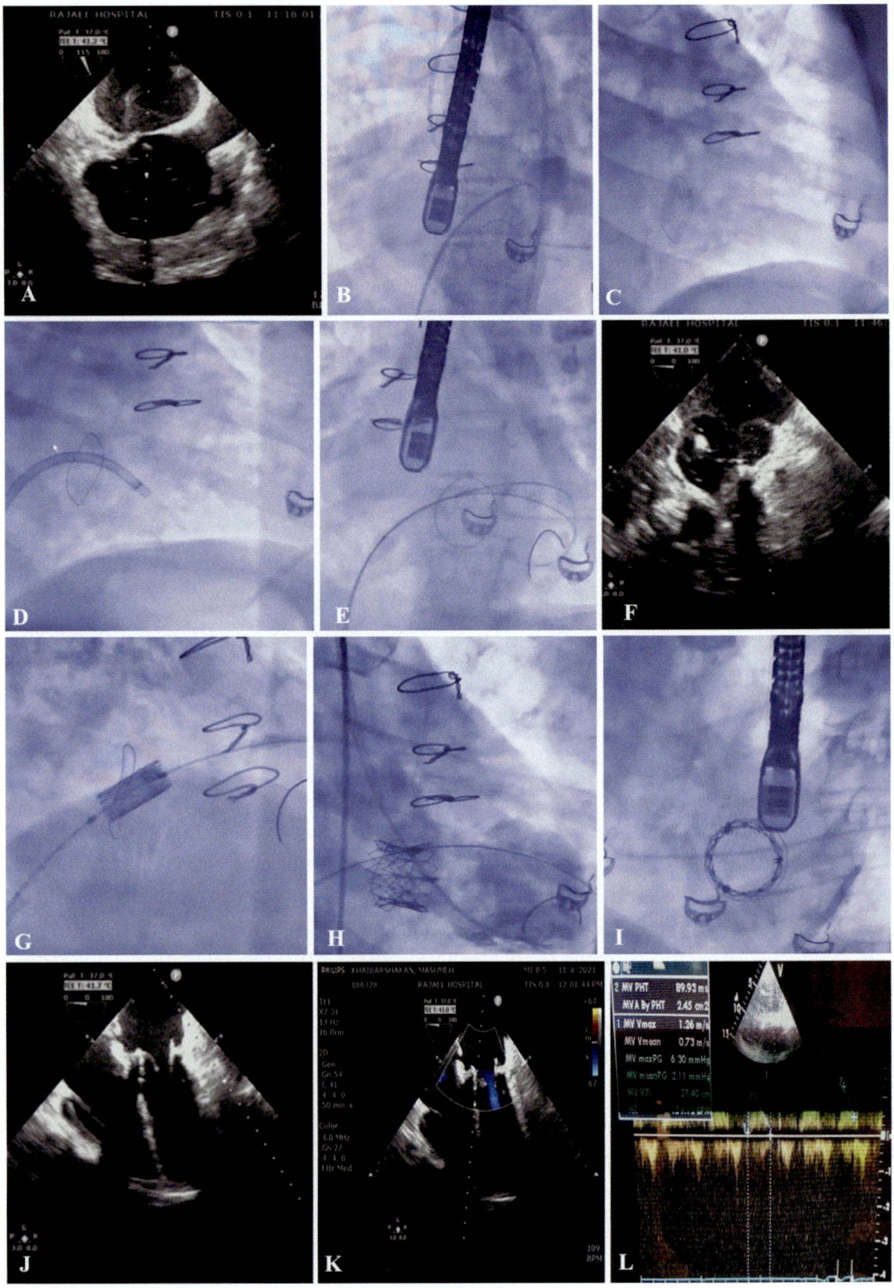
A
B
C
D
E
F
G
H
I
J
K
L
RAJAEI HOSPITAL
MV PHT
MVA By PHT
MV Vmax
MV Vmean
MV maxPG
MV meanPG

using a 0.035-inch hydrophilic guidewire, and the Agilis catheter was advanced over it (Fig. 19.3d). For additional support, two Super Stiff guidewires, 260 cm each (Amplatz extra-stiff and Lunderquist), were positioned in the LV apex (Fig. 19.3e). Septal dilatation was then performed using a 14 × 30 balloon. Due to the significant thickening and calcification of the SHV, predilation was conducted with a 23.30 balloon (Fig. 19.3f). Subsequently, a long delivery sheath (16 Fr) was introduced, and the chosen valve (Edwards SAPIEN 3–29) was mounted for antegrade delivery onto the delivery catheter with the Edwards E logo facing downward. Assembly and alignment of the valve were carried out in the inferior vena cava.

For optimal SAPIEN valve function, no more than 20% of the valve (3–5 mm) should be atrial during deployment. After the SHV was crossed under fluoroscopic and TEE guidance, the THV position was aligned so that the central marker of the balloon was positioned 3 mm below the opacified swing ring (Fig. 19.3g), and its position was confirmed by TEE. Thereafter, the valve was inflated slowly. During the partial inflation of the balloon (10%–15%) and before complete valve deployment, the position of the THV was reassessed using TEE. The final LV injection confirmed the correct position of the THV without MR (Fig. 19.3h, i). TEE verified an appropriate THV position with excellent leaflet motion, mild trans-valvular MR, and no PVL (Fig. 19.3j, k), along with hemodynamic data showing a mean pressure gradient of 2 mm Hg and a peak pressure gradient of 6 mm Hg (Fig. 19.3l).

After the procedure, the patient's hemodynamic condition was unstable, prompting the initiation of vasopressor agents. Following intensive care, within 24 hours, the CO increased to 4 liters/min/m^2, and the patient was ultimately discharged in good condition with warfarin.

Fig. 19.3 (**a**–**l**) Due to significant enlargement of the LA, the septostomy procedure took several minutes, during which TEE indicated impending layered clot formation in the LA (**a**). In contrast, the correct position of the Mullins catheter was confirmed through injection (**b**). The fluoroscopic appearance of the More valve is shown (**c**). A diagnostic multipurpose catheter (6 Fr) was telescoped into an 8.5 Fr large curve Agilis NxT steerable sheath (St. Jude Medical, St. Paul, Minnesota), allowing the SHV to be crossed with a 0.035-inch hydrophilic guidewire, over which the Agilis catheter was advanced (**d**). For additional support, two Super Stiff guidewires (260 cm) were positioned in the LV apex (**e**). Given the significant thickening and calcification of the SHV, predilation was performed using a 23.30 mm balloon (**f**). After crossing the SHV under fluoroscopic and TEE guidance, the THV was positioned so that the central marker of the balloon was located 3 mm below the opacified sewing ring (**g**). The final LV injection confirmed the proper positioning of the THV without MR (**h**, **i**). TEE demonstrated an appropriate THV position with excellent leaflet motion, mild transvalvular MR, and no PVL (**j**, **k**), along with hemodynamic data (**l**)

Discussion

Here are some detailed tips and tricks for transcatheter mitral ViV procedures to minimize complications and achieve the best results:

- As the most commonly used TVHs are the SAPIEN group (low profile), the trans-septal approach is the most frequently used method. TEE-guided septostomy is highly recommended, targeting the postero-inferior part of the inter-atrial septum, with at least a 3 cm distance between the puncture site and the mitral annulus.
- Another critical point is selecting the best deployment view. Some valves can be challenging to visualize, so it is essential to clearly identify the plane during this process. In most cases, a deep right anterior oblique projection provides a satisfactory view.
- The depth of implantation is crucial to attain the best neo-LVOT and avoid valve embolization. During deployment, no more than 20% of the THV (3–5 mm) should be atrial.

For more controlled implantation, it is advisable to inflate the balloon slowly and check the THV position using fluoroscopy and TEE. Once the appropriate position is confirmed, full inflation can be performed.

LVOTO is the Achilles heel of any transcatheter mitral intervention. LVOTO is defined as an LVOT peak gradient increase of ≥10 mm Hg post-transcatheter MVR. This occurs when the metal frame of the transcatheter valve pushes the anterior leaflet of the old bioprosthesis toward the interventricular septum, consequently narrowing the LVOT. The risk is more pronounced for pericardial valves compared with porcine valves and for taller valves compared with shorter ones. Factors influencing LVOTO after mitral ViV are depicted in Fig. 19.4.

Some techniques have been described to mitigate the risk of LVOTO. By way of example, the intentional transcatheter laceration of the anterior MV leaflet (LAMPOON) technique employs radiofrequency energy to lacerate the anterior mitral leaflet, preventing it from obstructing the LVOT after ViV-transcatheter MVR [14]. Pre-emptive alcohol septal ablation has also been described to increase the area of the predicted neo-LVOT in patients who would otherwise be excluded from transcatheter MVR due to the prohibitive risk of LVOTO [15].

Another important consideration is the risk of THV embolization. Correct oversizing using the True ID remains critical. To avoid embolization, it is important to avoid parallel deployment and achieve conical deployment [13].

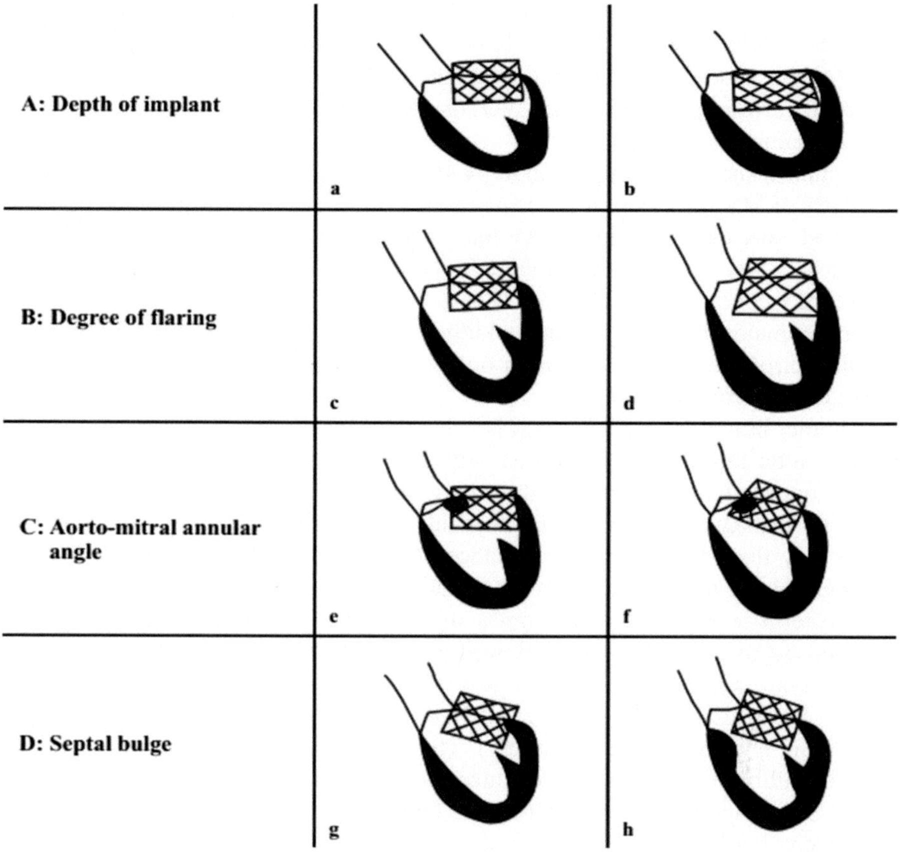

Fig. 19.4 (**a**) Depth of implant: a deeper THV results in a higher LVOTO (b > a); (**b**) Flare in the LV: an increase in flare leads to a higher LVOTO (d > c); (**c**) Aorto-mitral annular angle: a less obtuse angle increases the risk of LVOTO (f > e); (**d**) Septal bulge: a larger septal bulge raises the risk of LVOTO (h > g) [13]

Clinical Results

The Transcatheter Mitral Valve Replacement registry is a multicenter observational study that collected data on THV implantation in patients with failed MV bioprostheses, failed repair, and MAC [16]. Over a 9-year period (2009 through 2018), a total of 521 patients were enrolled, including 322 mitral ViV procedure patients and

141 MV intervention or repair patients. The patient cohort was considered high risk for surgery, with a mean STS score of 9%. A balloon expandable valve was used in 90% of the cases. Regarding procedural outcomes—conversion to surgery, need for a second valve implantation, and LVOTO—MV intervention or repair fared worse than the mitral ViV procedure.

Technical success was achieved in 94.4% of mitral ViV procedure cases but only in 80.9% of MV intervention or repair cases. Postprocedural MR (> moderate) was observed more frequently in the MV intervention or repair group (18.4% vs. 5.6%). All-cause 30-day mortality was higher in the MV intervention or repair cohort than in the MIV cohort (9.9% vs. 6.2%). There were no significant differences in terms of stroke, major bleeding, and acute kidney injury. The authors also demonstrated the impact of the learning curve on outcomes, as the latter half of the study had lower rates of complications.

Another important consideration is valve thrombosis. This event occurred in 10 patients after mitral ViV procedures, with 50% of cases occurring within the first three months after the procedure, three between three months and one year, and two after one year. Overall, at one year, the incidence was higher in patients on antiplatelet therapy than those on anticoagulation (60.6% vs. 1.6%; log-rank P = 0.019), highlighting the importance of anticoagulation in this subset.

Mid-term and long-term outcomes of mitral ViV procedures have also been reported by Ye et al. [17] who followed a cohort of 31 patients with failed mitral bioprosthetic valves who underwent ViV procedures with balloon expandable valves. The patients with degenerated MV prostheses had an STS score of 9.7% (indicating high risk for redo-surgery). The study demonstrated the overall safety and feasibility of the transcatheter intervention, with a low rate of major adverse events and periprocedural complications. Thirty-day mortality and disabling stroke rates were 1.4%, the rate of life-threatening bleeding was 4.1%, and no patient experienced more than mild paravalvular regurgitation. One mitral THV migrated aerially two months after the procedure, causing severe PVL. Two patients on postoperative dual antiplatelet therapy developed valve thrombosis, which was resolved with the use of oral anticoagulants. Although there was merely a trend in the improvement of pulmonary pressures and LV ejection fraction over the first 36 months, the authors noted a significant improvement in heart failure symptoms and NYHA functional class for up to eight years. Estimated survival rates were 88.9%, 79.5%, 69.8%, 61.9%, and 40.5% at one, two, three, four, and five years, respectively.

Kamioka et al. [18] conducted a retrospective study comparing mitral ViV procedures and redo surgery, analyzing 1-year echocardiographic and clinical outcomes. They found no difference in 1-year mortality and 30-day echocardiographic results between the two groups. Nevertheless, mitral gradients were higher in the mitral ViV procedure group (7.2 ± 2.7 vs. 5.5 ± 1.8 mm Hg), though this difference was not clinically significant. These findings suggest that good results can be achieved through less invasive transcatheter interventions, thereby avoiding high-risk redo surgery.

Conclusion

Advancements in mitral ViV technologies and a preference for less invasive approaches have enabled the rapid adoption of ViV transcatheter MVR among patients with a failed mitral bioprosthesis who are at high or prohibitive surgical risk. The reported outcomes indicate that this procedure is safe and feasible; however, further studies with long-term follow-ups are necessary.

References

1. Otto CM, Bonow RO. Valvular heart disease. In: Otto CM, Bonow RO, editors. Valvular heart disease: a companion to Braunwald's Heart Disease. Philadelphia: Saunders; 2013. p. 327–40.
2. Bonow RO, O'Gara PT, Adams DH, Badhwar V, Bavaria JE, Elmariah S, et al. 2020 focused update of the 2017 ACC expert consensus decision pathway on the management of mitral regurgitation: a report of the American College of Cardiology Solution set Oversight Committee. J Am Coll Cardiol. 2020;75(17):2236–70. https://doi.org/10.1016/j.jacc.2020.02.019. PMID: 32068084
3. Vongpatanasin W, Hillis LD, Lange RA. Prosthetic heart valves. N Engl J Med. 1996;335(6):407–16. https://doi.org/10.1056/NEJM199608083350607. PMID: 8676934
4. Bapat VN. Mitral valve-in-valve, valve-in-ring, and valve-in-mitral annular calcification: are we there yet? Circulation. 2021;143(2):117–9. https://doi.org/10.1161/CIRCULATIONAHA.120.051016. PMID: 33428465
5. Regueiro A, Granada JF, Dagenais F, Rodés-Cabau J. Transcatheter mitral valve replacement: insights from early clinical experience and future challenges. J Am Coll Cardiol. 2017;69(17):2175–92. https://doi.org/10.1016/j.jacc.2017.02.045. PMID: 28449778
6. Webb JG, Wood DA, Ye J, Gurvitch R, Masson JB, Rodés-Cabau J, et al. Transcatheter valve-in-valve implantation for failed bioprosthetic heart valves. Circulation. 2010;121(16):1848–57. https://doi.org/10.1161/CIRCULATIONAHA.109.924613. PMID: 20385927
7. Webb JG, Murdoch DJ, Boone RH, Moss R, Attinger-Toller A, Blanke P, et al. Percutaneous transcatheter mitral valve replacement: first-in-human experience with a new transseptal system. J Am Coll Cardiol. 2019;73(11):1239–46. https://doi.org/10.1016/j.jacc.2018.12.065. PMID: 30898201
8. Yoon SH, Bleiziffer S, Latib A, Eschenbach L, Ancona M, Vincent F, et al. Predictors of left ventricular outflow tract obstruction after transcatheter mitral valve replacement. JACC Cardiovasc Interv. 2019;12(2):182–93. https://doi.org/10.1016/j.jcin.2018.10.001. PMID: 30678796
9. Bapat V, Mydin I, Chadalavada S, Tehrani H, Attia R, Thomas M. A guide to fluoroscopic identification and design of bioprosthetic valves: a reference for valve-in-valve procedure. Catheter Cardiovasc Interv. 2013;81(5):853–61. https://doi.org/10.1002/ccd.24420. PMID: 22926957

10. Bapat VN, Attia R, Thomas M. Effect of valve design on the stent internal diameter of a bioprosthetic valve: a concept of true internal diameter and its implications for the valve-in-valve procedure. JACC Cardiovasc Interv. 2014;7(2):115–27. https://doi.org/10.1016/j.jcin.2013.10.012. PMID: 24556091
11. Bapat VN, Attia RQ, Condemi F, Visagan R, Guthrie M, Thomas M. Fluoroscopic guide to an ideal implant position for Sapien XT and CoreValve during a valve-in-valve procedure. JACC Cardiovasc Interv. 2013;6(11):1186–94. https://doi.org/10.1016/j.jcin.2013.05.020. PMID: 24139930
12. Conradi L, Silaschi M, Seiffert M, Lubos E, Blankenberg S, Reichenspurner H, et al. Transcatheter valve-in-valve therapy using 6 different devices in 4 anatomic positions: clinical outcomes and technical considerations. J Thorac Cardiovasc Surg. 2015;150(6):1557–65., 1567.e1–3; discussion 1565–7. PMID: 26432720. https://doi.org/10.1016/j.jtcvs.2015.08.095.
13. Pirelli L, Hong E, Steffen R, Taramasso M, Whisenant B, Babaliaros V, et al. Mitral valve-in-valve and valve-in-ring: tips, tricks, and outcomes. Ann Cardiothorac Surg. 2021;10(1):96–112. https://doi.org/10.21037/acs-2020-mv-14. PMID: 33575181
14. Babaliaros VC, Greenbaum AB, Khan JM, Rogers T, Wang DD, Eng MH, et al. Intentional percutaneous laceration of the anterior mitral leaflet to prevent outflow obstruction during transcatheter mitral valve replacement: first-in-human experience. JACC Cardiovasc Interv. 2017;10(8):798–809. https://doi.org/10.1016/j.jcin.2017.01.035. PMID: 28427598
15. Wang DD, Guerrero M, Eng MH, Eleid MF, Meduri CU, Rajagopal V, et al. Alcohol septal ablation to prevent left ventricular outflow tract obstruction during transcatheter mitral valve replacement: first-in-man study. JACC Cardiovasc Interv. 2019;12(13):1268–79. https://doi.org/10.1016/j.jcin.2019.02.034. PMID: 31202950
16. Yoon SH, Whisenant BK, Bleiziffer S, Delgado V, Dhoble A, Schofer N, et al. Outcomes of transcatheter mitral valve replacement for degenerated bioprostheses, failed annuloplasty rings, and mitral annular calcification. Eur Heart J. 2019;40(5):441–51. https://doi.org/10.1093/eurheartj/ehy590. PMID: 30358809
17. Ye J, Cheung A, Yamashita M, Wood D, Webb J, Chan F, et al. Transcatheter aortic and mitral valve-in-valve implantation for failed surgical bioprosthetic valves: an 8-year single-center experience. JACC Cardiovasc Interv. 2015;8(13):1735–44. https://doi.org/10.1016/j.jcin.2015.08.012. PMID: 26516002
18. Kamioka N, Babaliaros V, Morse MA, Frisbee J, Lerakis S, Lisko JC, et al. Comparison of clinical and echocardiographic outcomes after surgical redo mitral valve replacement and transcatheter mitral valve-in-valve therapy. JACC Cardiovasc Interv. 2018;11(11):1131–8. https://doi.org/10.1016/j.jcin.2018.03.011. PMID: 29880106

Chapter 20
Transcatheter Closure of Aortic Paravalvular Leakage

Ata Firouzi, Zahra Hosseini, and Ehsan Khalilipur

Abstract Aortic PVL is characterized by a regurgitation jet between the swing ring and the native valve tissue. The underlying cause of this side effect is dependent on the surgical timing and may be attributed to technical issues, endocarditis, or severe calcification of the surrounding tissue. The severity of patient symptoms can vary from asymptomatic cases to pulmonary edema, largely dependent on the size of the defects. While the standard treatment for severe leakage involves surgical repair, transcatheter device closure serves as a viable alternative approach for patients with high surgical mortality rates, provided that the anatomy is suitable and expert interventionists perform the procedure. Despite the lack of a dedicated device, various devices can be employed based on the shape and location of the defect as visualized through 3D TEE.

Keywords Aortic paravalvular leakage · Etiology · Symptoms · Surgical repair · 3D TEE · Device closure

Abbreviations

CT	Computed tomography
LV	Left ventricle
PVL	Paravalvular leakage
RV	Right ventricle
TEE	Transesophageal echocardiography

A. Firouzi · Z. Hosseini · E. Khalilipur (✉)
Cardiovascular Intervention Research Center, Rajaie Cardiovascular Institute, Iran University of Medical Sciences, Tehran, Iran

A. Firouzi et al. (eds.), *Mastering Skills in Complex Cardiovascular Interventions*, https://doi.org/10.1007/978-3-031-97321-5_20

Introduction

PVL is an infrequent but significant complication that may arise following prosthetic valve implantation. PVL is characterized by abnormal communication between the cardiovascular chambers neighboring a prosthetic valve. It affects individuals who have undergone surgical valve replacement, with an incidence rate of 2%–10% in the aortic position and 7%–17% in the mitral position [1]. While most PVLs are asymptomatic and typically follow a benign clinical course, approximately 1%–5% of PVL patients may experience serious clinical consequences [2]. PVL results from an incomplete seal between the sewing ring and annulus, which may develop due to abnormal pressure or traction forces exerted on the prosthesis post-surgery. Various factors are known to elevate the risk of PVL formation, including annular calcification, infection, suturing technique, and the size and shape of the prosthetic implant. Early PVLs are typically associated with technical aspects of surgical implantation. In contrast, late PVLs often stem from suture dehiscence resulting from endocarditis or the progressive resorption of inadequately debrided annular calcifications [3].

While most PVLs are minor and asymptomatic, larger PVLs can lead to severe clinical consequences, including heart failure, hemolytic anemia, or endocarditis. These complications are reported in 1%–5% of patients who have undergone surgical valve replacement. Symptomatic PVLs present with congestive heart failure caused by volume overload in 90% of cases and hemolytic anemia resulting from shear stress on red blood cells, affecting one-third to three-quarters of these patients [4].

Most patients presenting with congestive heart failure have a mean New York Heart Association functional class of III or higher. The severity of congestive heart failure symptoms is linked to the size and number of defects, as well as the degree of LV volume overload. Interestingly, hemolytic complications do not correlate with defect size, as they can occur even in small defects. Severe hemolytic anemia may also manifest as congestive heart failure. In addition, PVL is an essential pre-existing condition in the presence of bacteremia for the development of infective endocarditis, as it acts as an intracardiac defect creating significant turbulent flow.

Auscultation is commonly employed as a screening method for PVL, with cardiac murmurs raising concern when detected in patients with prosthetic valves. In cases of aortic PVL, a high-pitched decrescendo murmur can be detected at the left sternal border during diastole. Nonetheless, due to its lack of specificity in diagnosis, additional imaging modalities such as echocardiography should be promptly utilized to confirm or exclude the presence of PVL.

In cases where patients do not respond to medical therapy targeting congestive heart failure and/or hemolysis (including erythropoietic agents and blood transfusion), consideration should be given to either redo open-heart surgery or percutaneous PVL closure as alternative treatment options. For many years, surgical re-intervention has been regarded as the preferred treatment for symptomatic PVL patients. It has been demonstrated that the overall mortality rate is lower in patients who undergo surgical re-intervention than in those receiving conservative treatment. Surgery is considered the gold standard treatment for operable patients with intractable symptoms and severe PVL (Class I). Nevertheless, caution must be exercised when approaching re-operation, as redo surgery typically involves higher risks compared to the initial procedure, and PVL recurrence may be observed in over one-third of patients undergoing redo open-heart surgery for PVL [5]. In cases involving patients at high surgical risk, presenting with intractable symptoms, and demonstrating suitable anatomy, catheter-based closure performed by expert operators may serve as a viable alternative treatment option. Transcatheter PVL closure has surfaced as a less invasive, safe, and effective alternative to surgical re-intervention, offering long-lasting symptom relief for carefully selected patients. Consequently, percutaneous PVL closure has witnessed an increase in popularity in recent years. First documented in 1992, this procedure has evolved steadily and is now successfully executed in multiple centers with considerable experience in structural cardiac intervention [6].

Diagnostic Workup

Transthoracic echocardiography is the primary modality used for evaluating LV and RV size and function, assessing mechanical valve function, determining the severity of transvalvular leakage or PVL, and detecting any clots, pannus formation, or vegetation.

Three-dimensional TEE is the gold standard modality for PVL diagnosis, as it can accurately illustrate the location, shape, size, depth, and number of defects, dehiscence, and surrounding structures while distinguishing PVL from transvalvular regurgitations.

Although other imaging techniques, such as CT angiography and cardiac magnetic resonance imaging, can provide further information for a more comprehensive evaluation of defects, it is recommended to report the location of mitral PVL using a clockwise format from the surgeon's perspective or surgical view. Similarly, the location of aortic PVL should be described in a clockwise format (Figs. 20.1, 20.2, and 20.3).

In the context of the aortic valve, the left coronary cusp is situated between the 11 and 3 o'clock positions, the right coronary cusp lies between 3 and 7 o'clock positions, and the noncoronary cusp is positioned between 7 and 11 o'clock positions. Aortic PLV is most commonly found between 7 and 11 o'clock (46%), with a secondary peak incidence between 11 and 3 o'clock (36%).

Case

The patient was a 65-year-old man with a history of coronary artery bypass graft surgery, mechanical aortic valve replacement, and mechanical mitral valve replacement 15 years prior. He was admitted for refractory congestive heart failure, with five hospitalizations within the last six months. TEE revealed severe aortic PVL. His hemodynamics remained stable.

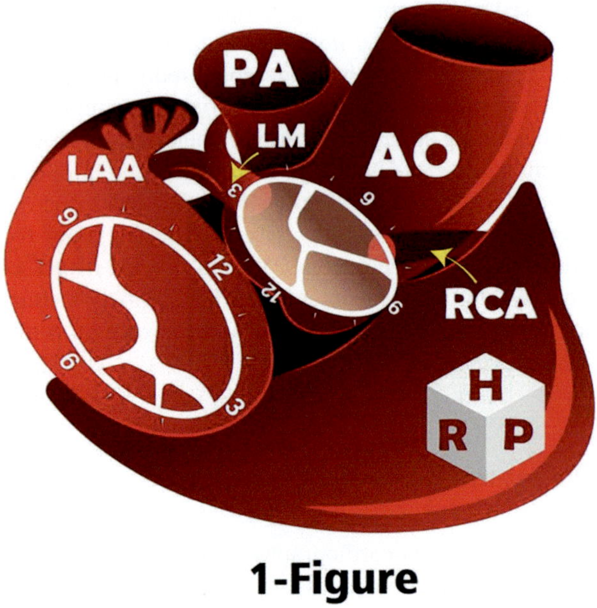

Fig. 20.1 The location of PVL using a clockwise format from the surgical view

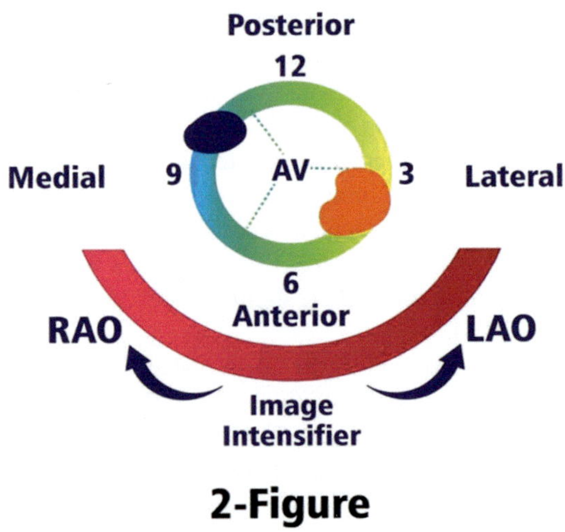

Fig. 20.2 The location of aortic PVL should be described in a clockwise format

On cardiac examination, a mid-diastolic murmur was audible at the right second intercostal space and along the left sternal border. Bilateral fine rales were detected in the lower third of the lungs, accompanied by diminished tactile fremitus. ECG met the criteria for LV hypertrophy

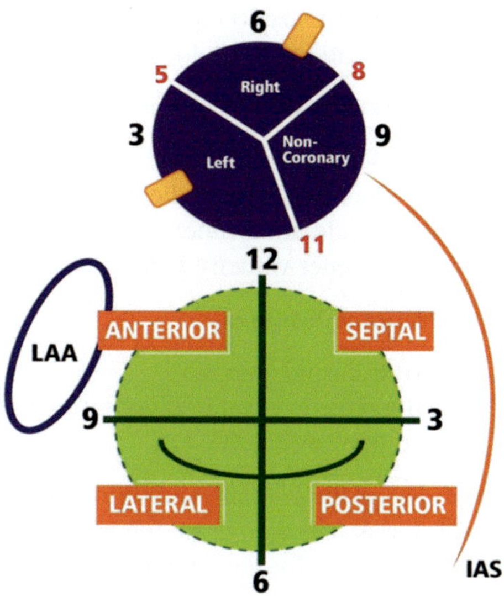

Fig. 20.3 In the left atrium, from the surgeon's perspective, leaks can be referenced in hours, similar to a clock, with the mitral-aortic curtain representing 12 o'clock and/or by positional segments. IAS: interatrial septum; LAA: left atrial appendage

- **3D TEE findings**
  - Severe LV enlargement with moderate systolic dysfunction (LV ejection fraction: 35%);
  - Moderate eccentric LV hypertrophy;
  - Mild RV enlargement with moderate dysfunction; and
  - Moderate left atrial enlargement; and
  - No clots.
- **Mitral valve:**
  - Mechanical bi-leaflet mitral valve;
  - Normal leaflet motion;
  - Mild transvalvular regurgitation;
  - No PVL;
  - No visible mass; and
  - Hemodynamic measurements:
    - Mean pressure gradient: 5 mm Hg;
    - Peak pressure gradient: 12 mm Hg;
    - Pressure half-time: 80 milliseconds; and
    - Doppler velocity index: 2.
- **Aortic valve:**
  - Mechanical bi-leaflet aortic valve;
  - Normal leaflet motion;
  - No transvalvular regurgitation;
  - Severe PVL:

    - Defect size: 6 mm and
    - Location: 10 o'clock at surgical view.
  - Hemodynamic measurements:
    - Mean pressure gradient: 15 mm Hg;
    - Peak pressure gradient: 25 mm Hg;
    - Acceleration time: <80 milliseconds; and
    - Doppler velocity index): >0.35.
- **Additional findings:**
  - No tricuspid stenosis;
  - Moderate tricuspid regurgitation with a gradient of 45 mm Hg; and
  - No pericardial effusion.

Considering the patient's condition and the echocardiographic findings, and given the high risk of reoperation, the heart team recommended transcatheter aortic paravalvular closure for the patient.

Procedural Technique

Under general anesthesia, a 6 Fr sheath was inserted into the right common femoral artery. An aortic root injection was performed in both left anterior oblique and right anterior oblique projections to assess the location and severity of the PVL (Fig. 20.4a).

Subsequently, under the guidance of 3D TEE and fluoroscopy, a retrograde approach was employed in the left anterior oblique projection. The defect, located at the 10 o'clock position, was crossed using a 0.035-inch straight hydrophilic wire (260 cm). A JR catheter was then advanced over the wire (Fig. 20.4b), and the wire was exchanged for an extra-support wire (Lunderquist). A kink-resistant hydrophilic delivery sheath (Epsilar 7 Fr) was advanced into the LV.

Based on the TEE measurements, which indicated a defect size of 6 mm, along with its shape and location, an 8 mm Muscular ventricular septal defect device (Occlutech) was selected. The device was loaded onto the delivery cable and passed through the delivery sheath. The distal disk was partially extruded in the LV (Fig. 20.4c). By retracting the entire system toward the defect, the distal disk was fully deployed on the ventricular side and anchored within the defect. After confirming proper positioning with TEE and angiography, the waist and proximal disk were sequentially deployed (Fig. 20.4d).

Prior to releasing the device, careful attention was given (using both TEE and fluoroscopy) to ensure there was no interference with the motion of the prosthetic valve leaflets (confirming free-moving leaflets) and no encroachment on the

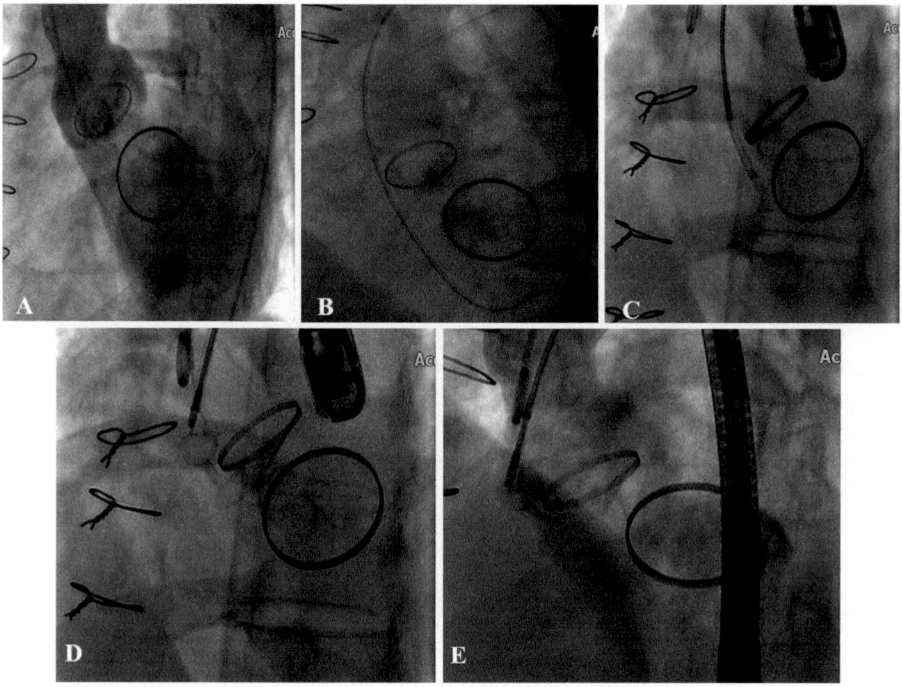

Fig. 20.4 (**a–e**) Aortic root injection revealed severe PVL from the medial side of the swing ring (**a**). Under TEE guidance, the defect was wired, and the JR catheter was advanced over it (**b**). The first distal disk was partially extruded into the LV (**c**). Subsequently, by retracting the entire system toward the defect, the distal dick was fully deployed on the ventricular side and anchored within the defect. After re-evaluating with TEE and angiography, the waist and proximal disk were deployed (**d**). The final injection indicated mild residual leakage, with no interference in leaflet motion and a stable position of the device (**e**)

coronary ostia (Fig. 20.4e). The final injection revealed mild residual leakage, no interference with the valve leaflets, and a stable device position.

In cases where the device is implanted near the noncoronary cusp, the anterior mitral valve leaflet should also be thoroughly assessed to rule out any potential interference.

Discussion

The safety and feasibility of transcatheter PVL closure have been validated in several registries and a meta-analysis [7]. Reported technical and clinical success rates range from 77%–86% and 67%–77%, respectively [8]. Moreover, successful transcatheter PVL closure has been linked to a lower cardiac mortality rate compared with unsuccessful closure attempts (260 patients; OR: 0.08, 95% CI: 0.01 to 0.90), as well as a positive trend toward lower all-cause mortality (311 patients; OR: 0.52, 95% CI: 0.09 to 1.74) and superior functional class improvement or improved hemolytic anemia compared with failed transcatheter PVL closure (267 patients; OR: 9.95, 95% CI: 2.10 to 66.73). Experienced centers report low complication rates for these procedures [9].

A recent single-center, non-randomized study demonstrated that percutaneous PVL closure yielded outcomes comparable to surgical treatment [10]. Furthermore, in a study by Calvert et al. [9] examining 308 PVL closure procedures in 259 patients across 20 centers, a technical success rate of 91% was reported, with significant improvements in PVL post-procedure ($P < 0.001$). At the final follow-up, PVL severity was classified as none (33.3%), mild (41.4%), moderate (18.6%), or severe (6.7%). The mean New York Heart Association class also showed improvement from 2.7 ± 0.8 pre-procedure to 1.6 ± 0.8 ($P < 0.001$) after a median follow-up of 110 (7–452) days.

Device Selection

Despite their widespread use, most devices employed for PVL closure today have not been specifically designed, tested, or approved for this purpose, leading to their "off-label" application. The variability in size and shape of PVLs, with many exhibiting crescentic and serpiginous features instead of a cylindrical form, poses a significant challenge for a single device to effectively address all types of PVLs. Among the available options, Amplatzer devices (St. Jude Medical, St. Paul, MN, USA) have emerged as the most frequently used devices for transcatheter PVL closure [11].

The Amplatzer Vascular Plug II is commonly employed in the United States, whereas the Amplatzer Vascular Plug III is the most frequently utilized device for PVL closure in Europe. Currently, the Occlutech PLD is the sole device approved by the European Commission specifically for transcatheter PVL closure [12].

Device selection for different PVL morphologies may vary, with Amplatzer Vascular Plug II or patent ductus arteriosus occluder being favored for small cylindrical PVLs. For oval-shaped PVLs, the Amplatzer Vascular Plug III occluder is often preferred. In cases involving small PVLs with substantial angulation and a small neck, which commonly occur after transcatheter aortic valve replacement, the Amplatzer Vascular Plug IV occluder may be a suitable option.

Device sizing typically relies on measurements obtained from 2D and 3D TEE. Angiography can also serve as a helpful tool for evaluating aortic PVLs. The PLD's rectangular-shaped design is particularly recommended for crescent-shaped defects as it facilitates effective coverage while minimizing interference with the valve (Table 20.1).

Table 20.1 Major complications associated with PVL closure

Complication	Percentage
Device embilisation	%4
Cardic perfoation	%4
Death	%2
Vascular complications	%2
Embolic stroke	%1.7
Emergency cardiac surgery for prosthetic impingement	%0.9
Intracranial haemorrhage	%0.9
Sepsis	%0.9

Conclusion

Despite its promising potential, percutaneous PVL closure still has limited clinical experience. The procedural success rate exhibits a range of 63%–95%, influenced by factors such as the PVL location and the availability of appropriate devices. Following successful closure, patients typically experience improvement in congestive symptoms within weeks. Still, hemolysis may persist or potentially worsen during the initial months post-procedure as the device undergoes endothelialization.

References

1. Hammermeister K, Sethi GK, Henderson WG, Grover FL, Oprian C, Rahimtoola SH. Outcomes 15 years after valve replacement with a mechanical versus a bioprosthetic valve: final report of the Veterans Affairs randomized trial. J Am Coll Cardiol. 2000;36(4):1152–8. https://doi.org/10.1016/S0735-1097(00)00834-2. PMID: 11028464
2. Davila-Roman VG, Waggoner AD, Kennard ED, Holubkov R, Jamieson WR, Englberger L, et al. Prevalence and severity of paravalvular regurgitation in the Artificial Valve Endocarditis Reduction Trial (AVERT) echocardiography study. J Am Coll Cardiol. 2004;44(7):1467–72. https://doi.org/10.1016/j.jacc.2004.06.060. PMID: 15464328
3. De Cicco G, Lorusso R, Colli A, Nicolini F, Fragnito C, Grimaldi T, et al. Aortic valve periprosthetic leakage: anatomic observations and surgical results. Ann Thorac Surg. 2005;79(5):1480–5. https://doi.org/10.1016/j.athoracsur.2004.10.010. PMID: 15854921
4. Rallidis LS, Moyssakis IE, Ikonomidis I, Nihoyannopoulos P. Natural history of early aortic paraprosthetic regurgitation: a five-year follow-up. Am Heart J. 1999;138(2 Pt 1):351–7. https://doi.org/10.1016/S0002-8703(99)70124-9. PMID: 10426849
5. Genoni M, Franzen D, Vogt P, Seifert B, Jenni R, Künzli A, et al. Paravalvular leakage after mitral valve replacement: improved long-term survival with aggressive surgery? Eur

J Cardiothorac Surg. 2000;17(1):14–9. https://doi.org/10.1016/S1010-7940(99)00346-0. PMID: 10748385
6. Ruiz CE, Jelnin V, Kronzon I, Dudiy Y, Del Valle-Fernandez R, Einhorn BN, et al. Clinical outcomes in patients undergoing percutaneous closure of periprosthetic paravalvular leaks. J Am Coll Cardiol. 2011;58(21):2210–7. https://doi.org/10.1016/j.jacc.2011.03.074. PMID: 22078425
7. Millan X, Skaf S, Joseph L, Ruiz CE, García E, Smolka G, et al. Transcatheter reduction of paravalvular leaks: a systematic review and meta-analysis. Can J Cardiol. 2015;31(3):260–9. https://doi.org/10.1016/j.cjca.2014.12.012. PMID: 25746024
8. Noble S, Jolicoeur EM, Basmadjian A, Levesque S, Nozza A, Potvin J, et al. Percutaneous paravalvular leak reduction: procedural and long-term clinical outcomes. Can J Cardiol. 2013;29(11):1422–8. https://doi.org/10.1016/j.cjca.2013.03.020. PMID: 23916751
9. Calvert PA, Northridge DB, Malik IS, Shapiro L, Ludman P, Qureshi SA, et al. Percutaneous device closure of paravalvular leak: combined experience from the United Kingdom and Ireland. Circulation. 2016;134(13):934–44. https://doi.org/10.1161/CIRCULATIONAHA.116.022684. PMID: 27578887
10. Angulo-Llanos R, Sarnago-Cebada F, Rivera AR, Elizaga Corrales J, Cuerpo Caballero G, Solis Martín J, et al. Two-year follow-up after surgical versus percutaneous paravalvular leak closure: a non-randomized analysis. Catheter Cardiovasc Interv. 2016;88(6):972–9. https://doi.org/10.1002/ccd.26459. PMID: 26856433
11. Shapira Y, Hirsch R, Kornowski R, Hasdai D, Assali A, Vaturi M, et al. Percutaneous closure of perivalvular leaks with Amplatzer occluders: feasibility, safety, and short-term results. J Heart Valve Dis. 2007;16(3):305–13. PMID: 17578052
12. Goktekin O, Vatankulu MA, Tasal A, Baykan AO, Sönmez K, Pamukcu B, et al. Transcatheter trans-apical closure of paravalvular mitral and aortic leaks using a new device: first in man experience. Catheter Cardiovasc Interv. 2014;83(2):308–14. https://doi.org/10.1002/ccd.24942. PMID: 23613352

Chapter 21
Transcatheter Closure of Mitral Paravalvular Leakage

Ata Firouzi, Zahra Hosseini, and Ehsan Khalilipur

Abstract Mitral PVL occurs more frequently than aortic valve leakage. The symptoms experienced by patients depend on the size of the defect and the severity of regurgitation. In cases of severe PVL, patients are typically symptomatic, which can ultimately lead to biventricular failure. Surgical repair is the recommended treatment. Nonetheless, for those at high surgical risk, transcatheter closure presents an alternative option. This intervention is highly complex and should be thoroughly evaluated by an experienced team to assess the anatomical features of the defect. Various devices are available for closure (off-label), depending on the shape and size of the defect.

Keywords Paravalvular mitral leakage · Biventricular failure · Surgical repair Complex procedure · Device

Abbreviations

LA	Left atrium
LV	Left ventricle
PLV	Paravalvular leakage
TEE	Transesophageal echocardiography
TTE	Transthoracic echocardiography
VSD	Ventricular septal defect

A. Firouzi · Z. Hosseini · E. Khalilipur (✉)
Cardiovascular Intervention Research Center, Rajaie Cardiovascular Institute, Iran University of Medical Sciences, Tehran, Iran

A. Firouzi et al. (eds.), *Mastering Skills in Complex Cardiovascular Interventions*, https://doi.org/10.1007/978-3-031-97321-5_21

Introduction

Despite recent advancements in valve replacement techniques, PVL remains a life-threatening complication for patients with prosthetic heart valves [1]. The incidence of mitral PVL after valve surgery is reported to be 2.2%, which may increase to 17% for mechanical valves and 9% for bioprosthetic valves during follow-up periods exceeding 10 years [2]. PVLs are twice as likely to occur with mitral prostheses compared to aortic prostheses, and the risk of PVL is greater with mechanical heart valves than with bioprosthetic valves. In addition, redo surgeries are associated with a higher rate of PVL recurrence and increased mortality compared to initial procedures. The prognosis worsens with repeat surgeries, with mortality rates of 13%, 15%, and 35% for the first, second, and third redo surgeries, respectively [3].

Each repeat operation is less likely to succeed, highlighting the need for percutaneous approaches. While most PVLs following surgical valve replacement are trivial to mild and typically have a benign course during the first year, moderate-to-severe PVLs can exacerbate hemolysis or lead to heart failure due to regurgitation. Consequently, patients may ultimately require either surgical closure or transcatheter closure of the defects. The mechanisms behind these leaks are not well understood. Incomplete alignment between the sewing ring and the annulus may occur due to significant annular calcification. Additionally, the tissue surrounding the valves can weaken as a result of chronic infection. Even the sutures themselves may fail to ensure adequate apposition of the valve to the annulus. All of these factors can contribute to significant PVL. The risk of PVL varies across different areas of the prosthetic ring. Several studies have shown that mitral PVLs occur more frequently at the anterolateral and posteromedial segments of the mitral valve annulus [4]. Multiple theories have identified predisposing factors for PVL development in these areas. One study indicated that collagen fibers in the mitral valve annulus are not distributed homogeneously [5]. The mitral annulus lacks a uniformly well-formed cord-like fibrous structure, particularly along the posterior segment. This characteristic may make the posterior annulus susceptible to mechanical injuries, potentially resulting in PVL [6]. Furthermore, the posterior annulus is longer than the anterior region, making it more susceptible to suture separation, and annular dilation particularly affects this area. The majority of PVLs are clinically insignificant and typically have a benign course in the absence of endocarditis. Early mild PVL may spontaneously resolve during the wound healing process, while new PVLs can develop as a late complication of valve surgery [7]. Large PVLs that are hemodynamically significant can lead to heart failure and increase the risk of infectious endocarditis. Additionally, large PVLs may cause clinically significant hemolytic anemia, necessitating repeated blood transfusions. PVL is the most common cause of hemolytic anemia in patients with prosthetic heart valves, particularly in those with mitral prostheses, due to the high velocity of the regurgitant jet resulting from the larger pressure gradient during the systolic phase. Turbulent flow through the PVL increases shear stress on red blood cells, leading to mechanical trauma and fragmentation of these cells. Notably, the severity of hemolysis does not always correlate with the size of the PVL. A small defect with a high-velocity jet can also lead to clinically significant hemolysis, particularly in the case of eccentric jets [8].

Diagnostic Workup

During a physical examination, a holosystolic murmur, typically heard at the left sternal border or in the mid-axillary line, may accompany the mechanical heart sound.

In TTE, measurements should include LV size and function, LA size, mechanical leaflet motion, and hemodynamic data (mean pressure gradient, pressure half-time, Doppler velocity index, and effective orifice area. Increased transmitral flow and gradient compared with LV outflow tract flow, in the context of a normal pressure half-time, may indicate mitral prosthetic valve regurgitation.

In TEE, accurate assessment of leaflet motion, any mass, differentiation between transvalvular and paravalvular regurgitation, and pulmonary artery pressure is essential. In 3D TEE, it is important to define the defect's location, shape, size, direction of the PVL, its relation to the prosthetic ring, and suitability for transcatheter closure. PVL is classified as "small" if the defect is <10% of the circumference of the annular ring, "moderate" if the defect is between 10% and 20%, and "large" if the defect exceeds 20%. Additionally, the rocking motion of the prosthesis with a large paravalvular defect exceeding 25% of the circumference of the annular ring indicates dehiscence. [8]

The location of PVL is best defined in relation to internal landmarks such as the LA appendage and aortic valve. On 3D imaging of mitral valve prostheses, the aortic valve is positioned anteriorly at the 12 o'clock position, the interatrial septum medially at the 3 o'clock position, and the LA appendage laterally at the 9 o'clock position. This type of description is commonly referred to as the "surgeon's view."

During fluoroscopy, anterolateral PVL is best approached in the posteroanterior view with cranial angulation, posteroseptal PVL in the right anterior view, and lateral PVL in the lateral view [8]. Cardiac computed tomography scans and cardiac magnetic resonance imaging are other modalities that can help estimate the severity, location, and size of PVL, as well as assess LV size and function [9]. These modalities can also define optimal fluoroscopic angles of the prosthetic annular plane for PVL transcatheter closure (Fig. 21.1).

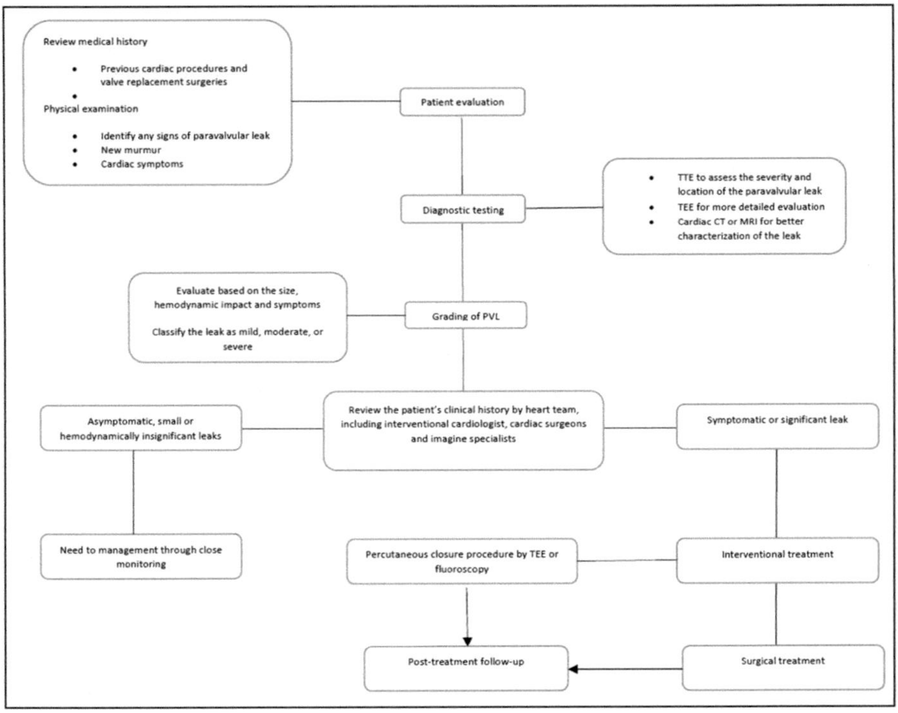

Fig. 21.1 The image presents an algorithm for the diagnosis and treatment of PVL

Transcatheter Mechanical Mitral PVL Closure

While surgical repair remains the gold standard for treating severe paravalvular regurgitation in patients with intractable symptoms (e.g., right and left-sided congestive symptoms and hemolytic anemia—Class I recommendation), transcatheter closure is a preferable option for high-risk patients (with a Society of Thoracic Surgeons risk score >8% or a >15% risk of mortality within 30 days) who experience refractory symptoms and are anatomically suitable for percutaneous intervention at a specialized center. According to the 2017 American College of Cardiology/American Heart Association Focused Guidelines Update for managing valvular heart disease, this approach (Class IIa) has demonstrated superior outcomes and fewer complications than open surgery. Success rates are highest for small, slit-like, crescentic, and single defects, while larger, irregular, sloped tunnel-shaped, and multiple defects tend to yield poorer results [10].

Case 1

The patient was a 70-year-old man with a history of mechanical mitral valve replacement performed 16 years prior. He presented with refractory dyspnea (New York Heart Association functional class III–IV) and orthopnea over the past three months, despite being on high-dose diuretic therapy. TEE revealed severe PVL on the medial side of the prosthetic ring.

On physical examination, the patient was hemodynamically stable but exhibited elevated jugular venous pressure. A pansystolic murmur, accompanied by metallic sounds, was auscultated along the left sternal border. Bilateral basilar rales were noted during lung examination. ECG demonstrated left axis deviation with an LV hypertrophy pattern.

- **3D TEE findings**
  - **LV:**
    - Moderate enlargement;
    - Mild systolic dysfunction (LV ejection fraction: 45%); and
    - Moderate eccentric LV hypertrophy.
  - **Right ventricle:**
    - Mild enlargement and
    - Moderate dysfunction.
  - **LA:**
    - Moderate enlargement and
    - No evidence of clots.
  - **Aortic valve:**
    - Tricuspid aortic valve;
    - No aortic stenosis; and
    - Moderate aortic insufficiency.
  - **Mitral valve:**
    - Bileaflet mechanical valve with normal leaflet motion;
    - Hemodynamic study:
      - Mean pressure gradient: 7 mm Hg;
      - Peak pressure gradient: 12 mm Hg;
      - Pressure half-time: 70 milliseconds; and
      - Doppler velocity index: 2.5.
    - No evidence of thrombus or mass on the leaflets;
    - No transvalvular regurgitation; and
    - Severe PVL (1 crescent-shaped defect at 2 o'clock of the swing ring).

- **Tricuspid valve:**
  - No tricuspid stenosis and
  - Moderate tricuspid regurgitation.
- **Pulmonary hypertension:**
  - Moderate (systolic pulmonary artery pressure: 65 mm Hg).
- **Other:**
  - No evidence of pulmonary embolism.

Based on the patient's symptoms and echocardiographic findings, he was deemed a good candidate for transcatheter PVL closure.

Procedural Technique [11]

Under general anesthesia, the right common femoral artery and vein sheaths (6 Fr) were inserted, followed by an LV injection in the left and right anterior oblique projections to identify the defect site and assess the severity of PVL. Guided by 3D TEE, a comprehensive evaluation of the mechanical valve was conducted, including the degree of transvalvular and paravalvular regurgitation; the number, shape, and location of defects; and the exclusion of any thrombus, vegetation, or dehiscence.

Based on the defect's location (designated by clock numbers), either a transseptal or transapical approach is employed. In this case, the largest defect was crescent-shaped, measuring 6 mm, and located at the 2 o'clock position (corresponding to the mitral-aortic fibrous continuity). For defects situated between 6 and 11 o'clock, a transapical approach is preferred. When defects are positioned away from the septum, the puncture location is less critical. Nevertheless, for medial defects near the interatrial septum, a posterior and slightly superior puncture is recommended to achieve the appropriate working height within the LA.

After performing a septostomy using a Brockenbrough needle and injecting contrast to confirm entry into the LA (Fig. 21.2a), intravenous heparin was administered to achieve an activated clotting time of >250 seconds. A spiral wire was then inserted into the LA for additional support and to help localize the LA borders, followed by septal dilation. Using a 125 cm 6 Fr multipurpose diagnostic catheter telescoped through a steerable transseptal sheath (Agilis sheath), several attempts were made to manipulate a 0.035-inch straight hydrophilic wire under the guidance of TEE and fluoroscopy to cross the defect. The wire was advanced into the ascending aorta, after which the catheter was passed over the wire (Fig. 21.2b). Subsequently, the wire was exchanged for an Amplatzer Super Stiff guidewire, and a hydrophilic kink-resistant delivery sheath (Epsilar-7 Fr) was carefully advanced over the wire into the LV. If significant difficulty arises in crossing the delivery

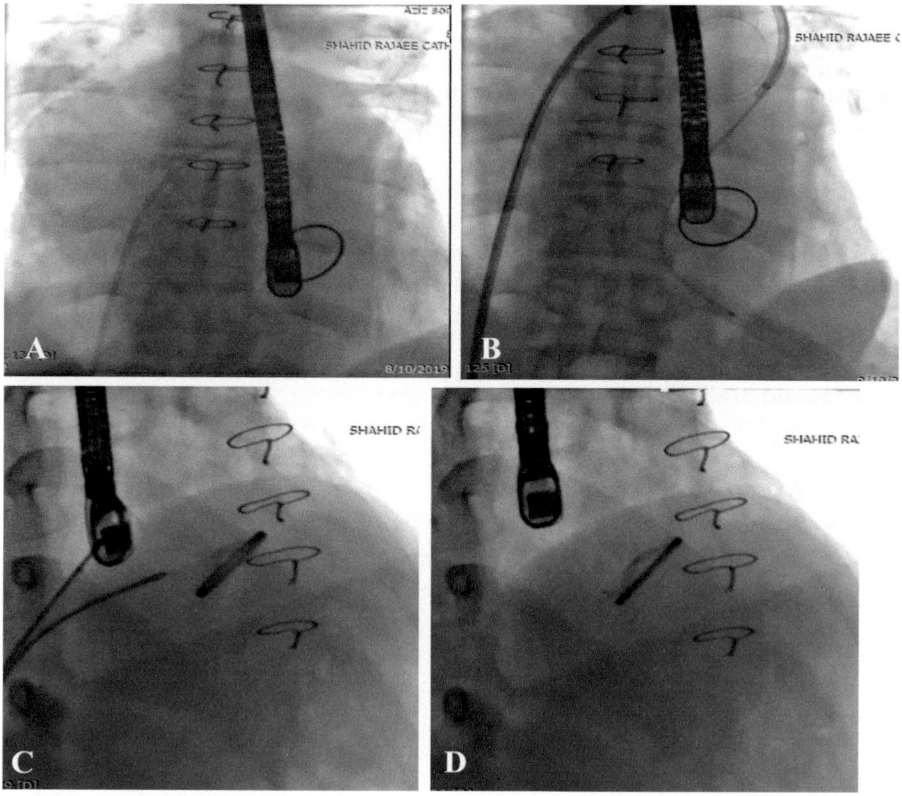

Fig. 21.2 (**a**–**d**) Following a septostomy (**a**), wiring of the defect at the 2 o'clock position was performed using a hydrophilic guidewire through the Agilis sheath (**b**). Based on the characteristics of the defect, an 8 mm Occlutech Muscular VSD Occluder was selected. Under the guidance of TEE and fluoroscopy, the distal disk was partially exposed, and the entire system was retracted, allowing for the full deployment of the distal disk in the LV. After confirming the correct position of the distal disk via TEE and LV injections, the waist and proximal disk were deployed in the LA (**c**). Before releasing the device, TEE confirmed that the mitral leaflets were mobile, the pulmonary veins were open, there was no residual leakage, and the device was stable (**d**)

sheath, additional support can be achieved by creating an arteriovenous rail. In this instance, based on the characteristics of the defect, an 8 mm Occlutech Muscular VSD Occluder was selected. After loading the device and connecting it to the delivery cable, the device was advanced over the delivery sheath under the guidance of TEE and fluoroscopy. Once the distal disk was partially exposed, the entire system was retracted, allowing the distal disk to fully deploy in the LV. Following confirmation of the correct position of the distal disk via TEE and LV injections, the waist and proximal disk were deployed in the LA (Fig. 21.2c). Before releasing the device, TEE should verify that the mitral leaflets are mobile, the pulmonary veins are open, any residual leakage is assessed, and the stability of the device is confirmed.

Case 2

The patient was a 75-year-old man with a history of mechanical mitral valve replacement performed 10 years prior. He presented with complaints of resting dyspnea and orthopnea.

On physical examination, the patient was hemodynamically stable (SpO_2: 90%). A pansystolic murmur was auscultated at the left sternal border, radiating to the apex. Lung examination revealed bilateral diffuse fine rales extending up to the apex. ECG showed atrial fibrillation with a rapid ventricular response.

Diuretic and vasodilator therapies were initiated immediately, to which the patient responded significantly.

- **TTE and 3D TEE findings (after 24 hours)**
  - **LV:**
    - Severe enlargement;
    - Moderate systolic dysfunction (LV ejection fraction: 35%); and
    - Moderate concentric LV hypertrophy.
  - **Right ventricle:**
    - Moderate enlargement and
    - Severe dysfunction.
  - **LA:**
    - Severe enlargement and
    - No evidence of clots.
  - **Aortic valve:**
    - Tricuspid aortic valve;
    - Mild aortic stenosis (aortic valve area: 1.7 cm^2); and
    - Moderate aortic insufficiency.
  - **Mitral valve:**
    - Bi-leaflet mechanical valve with normal leaflet motion;
    - Hemodynamic data:
      - Mean pressure gradient: 6 mm Hg;
      - Peak pressure gradient: 14 mm Hg;
      - Pressure half-time: 65 milliseconds; and
      - Doppler velocity index: 2.2.
    - No transvalvular regurgitation; and
    - Severe PVL (single 6 mm oval-shaped defect at 1 o'clock of the swing ring).
  - **Tricuspid valve:**
    - No tricuspid stenosis and

 - Moderate tricuspid regurgitation.

- **Pulmonary hypertension:**

 - Severe (systolic pulmonary artery pressure: 75 mm Hg).

- **Other:**

 - No evidence of pulmonary embolism.

Based on the patient's clinical condition and echocardiography findings, he was deemed high-risk for a redo-operation but suitable for transcatheter PVL closure. Following a thorough discussion with the heart team, the decision was made to proceed with trans-septal PVL closure.

Procedural Technique

Under conscious sedation, after the insertion of the right common femoral artery and common femoral vein sheaths (6 Fr and 8 Fr, respectively), an LV injection was performed in the right and left anterior oblique projections to characterize the severity of the PVL and the defect location. Guided by 3D TEE, a complete assessment of the mechanical valve was conducted. The evaluation included a review of the degree of transvalvular and paravalvular regurgitation, as well as the number, shape, and location of the defect while ruling out any thrombus, vegetation, or dehiscence. Following a septostomy using a Brockenbrough needle, positioned more posteriorly and inferiorly, contrast was injected to confirm entry into the LA. Subsequently, intravenous heparin was administered to achieve an activated clotting time of >250 seconds, and a spiral wire was inserted into the LA for additional support. Thereafter, septal dilatation was performed using a 125 cm 6 Fr multipurpose diagnostic catheter, telescoped through a steerable trans-septal sheath (Agilis sheath). After several attempts, a 0.035-inch straight hydrophilic wire was carefully manipulated under TEE and fluoroscopy guidance to cross the defect. Following this, the Agilis catheter was advanced over the wire, and the multipurpose catheter was exchanged for a JR guiding catheter. To provide additional support, two Super Stiff guidewires were inserted into the LV apex. A kink-resistant delivery sheath (LIFESCIENCE sheath-10 Fr) was gently advanced over the wire. Based on the characteristics of the defect, an 8 mm Occlutech Muscular VSD Occluder was selected for the procedure. After loading the device and connecting it to the delivery cable, the device was passed over the delivery sheath. Under the guidance of TEE and fluoroscopy, the distal disk was partially exposed. The entire system was subsequently pulled back, allowing the distal disk to fully deploy in the LV. In the next step, the correct position of the distal disk was confirmed using TEE and LV injections, and the waist and proximal disk were deployed in the LA. Before releasing the device, TEE confirmed that the mitral leaflets were mobile, the pulmonary veins were open, there was no residual leakage, and the device was stable (Fig. 21.3).

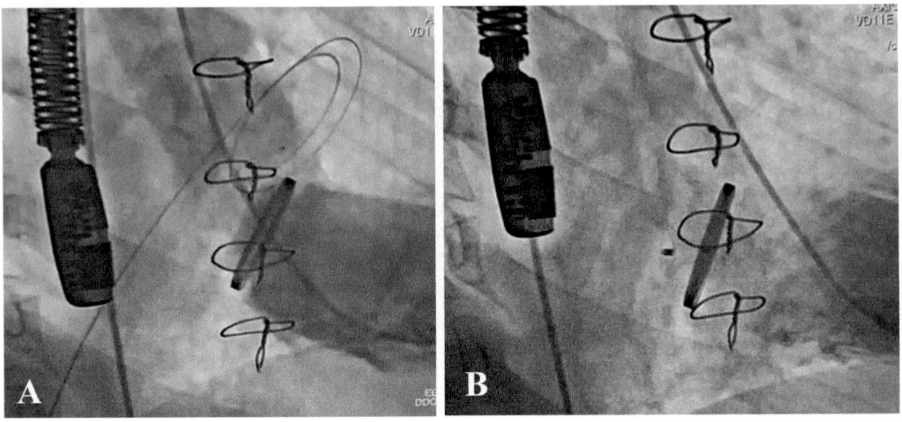

Fig. 21.3 (**a**, **b**) LV injection in the right anterior oblique projection revealed severe PVL, with the spiral wire visible in the LA (**a**). Following the deployment of the occluder device, the final injection indicated the proper positioning of the device, normal leaflet motion, and no residual PVL (**b**)

Discussion

Various techniques are available for transcatheter mitral PVL closure, including antegrade (trans-septal, particularly for posterior and septal defects), retrograde aortic, and apical approaches. The trans-septal puncture is generally the preferred method for mitral PVL closure. Optimal trans-septal approach puncture location depends on PVL localization, with TEE playing a crucial role in guiding the puncture process for mitral valve interventions. For instance, lateral defects are amenable to high puncture on the interatrial septum, while more posterior and lower punctures facilitate direct access to the origin of posteriorly and medially localized PVLs [12]. The trans-apical approach, offering the shortest route, may be a suitable alternative for patients with posterior or septal PVL or multiple PVLs at various locations [13]. Be that as it may, this procedure may lead to challenges in other steps of transcatheter closure, potentially prolonging the overall procedural time. Crossing the defect stands out as the most demanding aspect of the procedure, requiring considerable expertise, particularly when addressing posterior defects through the trans-septal approach. In certain instances, crossing the defect with a guidewire is feasible; however, it may still be impossible to cross with an available delivery sheath or catheter. While the defect may be crossed and closed, the closure device could interfere with the prosthetic valve's function, preventing device release [14]. Moreover, a concomitant aortic prosthesis can further complicate the transcatheter closure of the mitral prosthesis. For experienced operators, alternative approaches such as trans-apical, retrograde, or combined retrograde/antegrade may also be viable options [15]. Given these potential challenges, thorough periprocedural planning and effective communication among experienced members of the structural heart team are essential for procedural success.

Irregular defect morphology and complex anatomy of the surrounding tissue may contribute to incomplete PVL sealing. While partial transcatheter closure of

PVL can reduce regurgitant volume and alleviate heart failure symptoms, hemolysis may remain unaffected or potentially worsen. Hence, the availability of dedicated devices in various sizes and shapes is critical for achieving complete PVL sealing [16].

An ideal device for transcatheter PVL closure should be retrievable and repositionable, with a size larger than the defects to minimize the risk of embolization [17]. Nonetheless, larger devices and those placed near the hinge point may interfere with prosthetic function, potentially causing the sewing ring to fracture from the annulus and exacerbating PVL.

Several complications may occur during transcatheter PVL closure procedures, including bleeding, pericardial effusion, hemothorax (in trans-apical procedures), transient hemolysis, transient ischemic events, or stroke [18].

Numerous devices have been employed for transcatheter PVL closure; still, outcomes have often been suboptimal due to device limitations and the crescentic nature of most PVLs. Most commonly used devices are from the Amplatzer family (St. Jude Medical [now Abbott] and St. Paul, MN, USA), which includes the Amplatzer Vascular Plug (AVP) series (AVP II, AVP III, and AVP IV), Amplatzer Duct Occluder (ADO I and ADO II), Amplatzer Atrial Septal Occluder, and Amplatzer Muscular VSD Occluder. A round or long tunnel-shaped PVL can be effectively sealed using a Vascular Plug II device, while a large crescentic leak may be managed with a larger device such as a VSD Occluder [19]. However, the use of a large device like the VSD Occluder carries an inherent risk of hemolysis. Consequently, employing a "nesting" technique with multiple devices may prove more successful in closing a large PVL defect than using a single large device, such as the VSD Occluder (Fig. 21.4).

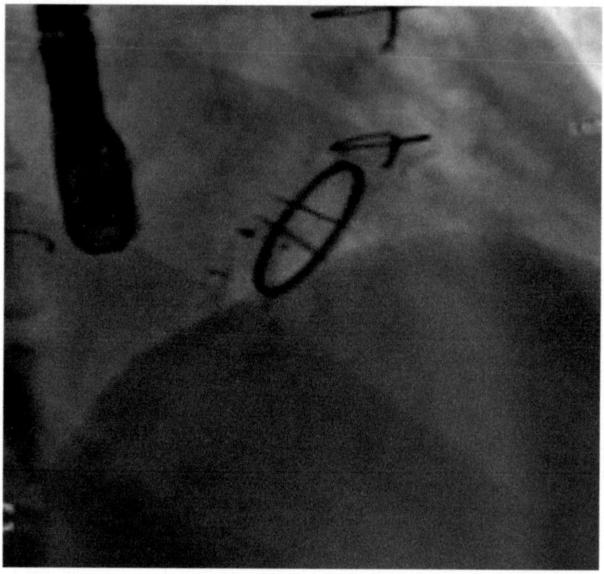

Fig. 21.4 Due to the large size of the defect, two muscular VSD occluder devices (6 mm and 8 mm) were implanted instead of a single device

According to the Spanish registry, which encompassed 514 procedures in 469 patients across 19 centers (70.2% involving mitral prostheses), technical and procedural success was achieved in 86.6% and 73.2% of patients, respectively [19]. In addition, 80.2% of patients were complication-free, with the most common complication being minor vascular access-related bleeding (8.6%). Notably, this complication did not result in significant clinical consequences. The 30-day incidence of major complications (death, stroke, or need for emergency surgery) was reported to be 5.6%. The relatively high procedural success rate and low complication rate in this registry can be attributed to the utilization of specifically designed devices, dedicated catheter delivery techniques, and advanced echocardiographic imaging modalities, such as real-time 3D TEE. Interestingly, the registry data also demonstrated a correlation between operator experience and procedural success rates for mitral interventions.

The registry from the United Kingdom and Ireland, which included 259 patients from 20 different centers, revealed similar findings and indicated a hospital mortality rate of 2.9% for elective procedures. The authors strongly suggested transcatheter closure as a viable alternative to repeat surgery [20]. In 2018, Busu et al. [21] conducted a meta-analysis comparing surgical and transcatheter closure treatment strategies, providing significant insights. The analysis encompassed 22 studies: 17 single-arm retrospective studies and five retrospective studies directly comparing the two approaches. A total of 2373 patients were enrolled, with 862 (36.3%) in the surgical group and 1511 (63.7%) in the transcatheter group. The primary indications for transcatheter closure were symptomatic heart failure (65.9%) and hemolytic anemia (21.9%). The mitral position was the most common site of PVL (74.4%), and the Amplatzer Vascular Plug II was the most frequently used device (71%). In the surgical closure group, prosthetic reimplantation and patch or pledgeted suture repair were performed at similar rates. Surgical closure demonstrated higher technical success rates (96.7% vs. 72.1%), attributed to the nature of the intervention. Nevertheless, it was associated with higher upfront costs, including increased 30-day mortality (8.6% vs. 6.8%), stroke rates (3.3% vs. 1.4%), and longer hospitalization durations. At one-year follow-up, no significant differences were observed between the two strategies in terms of mortality (17.3% vs. 17.2%), reoperation rates (9.1% vs. 9.9%), New York Heart Association functional class, or heart failure readmissions.

Device Selection

Most devices used for this procedure are employed off-label. PVLs vary in size and shape (crescentic, oval, serpiginous, or cylindrical), making a single device unsuitable for all PVLs. Typically, for small cylindrical PVLs, AVP II or PDA occluders are utilized; for oval-shaped PVLs, AVP III is recommended; and for small, angulated PVLs, AVP IV occluders may be considered.

Conclusion

Transcatheter closure of PVLs in symptomatic patients presents a less invasive alternative to surgery, associated with reduced morbidity and mortality. The most common complications include device embolization, cardiac perforation, vascular issues, prosthetic valve impingement, and stroke. In experienced centers, transcatheter closure of severe mitral PVLs can be considered an initial treatment option, as this approach does not exclude the possibility of subsequent surgical intervention, thereby facilitating a stepwise strategy for managing this complex clinical condition. Furthermore, advancements in specialized devices, catheters, surgical techniques, and materials, along with multimodal imaging guidance, are expected to enhance the success rates of interventional procedures. It is essential to adopt an integrated approach to determine the most suitable treatment for each patient. Prospective registries and further randomized studies are likely to help fill the existing gaps in evidence.

References

1. Ruiz CE, Hahn RT, Berrebi A, Borer JS, Cutlip DE, Fontana G, et al. Clinical trial principles and endpoint definitions for paravalvular leaks in surgical prosthesis: an expert statement. J Am Coll Cardiol. 2017;69(16):2067–87. https://doi.org/10.1016/j.jacc.2017.02.038. PMID: 28330792
2. Duncan BF, McCarthy PM, Kruse J, Andrei AC, Li Z, Russell HM, et al. Paravalvular regurgitation after conventional aortic and mitral valve replacement: a benchmark for alternative approaches. J Thorac Cardiovasc Surg. 2015;150(4):860–8. https://doi.org/10.1016/j.jtcvs.2015.06.047. PMID: 26210274
3. Genoni M, Franzen D, Vogt P, Seifert B, Jenni R, Künzli A, et al. Paravalvular leakage after mitral valve replacement: improved long-term survival with aggressive surgery? Eur J Cardiothorac Surg. 2000;17(1):14–9. https://doi.org/10.1016/S1010-7940(99)00346-0. PMID: 10748385
4. De Cicco G, Russo C, Moreo A, Beghi C, Fucci C, Gerometta P, et al. Mitral valve periprosthetic leakage: anatomical observations in 135 patients from a multicentre study. Eur J Cardiothorac Surg. 2006;30(6):887–91. https://doi.org/10.1016/j.ejcts.2006.09.017. PMID: 17064949
5. Kunzelman KS, Cochran RP, Murphree SS, Ring WS, Verrier ED, Eberhart RC. Differential collagen distribution in the mitral valve and its influence on biomechanical behaviour. J Heart Valve Dis. 1993;2(2):236–44. PMID: 8269138
6. Angelini A, Ho SY, Thiene G. Anatomy of the mitral valve. In: Boudoulas H, Wooley CF, editors. Mitral valve: floppy mitral valve, mitral valve prolapse, mitral valvular regurgitation. 2nd ed. New York: Futura Publishing Company; 2000. p. 5–30.
7. Eleid MF. Interventional management of paravalvular leak. Heart. 2018;104(21):1797–802. https://doi.org/10.1136/heartjnl-2017-312424. PMID: 29871966
8. Gürsoy OM, Astarcıoğlu MA, Gökdeniz T, Aykan AÇ, Bayram Z, Çakal B, et al. Severe mitral paravalvular leakage: echo-morphologic description of 47 patients from a real-time three-dimensional transesophageal echocardiography perspective. Anatol J Cardiol. 2013;13(7):633–40. https://doi.org/10.5152/akd.2013.4575. PMID: 24064140

9. Suh YJ, Hong GR, Han K, Im DJ, Chang S, Hong YJ, et al. Assessment of mitral paravalvular leakage after mitral valve replacement using cardiac computed tomography: comparison with surgical findings. Circ Cardiovasc Imaging. 2016;9(6):e004153. https://doi.org/10.1161/CIRCIMAGING.115.004153. PMID: 27282823
10. Ozkan M, Gürsoy OM, Astarcıoğlu MA, Wunderlich N, Sievert H. Percutaneous closure of paravalvular mitral regurgitation with Vascular Plug III under the guidance of real-time three-dimensional transesophageal echocardiography. Turk Kardiyol Dern Ars. 2012;40(7):632–41. https://doi.org/10.5543/tkda.2012.78000. PMID: 23363896
11. Firouzi A, Hosseini Z. Transcatheter closure of mitral paravalvular leakage (PVL). In: Maleki M, Alizadehasl A, editors. Case-based clinical cardiology. London: Springer; 2021. p. 701–8. https://doi.org/10.1007/978-1-4471-7496-7_77.
12. Krishnaswamy A, Kapadia SR, Tuzcu EM. Percutaneous paravalvular leak closure—imaging, techniques and outcomes. Circ J. 2013;77(1):19–27. https://doi.org/10.1253/circj.CJ-12-1289. PMID: 23220796
13. Kumar R, Jelnin V, Kliger C, Ruiz CE. Percutaneous paravalvular leak closure. Cardiol Clin. 2013;31(3):431–40. https://doi.org/10.1016/j.ccl.2013.05.004. PMID: 23931104
14. Komoda T, Hetzer R, Siniawski H, Oellinger J, Felix R, Uyama C, et al. Effects of prosthetic valve placement on mitral annular dynamics and the left ventricular base. ASAIO J. 2001;47(1):60–5. https://doi.org/10.1097/00002480-200101000-00014. PMID: 11199317
15. Zhou D, Pan W, Guan L, Qian J, Ge J. Retrograde transcatheter closure of mitral paravalvular leak through a mechanical aortic valve prosthesis: two successful cases. Tex Heart Inst J. 2016;43(2):137–41. https://doi.org/10.14503/THIJ-14-4898. PMID: 27127431
16. Onorato EM, Muratori M, Smolka G, Malczewska M, Zorinas A, Zakarkaite D, et al. Midterm procedural and clinical outcomes of percutaneous paravalvular leak closure with the Occlutech Paravalvular Leak Device. EuroIntervention. 2020;15(14):1251–9. https://doi.org/10.4244/EIJ-D-19-00517. PMID: 31638576
17. Bernard S, Yucel E. Paravalvular leaks—from diagnosis to management. Curr Treat Options Cardiovasc Med. 2019;21(11):67. https://doi.org/10.1007/s11936-019-0776-0. PMID: 31673867
18. Lampropoulos K, Aggeli C, Megalou A, Barbetseas J, Budts W. Diagnosis and treatment of left-sided prosthetic paravalvular regurgitation. Cardiology. 2016;133(1):27–34. https://doi.org/10.1159/000439169. PMID: 26421997
19. García E, Arzamendi D, Jimenez-Quevedo P, Sarnago F, Martí G, Sanchez-Recalde A, et al. Outcomes and predictors of success and complications for paravalvular leak closure: an analysis of the SpanisH real-wOrld paravalvular LEaks closure (HOLE) registry. EuroIntervention. 2017;12(16):1962–8. https://doi.org/10.4244/EIJ-D-16-00119. PMID: 28191876
20. Calvert PA, Northridge DB, Malik IS, Shapiro L, Ludman P, Qureshi SA, et al. Percutaneous device closure of paravalvular leak: combined experience from the United Kingdom and Ireland. Circulation. 2016;134(13):934–44. https://doi.org/10.1161/CIRCULATIONAHA.116.022684. PMID: 27578887
21. Busu T, Alqahtani F, Badhwar V, Cook CC, Rihal CS, Alkhouli M. Meta-analysis comparing transcatheter and surgical treatments of paravalvular leaks. Am J Cardiol. 2018;122(2):302–9. https://doi.org/10.1016/j.amjcard.2018.04.009. PMID: 29861049

Index

A. Firouzi et al. (eds.), *Mastering Skills in Complex Cardiovascular Interventions*, https://doi.org/10.1007/978-3-031-97321-5

MIX
Papier aus verantwortungsvollen Quellen
Paper from responsible sources
FSC® C105338

If you have any concerns about our products,
you can contact us on
ProductSafety@springernature.com

In case Publisher is established outside the EU,
the EU authorized representative is:
Springer Nature Customer Service Center GmbH
Europaplatz 3, 69115 Heidelberg, Germany

Printed by Libri Plureos GmbH
in Hamburg, Germany